SCIENCE AND ETHICAL RESPONSIBILITY

Proceedings of the U.S. Student Pugwash Conference
University of California, San Diego, June 19-26, 1979

SCIENCE AND ETHICAL RESPONSIBILITY

Proceedings of the U.S. Student Pugwash Conference
University of California, San Diego, June 19-26, 1979

Edited by
Sanford A. Lakoff

With the assistance of
Jeffrey Leifer, Ronald Bee, Eric Markusen

Foreword by
Bernard T. Feld
Massachusetts Institute of Technology
Cambridge, Massachusetts

1980
ADDISON-WESLEY PUBLISHING COMPANY
ADVANCED BOOK PROGRAM
Reading, Massachusetts

London • Amsterdam • Don Mills, Ontario • Sydney • Tokyo

Science and Ethical Responsibility
Proceedings of the U.S. Student Pugwash Conference, University of California, San Diego, June 19–26, 1979

This book was prepared with the support of NSF Grant OSS-7824990. However, any opinions, findings, conclusions, and/or recommendations herein are those of the contributors and do not necessarily reflect the views of the National Science Foundation.

Library of Congress Cataloging in Publication Data

U. S. Student Pugwash Conference, University of
    California, San Diego, 1979.
    Science and ethical responsibility.

    Bibliography: p.
    Includes index.
        1.   Science--Social aspects--Congresses.
    2.   Disarmament--Congresses.   3.   Research--Social
    aspects--Congresses.   4.   Science and ethics--
    Congresses.   I.   Lakoff, Sanford A.   II.   Title.
    III.   Title: Pugwash Conference.
    Q175.4.U54   1979   174'.95   80-25461
    ISBN 0-201-03993-1

Published simultaneously in Canada

Published by Addison-Wesley Publishing Company, Inc., Advanced Book Program, Reading, Massachusetts, 01867, U.S.A.

Manufactured in the United States of America

Contents

Contributors

CURT BIREN is completing graduate work in Science, Technology, and Public Policy at George Washington University.

EDMUND G. BROWN, Jr., is serving his second term as Governor of the State of California.

ROSEMARY A. CHALK is Staff Officer of the Committee on Scientific Freedom and Responsibility, the American Association for the Advancement of Science.

CHRISTOPHER CHYBA is completing his undergraduate studies, majoring in Physics at Swarthmore College.

BERNARD T. FELD is Professor of Physics at The Massachusetts Institute of Technology and a member of the executive committee of the international Pugwash conferences.

CRAIG B. GLIDDEN is a graduate student and a member of the Defense Workshop at Tulane University.

G. ALLEN GREB is Post-Graduate Research Historian with the Program in Science, Technology and Public Affairs at the University of California, San Diego.

CLIFFORD GROBSTEIN is Professor of Biological Science and Public Policy and was founding Dean of the Medical School at the University of California, San Diego.

FRASER HOMER-DIXON is an undergraduate at Carleton University, Ottawa, Canada.

GEORGE H. KIEFFER is Professor of Life Sciences at the University of Illinois, Urbana.

GEORGE B. KISTIAKOWSKY is Professor of Chemistry at Harvard University and served as Special Assistant for Science and Technology to President Eisenhower.

SANFORD A. LAKOFF is Professor of Political Science at the University of California, San Diego.

DIANE LEFEBVRE is completing her undergraduate studies at Michigan State University.

JEFFREY R. LEIFER is enrolled in the Graduate School of Management and Organization at Yale University.

HANS-GERD LOEHMANNSROEBEN is a graduate student at the University of North Carolina, Chapel Hill.

ALEXANDER MORIN is Director of the Office of Science and Society, the National Science Foundation.

JAMES M. MURTAGH is studying medicine at the University of Michigan.

BARRY L. PRICE is a graduate student in Political Science at Indiana University, Bloomington.

JOEL PRIMACK is Associate Professor of Physics at the University of California, Santa Cruz.

STEPHEN L. REISS is completing undergraduate work at the Woodrow Wilson School, Princeton University.

ROGER REVELLE is Professor of Science and Public Policy in the Department of Political Science and the Program in Science, Technology and Public Affairs at the University of California, San Diego. He is the former Director of the Center for Population Studies at Harvard University and previously of the Scripps Institution of Oceanography.

JONAS SALK is a Fellow and founding Director of the Salk Institute for Biological Research, La Jolla, California.

HERBERT SCOVILLE, JR., is President of the Arms Control Association and former Assistant Director of the Arms Control and Disarmanent Agency.

CRAIG SHAPIRO is a student at the Harvard Medical School.

JEREMY J. STONE is Executive Director of the Federation of American Scientists and earned a doctorate in Mathematics.

WILLIAM VAN CLEAVE is Professor of International Relations at the University of Southern California and Co-Director of the Committee on the Present Danger.

HERBERT F. YORK, JR., is Professor of and Director of the Program on Science, Technology and Public Affairs at the University of California, San Diego. He is currently on leave as U.S. Ambassador to the Comprehensive Test Ban Talks in Geneva and was previously Director of the Livermore Laboratories and Director of Defense Research and Engineering during the Eisenhower administration.

Foreword

It is with a sense of growing excitement and satisfaction that I have observed the germination and development of the ideas for a student version of the Pugwash Conferences on Science and World Affairs. The parent conferences—stemming directly, as they do, from the testamentary appeals of Albert Einstein and Bertrand Russell and supported by such great figures of the World War II generation as Leo Szilard, Eugene Rabinowitch, Frédéric Joliot-Curie, Max Born, Igor Tamm, and Alexander Topchief—have, from the beginning, been struggling with the problem of how to prevent the average age of their participants from increasing by one year with each passing year. As much as we have tried to involve younger scientists increasingly in the planning for and participation in the conferences, the very concept (involvement of scientists with the possibility of a more-or-less direct access to their governments) and the practice of defining the Pugwash movement in terms of the alumni of past conferences and symposia have tended to delay the inevitable takeover of the movement by the younger generation.

Nevertheless, those of us who have been entrusted with guiding the future of Pugwash have been seeking anxiously for a solution to this geriatric dilemma. Viewing the results of the Student Pugwash Conference on Science and Ethical Responsibility, as indicated by this stimulating collection of papers, I am immensely encouraged. Foremost among my reactions, in fact, is the regret that circumstances did not permit my personal participation.

The approach of the organizers was somewhat different from that found in the parent conferences. Rather than directly confronting the current set of techno-logically or scientifically based social dilemmas, and looking for "quick fixes," they have approached today's major science-society issues from the point of view of trying to understand the roles and responsibilities of scientists in contributing to our current predicaments. As a consequence, the discussion of possible solutions is from a rather longer-range perspective than usual, stressing much less the "quick fix" than an attack on the fundamentals. This is, of course, an example of the freshness of the approach of young, unfettered minds that has long been the hallmark of scientific innovation.

For their successful efforts in organizing this conference, as well as for the excellence of its output, the organizers—in particular Jeffrey Leifer—have earned our highest plaudits. This was an enterprise demanding a rare combination of idealistic vision and entrepreneurial acumen, and those responsible have carried it out with consummate skill and enthusiasm. We Pugwash veterans are feeling considerably more confident in the future of the movement—and, indeed, of the human race—now that we have been able to observe first-hand the competence and idealism of the representatives of the takeover generation. We look forward to their continuing and spreading participation in the Pugwash movement.

BERNARD T. FELD
Chairman, Executive Committee,
Council of the Pugwash Conferences
on Science and World Affairs

Preface

Assembled in this volume are the proceedings of a week-long conference held at the University of California, San Diego, in June, 1979. The volume includes a selection of the student papers submitted to the conference workshop, as well as contributions by senior participants. Some of the senior contributions (by Morin in Part I, Scoville, Van Cleave, and Stone in Part II, Kieffer in Part III, and Primack, Stone, Chalk, and Brown in Part IV) are revised versions of oral presentations. Since Dr. Salk's oral presentation drew upon a chapter in an earlier book, the chapter is reprinted here. All the other senior contributions were written expressly for publication in this volume. The select bibliography, which draws on references in the prepared papers as well as on other sources, was prepared by the Editors.

Warm appreciation is due Dorothea Thorburn for her most helpful copy-editing and to Lore Henlein, Director of the Addison-Wesley Advanced Book Program, as well as to her associates, for guiding this project into print with an adroit balance of patience and firmness. We are grateful to all the contributors for their cooperation during and after the conference. In the spirit of the conference, and especially of its primary aim of encouraging younger people to concern themselves directly with the social context of science and technology, this volume is very much the result of a collaborative effort between faculty and students.

SANFORD A. LAKOFF

SCIENCE AND ETHICAL RESPONSIBILITY

I. Science, Ethics, and the Aims of the Student Pugwash Conference

1.   The Origins and Objectives of Student Pugwash

Jeffrey R. Leifer

Introduction

The First Student Pugwash Conference, which convened in June 1979, had a three-year gestation. The original concept was simple enough, but getting from there to here involved a complex interplay of seemingly coincidental factors. Since this conference could be an important beginning, I should like to offer a brief description of the origins, objectives, and development of the Student Pugwash idea, along with some retrospective commentary.

As an undergraduate at the University of California, San Diego, where I was active in student government, I had come to believe that many university students have the dedication and intelligence to justify a larger role than they actually play in the continuing search for solutions to major national and world problems. They are a resource that is usually ignored except when candidates for office need campaign workers, or when there is some burning issue that leads to mass rallies on campus and demonstrations elsewhere. There is practically no effort, however, to engage students in long-term projects dealing with major social issues in such a way that they can develop thoughtful foundations for energetic efforts. One reason for this is that the educational system does a less than adequate job of sensitizing students to the major issues they will face as citizens and leaders.

It may seem pretentious to suggest that undergraduate and graduate students can help solve the problems of the world. There can be little argument, however, with the suggestion that well-designed forums, in which students can come to grips with real world problems and their proposed solutions, can provide useful preparation for future leadership.

Although my thinking about these issues had been developing for some time, the germ of the Student Pugwash idea arose from a confluence of two specific insights. Conversations with my father (who is a physicist) led me to recognize the

3

importance of workshop-type conferences as an educational tool in the professions. It seemed that the same approach could provide an even more important educational tool for students. Then, during my junior year, I became familiar with the international Pugwash conferences through discussions with Professors Herbert York and Roger Revelle. Suddenly, it became clear to me that one effective response, both to the student need for forums to confront major problems and to the failure of higher education to satisfy this need, could be found in the use of workshop-conferences modeled upon the international Pugwash conferences.

But what is international Pugwash, and why have so few students heard of it? The Pugwash movement began as a response to a deep concern over the threat of a nuclear holocaust. Bertrand Russell's "Pugwash Manifesto" in 1955 called upon the world—and particularly scientists—to accept the moral responsibility of seeking solutions to global problems.

It is sobering to recognize that the dominant threats to civilization today, particular the threat of nuclear wars, are the same as those grappled with at the first Pugwash Conference in 1957.* Yet the results of these conferences have shown that, despite Cold War tensions, scientists from the east and the west can reason together in an effort to resolve critical issues and that their work can be fruitful.

The subjects treated in early Pugwash conferences seemed most appropriate for physical scientists, although life scientists and social scientists participated and contributed even to the first meeting. With time it became apparent that other world problems would lend themselves well to the Pugwash format. Over the years, the focus of Pugwash conferences has expanded to include problems of world food and energy distribution, the social impact of technology, and other "North-South" concerns.

Although international Pugwash has helped to create conditions for concrete achievements, such as the conclusion of the Partial Test Ban Treaty, its role has been so much in the background that too often it goes unrecognized. Its low profile is in part the result of an early, conscious decision to avoid becoming a propaganda platform. Instead of seeking publicity, Pugwash members have sought to exert influence individually by carrying policy messages back to their respective governments. This unusual approach has probably increased their effectiveness to a level beyond that which would have been achieved by better-publicized conferences.

Although international Pugwash provided a stimulus and an ideal format, Student Pugwash was never intended to be simply a youthful replica. In addressing world problems, it would be unrealistic to suppose that a week of open give-and-take among a group of bright students could approximate two and a half decades of continuous effort by some of the world's leading minds

---

*The first conference was organized under the direction of Dr. Joseph Rotblat of Great Britain, an internationally respected scholar who continues to play a major role in the conferences.

and personalities. Student Pugwash is also distinguished by different objectives. International Pugwash seeks to formulate policy by consensus and to influence governments to cooperate in problem-solving. Student Pugwash seeks to provide information about those problems, to stimulate interest in them, and to foster dialogue about them. It expectations are twofold: (1) that participants will become dedicated to solving major problems, and (2) that their efforts toward this goal will have an impact beyond that of the student conferences themselves. In spite of this difference in focus, however, the subject matter, the format, and the mixture of regional, national, and global contexts are all derived from international Pugwash, and the long-term goals of both are identical.

I leave it to the participants and the papers* to demonstrate the degree to which these goals have been met by the first Student Pugwash Conference.

The Path to the Conference

Although the concept of a workshop-style conference based on the international Pugwash agenda came to mind early, establishing a format that would convince a sponsor to support the idea and then actually attracting sponsorship was a two-year process.

My own university community proved very supportive. A second important source of support came from international Pugwash itself. With help from Professors York and Revelle at UCSD, Professor Bernard T. Feld at the Massachusetts Institute of Technology, and Dr. Martin M. Kaplan, International Pugwash Director-General, I was able to attend Pugwash conferences in Munich, West Germany (1977), and Varna, Bulgaria (1978). The Munich Conference offered inspiration and information. At Varna, with enthusiastic support from the United States delegation, I received formal endorsement of the concept of a Student Pugwash Conference.

During the crucial search for financial support, I found that my own program objectives meshed with those of the National Science Foundation's Ethics and Values in Science and Technology Program, as well as with a similar program of the National Endowment for the Humanities. When Professor Sanford Lakoff of UCSD submitted a proposal outlining such a conference, both organizations decided that the concept had merit and agreed to cosponsor it. Their support has been flexible, encouraging, and nonpolitical, for which I am most grateful.

With the help of a faculty advisory committee, Professor Lakoff and I proceeded to develop the conference format. We recognized that the conference would succeed or fail on the basis of its structure and on the quality of the participants, both senior and junior. Several structural considerations emerged as vital during the search for an effective format:

---

*This volume represents only a selection from the more than 100 papers presented during the conference.

1. It was important that the subject matter be focused on major national and international problems of the day. Five topics were selected on the basis of their importance, interdisciplinary character, and applicability to the conference objectives: arms control and disarmament; science and values; scientists and politics; technology and development; and the implications of biomedical research.

2. A theme was needed to provide a bond common to all topics. Such a theme was the ethical implications of each problem area. This theme was to serve as the link between subjects as well as a way of treating the particular aspects of each topic.

3. The conference had to have a multidisciplinary scope in recognition of the interdisciplinary character of the problems. I am told that interdisciplinary approaches are ordinarily more successful in grant applications than in execution. Candor compels me to admit that this conference generated less interdisciplinary exchange than had been hoped for, but it did succeed modestly and established that an interdisciplinary approach is feasible.

4. The conference format was to be a mixture of plenary and workshop sessions. As an unexpected benefit, the plenary sessions, which were open to the public, were attended by many people in the La Jolla area.*

5. Each student participant was to present to one of the workshops an original paper on a theme addressing the workshop subject. Student participants were selected on the basis of background, interest, experience, and submission of a detailed outline of the paper to be presented. Although the conference announcements went out much later than planned, the response rate was good and the caliber of the applicants generally high. The selection committee reviewed every individual application and outline in detail, most of them more than once. We were delighted with our choices but probably would have been equally pleased with most of those whom we could not accommodate.

6. A large staff of senior experts in the many disciplines served as "tutors" and advisors. They were selected not only for their expertise but also for their dedication to the objectives of the conference. The contribution of these distinguished senior participants played an invaluable role in the success of the conference.

7. All the students and many of the senior participants were to live and eat together. This component of the program was very successful. Every meal and every evening was a workshop extension, even the social gatherings. Discussions frequently continued late into the night.

8. Finally, a working staff was needed to guarantee that the conference would proceed smoothly. We opted for an all-student working staff—only partially as an economy measure. The conference worked as well as it did because the staff understood and believed in its goals. Most of the participants volunteered compliments about the conference organization.

---

*Evening classes were offered during the week prior to the conference to prepare the audience, and afterward for the purpose of encouraging reflection.

The Conference

Judging from the favorable comments of the more than 100 student and senior participants during and after the conference, our major objectives were achieved. The daily morning plenaries were effective because the senior participants generally made thoughtful and provocative presentations. A highlight of the conference was the appearance of California's Governor Edmund G. Brown Jr., to make a major speech on arms control. In general, the plenaries were noteworthy for their provocative and wide-ranging exchanges. Question periods provided a few surprises and a good deal of intellectual exercise and further debate.

It was in the workshops, however, that the conference proved most valuable. The students were enthusiastic, interested, and usually well prepared; the senior convenors and commentators were patient and enthusiastic but not uncritical. The seniors tried to channel and focus student enthusiasm to maintain continuity, and to expose all sides of the issues. The students gained a great deal from the process. They learned that, although some world problems can be solved, others simply become compounded at a rate that calls for increased effort, concentration, imagination, and determination. Hours were dedicated to summarizing issues, listening to responses, and learning methods of analysis. Minds were opened and views were changed. It is hard to describe the energy, the sense of purpose, and the magic of enlightenment that engulfed the students. I doubt that any of us reached the end of the conference unmoved or unchanged intellectually and emotionally. I am sure that the participants will not forget what they learned, nor will they permit their new understanding to lie fallow.

Observations

It may be that problems do not change with time, only proposed solutions do; but that does not mean that problems are never successfully addressed. People and institutions are developing the capability to do good, and harm, at an ever-increasing pace. It is not surprising that the complexity and consequences of world problems are also growing rapidly. The inference to be drawn is that the dedication and effort required to deal with them must also grow. The problems often require fresh thinking by new generations of leaders. Conferences like this could help to nurture such leadership.

Plans are under way for a second Student Pugwash Conference in mid-1981, which, it is hoped, will stimulate the same enthusiastic support given this one from so many different quarters.

Future Considerations

Some have hinted to me that they consider Student Pugwash an "elitist" activity. I presume that this comment refers to the commitment of substantial resources to the education of a select few. I would reply that it is impractical to achieve the objectives of the conference by any other means. True, the money spent on the conference could, for example, have supported a televised lecture, reaching thousands of people. The same alternative, on an even grander scale, would supplement all higher education. Would the results be as good or better? I doubt it.

The Student Pugwash Conference brought together a group preselected for their interest and ability, and gave them an opportunity to work together in a short but intensive experience. The students had a no-holds-barred opportunity to try out their ideas in the presence of senior specialists in their areas of interest. They had a chance to learn how to think more clearly about very difficult problems, in the only way that such methodologies can be learned. They probably left with more information but also with a better feel for the relevant modes of inquiry and with a renewed enthusiasm for entering the fray. Solutions to problems are not developed in crowds or by isolated individuals watching television programs, as informative as they can be.

Another concern some have raised is that the conference participants were too homogeneous a group—in particular that ethnic minorities were not properly represented. The selection process itself was certainly fair and open. Participants included postdoctoral, graduate, and undergraduate students from forty-five universities and fifteen disciplines. Nevertheless, the conference would have been enriched by better minority representation. Future preparations should include plans for a more vigorous effort to solicit minority applicants. Because approval to go ahead with the first conference was so long delayed, the announcements were not sent out as early as planned. With more lead time, it should be easier to attract a more diverse group of applicants. The conference would also benefit from an effort to broaden its base by increasing the variety of disciplines represented.

Should Student Pugwash be institutionalized? For the time being, it would probably be best to delay a decision. Institutionalizing Student Pugwash suggests creation of a bureaucracy, with all the panoply of an administration, board of directors, secretariat, etc. That troubles me. A student organization is by its nature hard to institutionalize. Being a student is not, normally, a life's work, and therefore all student organizations are plagued by a lack of continuity. By design, student activities are influenced by the inherently transient nature of their leadership. In some organizations, lack of continuity can be a blessing; in the case of Student Pugwash, I fear it would be a disaster, a death knell. These negative comments hardly answer the question, but they may set the context within which we must search more carefully for a long-term solution. For myself, although I am uncertain about seeing Student Pugwash institutionalized, I am strongly in favor of its becoming an institution, with exceptional leaders, participants, and staff creating a context that succeeds in its objectives each time a conference is convened.

The question of whether Student Pugwash should become international has also arisen. My dream from the very beginning has been to include participants from many countries. Giving the conference an international character would contribute dramatically to achieving its objectives. In the first conference, the burden of presenting the perception of world problems as seen from a foreign cultural and political perspective had to be borne by the few participants who were foreign students in the United States, and, of course, by the senior experts. Although the foreign students performed their task well, my conversations with them suggest that a working relationship among a mix of students from throughout the world would have been even more effective. It was impractical to make more than a minimal effort to give this conference an international character, and I feel that we should proceed very slowly to interest individual governments and institutions in the idea. At the last International Pugwash Conference in Mexico City (1979) I offered a resolution (an idea conceived in conversation with Jeremy Stone) to the full assembly of delegates, calling upon all participating nations to sponsor a "student observer" to the next annual international conference as a start toward the founding of an international student conference. The idea was well received, and the Director-General has indicated that at least some countries will respond to the proposal for the 1980 conference. Now, while we wait, we should continue to seek support and expansion of the participation. With luck, it may happen.

Expanding Activities

In the aftermath of this first Student Pugwash Conference, the San Diego Public Broadcasting station (KPBS) produced a 90-minute film of the SALT II debate/plenary. This has been aired in the three largest cities in California, has been seen in four other states, and has been requested by twenty-five other stations. A second film produced by a community video organization has been transmitted via satellite to community stations nationwide. Ten major newspapers (including *The New York Times*, the *Los Angeles Times*, and the *Washington Post*) covered the conference, and some of the student papers presented at the conference have appeared in newspapers throughout the nation.

The conference has already been a source of inspiration and material for academic programs—for example, an extension course at UCSD* that explored the moral content, contradictions, and impact of world problems. Colloquia based on this same model have been held at Stanford University, the University of Minnesota, and the University of California, Berkeley. We are now considering the possibility of melding curricula, based on Student Pugwash concepts, into interdisciplinary study programs at universities across the country. We are, of course, also hopeful that this volume of papers will prove useful in such courses at universities, colleges, and high schools.

At the international level, one of our "alumni," Fraser Homer-Dixon, has

---

*In teaching this class, Eric Markusen and I used the conference material as model and content.

taken on the ambitious task of organizing a Canadian Student Pugwash in conformity with our objectives. Other foreign alumni are exploring similar goals in England, Italy, and the Federal Republic of Germany.

A new forum for participation by American students will be opening soon as a result of the conference in the form of a new section of *The Bulletin of the Atomic Scientists*, which will be open to the views of students on today's major problems.

The Role of the Conference Delegates

The student delegates to this first conference have become keenly aware that the great problems facing modern, technologically advanced societies can be held to manageable proportions only by concentrated and thoughtful effort. As they move into responsible positions, they cannot forget that these problems have a claim on their time and energy—that there is hope, but that work is required. They must show by word and deed that these problems can be made better known and can be managed, if not completely resolved. Since the problems we face are modern versions of age-old difficulties, it is as though our generation has been passed a set of historical and intellectual imperatives. Leo Szilard said it well: "Let your acts be directed toward a worthy goal, but do not ask if they will reach it; they are to be models and examples, not means to an end."

Acknowledgements

My three-year effort to bring the concept of a Student Pugwash Conference to fruition would have been impossible without a great deal of help. The evolution of this conference was a collaborative effort, and the list of individuals to whom I am indebted has become so long that I cannot, as a practical matter, list them all. I ask the many I must omit to accept my sincere thanks and think fondly of their contribution to the success of the conference.

Some of those who were helpful deserve particular recognition, however. I would begin by expressing my great respect for Cyrus Eaton, who died recently at the age of 95. An industrialist with great insight and moral fiber, Mr. Eaton recognized the merit of international dialogue dedicated to seeking solutions to world problems. It was he who provided support for the first conference and the site, Pugwash, Nova Scotia, his birthplace, from which the conference takes its name. It was an honor to dedicate this first Student Pugwash Conference to his memory.

I am also especially grateful to Professors Roger Revelle and Herbert York, and to Professor Bernard Feld and Dr. Martin Kaplan, all of whom responded with enthusiasm and offered immediate support when the idea for the conference was first brought to their attention.

Several UCSD administrators were invaluable in making the project feasible, notably Chancellor William McElroy and his staff, especially David Ryer,

Bernard Sisco, David Ernst, and Kenneth Bowers. Two members of the UCSD faculty, Professors Walter Kohn and Clifford Grobstein, took pains to criticize and refine many of the details of the conference. Professor Ronald Berman was also helpful in practical and intellectual matters.

Everyone over the age of 35 was surprised at the student staff's efficiency in running the conference. The secret of their success is very simple: The staff was highly motivated. They performed as competently and creatively as any professional group could ever hope to. Their names are listed elsewhere, and each and every one deserves warm appreciation for outstanding effort. A special note of thanks should go to the staff of the Department of Political Science, who dealt with the incessant stream of detail, kept us honest and solvent, and put up with my often peculiar office behavior.

I am grateful to the National Science Foundation, and especially to Alex Morin for his helpful suggestions, and to Bill Blanpied for his unwavering support and guidance. Thanks also are due Anne Eaton and Stanley Sheinbaum for their generosity and encouragement.

To Professor Sanford Lakoff I am indebted in a very special way. Without his help, none of my hopes would have been realized. Even though he may edit out most of what I would like to say, I must somehow thank him for the special relationship we developed and continue to share, for it was under his tutelage that I grew to understand the meaning of discipline.

My heartfelt thanks go to Lisa Endlich and to my family, particularly my mother, for their sensitivity, understanding, and indispensable assistance during so long a time; I needed their support, and they never failed to provide it. I owe more than I could ever record here to my father, Herbert, not only for his direct contributions to the project, but for his perceptions, spirit, and commitment to principle. He deserves to be considered the father of Student Pugwash.

2. Encouraging Scientific Responsibility*

Alexander Morin

The sponsorship of this conference by the National Science Foundation reflects a growing recognition within the Foundation, and more significantly within the scientific community as a whole, of the importance of addressing the ethical responsibilities of scientists. This particular meeting recognizes the importance of speaking and listening to young scientists and engineers—those who represent the changing perceptions of the extent and nature of such responsibilities, and in whose hands lies the possibility of dealing with these problems more effectively in the future.

What are the special responsibilities of scientists? We are good citizens, and like other good citizens we acknowledge that our behavior should be guided by ethical and moral codes. We participate—or should participate— in the making of decisions that affect our destinies and those of our society. We try to see that our institutions are responsive to the needs of the times. We do all the things that the civics textbooks teach us to do. But what is so special about science in these matters? What distinctive responsibilities do we have, if any?

One way to consider this question is to look at the de facto answers that have been given, in the form of the programs for which I am responsible at NSF. These programs reflect the changing attitudes toward the nature of scientific responsibility and the relationship of the scientific community to the general public over the past fifteen years or so. The Office of Science and Society, which I head, is a quasi-accidental agglomeration of programs that developed at different times and were designed to meet different needs. Historically, and by their nature, they tell us something about the problems we are dealing with.

The oldest of the programs is the Public Understanding of Science Program, which was originally conceived essentially as a sophisticated public relations effort for science. Through this program the scientists were to tell the public what

*The opinions expressed in these comments are those of the author and do not necessarily reflect the view of the National Science Foundation.

13

science is and what it does, with two motives: One was that it is good for people to know these things; people should understand the scientific bases of the world in which they live. The other motive was somewhat more self-serving: to sell science and to get greater support for scientific research from Congress and from the White House. Now these are not bad motives, if you believe in science. If you are persuaded (as I am) that the scientific enterprise is indeed very important, then it becomes necessary to inform the public of its content.

This approach is, however, a limited way of defining scientific responsibility. Over the years the program has changed, along with the changes that have occurred in the larger society. For example, administratively the program was separated from the public relations effort of the NSF and made an independent program, with independent peer review systems and advisory committees. It was given the new and more credible mandate of conveying an objective, balanced perspective on scientific development and its consequences—including the limitations of scientific knowledge, and the adverse consequences of techno-logical applications as well as the more positive ones. We require now of these activities that they not be "sales jobs," that they address policy issues, and that they cover the full range of values and points of view about these issues. There has been a substantial shift in our perception of what is responsible communication of science to the public.

A second program, started six or seven years ago, is directly responsible for supporting this conference: the Ethics and Values in Science and Technology Program. This program was initiated by a group of biologists within and outside the NSF who, like the atomic physicists in the late 1940s and early 1950s, were concerned about the moral implications of their own actions and those of their employers. The biologists considered two kind of issues that struck them as important and that needed study. One set of issues related to the public health impact of their work—viral research, toxic substance research, and DNA research (at a somewhat later date). The other was a set of moral issues evolving from their concern about meddling with nature. These scientists were deeply disturbed by the disruption of ecological chains, for example. These two sets of issues merged with a third set, involving the problems arising from the impact of science on personal rights and autonomy, such as the issues that arose because of the development of computer technology and research on human subjects.

The Ethics and Values Program is primarily intrascientific. It does not venture much outside the scientific community, but within the scientific community it responds to a sense of the possible dangers that are implicit in the research enterprise and its applications. It asks what can be understood about them, how they can be dealt with, how the political issues arise, and how they can be handled in a way that meets social needs. This program thus deals with a different level of concern about the relationship of science and society and a more sophisticated and contemporary view of the issues involved.

The third program—the most recent and most sensitive one from a political standpoint—is called Science for Citizens. This program explores yet another dimension of scientific responsibility, because it addresses problems of equity. It reflects the general spread of a popular mistrust of authority in our society, which affects science as well as all other institutions and forms of expertise. It rests on the assumption that scientific and technical knowledge is a scarce resource. Some groups of people have much more of it than others—notably, big business and industry, governmental institutions, and other large enterprises. When it comes to determining policies, these institutions, because of their access to scientific and technological expertise, have enormous advantages over the more diffuse interests of other groups in our society, whether they are environmentalists or taxpayers or minorities or poor people. The Science for Citizens Program was created to help redistribute this scarce resource, to see to it that the groups that lack access to scientific and technical expertise are able to obtain more of it. Obviously, there are profound ethical implications in this attitude. To what extent is it the responsibility of the scientific community to assure that its expertise is widely available to disadvantaged groups? How is this responsibility to be defined, and what is the proper role of government in meeting it? A host of complicated problems is involved here.

All these programs, and the problems lying behind them, pose ethical issues that reflect and affect the way in which science is perceived, both within the scientific community and by the public outside that community. The resulting issues can be expressed at various levels. One level relates to questions of personal conduct. We have always had a conventional scientific morality, enjoining the scientist to be objective, to base conclusions on evidence, to avoid plagiarism or falsification of data, to use citations properly. An enormous amount of institutional and organizational support is found in the universities and in the professional associations for this kind of morality—for ethics concerned with the appropriate relationship between the scientist and his or her peers and colleagues.

A second level of issues of conduct that has been evidenced in the programs of the NSF and also in the Pugwash meetings involves the question of personal responsibility for the social impact of scientific discoveries and applications. Some scientists refuse as a matter of principle to work on projects that they believe may be bad for people. They insist that they will use their skills only in ways that will help to achieve peace or other desirable social goals. Some social scientists refuse to work for the CIA, on the ground that this agency does not contribute to human welfare. These are decisions that individuals must make for themselves, based on their judgment. For this kind of decision there are no institutional codes, and no support within the scientific community as such. Conscientious objection is recognized as legitimate in our society and in the scientific community too, but these are personal decisions, not institutional or social ones.

And finally, there is a third level of personal conduct, sometimes called whistle blowing. This refers to people (most obviously the scientific technicians) who are employed to accomplish particular ends and who discover that the results of their efforts will have adverse consequences; they therefore "blow the whistle." Such ethical issues are beginning to become salient and to be recognized institutionally in codes of ethics, in the activities of professional associations, and to some degree even in legislation.

What should interest us most is the extent to which the scientific community, as it is organized in departments within universities, in associations and societies, and through its journals, has begun to develop appropriate institutions to deal more effectively with these ethical issues. This is a different matter from our concern with public understanding of science. We may talk about the informed consent of subjects for experimentation; what about the informed consent of the public to the consequences of our activities? Have we some special responsibility to provide an estimate of the results of our research and development activities and to secure general informed consent? To what extent should it become an institutional responsibility of science to invite greater public participation in decisions to undertake research affecting human welfare? To what extent should scientific institutions encourage and support the active participation of scientists and engineers in the political arena, not merely in the defense of science but in the defense of broader social objectives as well? There is little institutional support for this kind of activity. The reward system does not encourage it. You do not win a Nobel Prize, you do not get promoted, you do not establish a good publication record, and you do not receive the esteem of your scientific peers by engaging in social or political activities, even when they are closely coupled to your special field of research.

If these attitudes are to be changed, it will require a substantial shift in the balance of support within the scientific community for such expressions of professional responsibility. And if such a shift occurs, it will be the result of the activities of younger scientists determined to bring it about. But let me add a word or two of caution.

When we question the special responsibilities of scientists and engineers, two common but untrustworthy answers are given. One is the assumption often made by people trained as we are that there are rational solutions to everything—that you can mathematize ethics, that you can provide environments and institutions that will do away with conflicts in political and social values. We have seen a widespread public rejection of decision-making processes that rely only on technical expertise. Proposals to establish a science court, for example, are designed to depoliticize issues and to treat decisions that must be made about important issues such as SALT, and control of freedom of research, as though they were mainly technical matters to be dealt with by experts. We should be wary of such proposals. They are undemocratic and they do not work.

A second assumption of scientists is that, because we have a range of expertise on some highly technical and important subjects, our authority extends well beyond that range. An example is the Nobel Prize winner who is an expert in his own subject but who thinks that the award permits him to speak with great authority on every subject. One of the difficulties we face when dealing with issues of ethical responsibility is that everybody considers himself an expert. In politics and in ethics, there is no such thing as an admitted non-expert. The question is, What special authority do we scientists have as experts in these matters? I should like to point out that these questions have created a substantial body of thought and study that should not be ignored. Discussions of the appropriate relationship of the citizen to the state, and of the citizen to his peers, have taken place over hundreds of years. Philosophers have devoted considerable time to these matters. In spite of the weaknesses of contemporary social sciences, social scientists have also learned a good deal about organizations and institutions. These are not bodies of trivial data, and they must not be ignored if we are to discuss such questions seriously.

The importance of this conference hinges on these issues. It has been undertaken to introduce them to the young people who are coming into science, as early as possible; to introduce them to the best thinking of those who have already considered these matters; and to encourage fresh exploration of the various issues as they affect the lives and conduct of the younger generation and, more important, the conduct of the scientific enterprise as a whole.

3.　Ethical Responsibility and the Scientific Vocation

Sanford A. Lakoff

In 1918, at a time of great skepticism and demoralization, the German sociologist Max Weber issued a plea to students on behalf of the scientific vocation.[1] He did not promise that it would bring happiness, ethical guidance, or even the consolations of philosophy. On the contrary, he pointed out that the essential difference between the modern and the ancient pursuit of knowledge was that in the form of modern science the pursuit of truth had become detached from the quest for the meaning of life. When Plato had urged the youth of Athens to abandon the illusions of the cave for the sunlight of science, knowledge and values had been considered inseparable. By attaining a philosophic knowledge of the true forms, it was thought possible to understand not only the workings of the natural universe but also the meaning of the beautiful and the good; to learn "how to act rightly in life, and above all, how to act as a citizen of the state." For Plato and the ancient Greeks it was "for these reasons one engaged in science."[2]

The modern scientist, Weber suggested, must be prepared to accept a much more limited intellectual opportunity, because "the fate of our times is characterized by rationalization and intellectualization and, above all, by the 'disenchantment of the world.' Precisely the ultimate and most sublime values have retreated from public life either into the transcendental realm of mystic life or into the brotherliness of direct and personal human relations."[3] Who today still believes, he asks rhetorically, "that the findings of astronomy, biology, physics, or chemistry could teach us anything about the *meaning* of the world?"[4] Whatever

---

[1] "Science as a Vocation" (1918), *From Max Weber: Essays in Sociology*, translated and edited by H.H. Gerth and C. Wright Mills, London, Kegan Paul, Trench, Trubner & Co., 1947.

[2] *Ibid.*, p. 141.

[3] *Ibid.*, p. 155.

[4] *Ibid.*, p. 142.

unifying visions may have animated the earlier pioneers of scientific discovery, the commanding fact about the state of the scientific vocation now is that it is "organized in the service of self-clarification and knowledge of interrelated facts. It is not the gift of grace of seers and prophets dispensing sacred values and revelations, nor does it partake of the contemplation of sages and philosophers about the true meaning of the universe."[5]

If, while recognizing the divorce of scientific inquiry from the quest for the meaning of life, this acute analyst of modern social history nevertheless urged the young of his time to pursue careers in science, it was for a reason he left largely unspoken. Implicit in Weber's own commitment to the scientific vocation was the belief that, although it could not promise salvation (for those who "cannot bear the fate of the times like a man," he observed, "the arms of the old churches are opened widely and compassionately..."[6]), it could satisfy the profound modern craving for consciousness, for freedom from illusion. Insofar as Weber's commitment to the scientific vocation can be called an ethical commitment, it was very much the same sort of belief that Philip Rieff has characterized, in the case of Freud, as the "ethic of awareness."[7]

Such an ethic, however, offers only a limited guidance with respect to the responsibilities that may be inherent in or especially associated with the scientific vocation. At most, it suggests that scientists should be concerned about threats to their freedom of inquiry, and perhaps by extension to all constraints upon freedom of thought and expression. It does not indicate at all whether and in what respects scientists have an obligation to concern themselves with the uses to which their discoveries are put. It is this issue, however, more than any other, that today must be in the foreground of any discussion of the ethic of science.

It is understandable that Max Weber should not have been as troubled about it as we are now. In his time, science was still an activity largely pursued in isolation from the concerns of everyday life and even more from affairs of state. Even then, it is true, developments were in progress that were to end this isolation forever. The most dramatic example was the growing importance of science in warfare. The role of improved explosives and the introduction of poison gas in the First World War has led some historians, perhaps exaggeratedly, to refer to it as "the chemists' war." Few of the scientists who participated in this war, however, are known to have given much thought to their role in providing new and improved means of destruction, possibly because their contributions were not decisive, and any misgivings they might have expressed would not have been taken very seriously. Some, whose inventions had military application, like Alfred Nobel, overcame any pangs of conscience by adopting the naive view that inventions that

[5]*Ibid.*, p. 152.

[6]*Ibid.*, p. 155.

[7]Philip Rieff, *Freud: The Mind of the Moralist*, New York, Viking Press, 1959.

made warfare more destructive would contribute to peace by making conflict unthinkable.[8]

Developments during and since the Second World War have completed the virtually total transformation of warfare from its beginnings in hand-to-hand combat to the present stage, where in its most potentially destructive forms it consists largely in the remote manipulation of sophisticated technology of all sorts. The modern definition of a great superpower is not, as it once was, a state capable of fielding and equipping great masses of disciplined and armored foot-soldiers and sailors. Now, what is crucial is the capacity to deploy over great distances such numbers and tonnages of nuclear weapons, in such a variety of forms, as to deter any potential opponent and—according to a diminishing number of analysts—as to prevail over an opponent in an actual conflict. Plainly, warfare today is very much a matter of advanced science and technology, and not only of the more conventional factors of manpower, economic capacity, and morale.

For this reason, it has been as contributors to national war efforts that scientists have had their initial experience with the problems of ethical responsibility. Fear that the atomic bomb would be acquired first by Nazi Germany moved the nuclear physicists on the Allies' side to impose self-censorship on their publications and to urge the British, Canadian, and American governments to organize a major effort to develop the weapon first. Later, when the Germans were all but defeated, fear that the use of the bomb against Japan would set a terrible precedent prompted some of the same scientists vainly to petition the American government to avoid introducing the bomb as a military weapon. After the war, this early experience led, especially in the United States, to a wider "scientists movement" involving scientists in all disciplines, dedicated to warning the public of the dangers of nuclear weapons and to working toward the objective of achieving disarmament and structures of world peace. The world-wide Pugwash movement, created in response to a call from Albert Einstein and Bertrand Russell, is one manifestation of this campaign.

More recently, other applications of knowledge have drawn scientists and technologists into a concern for the impact of their work. Advances in biology, chemistry, and medical technology have had extraordinary effects upon preventive, diagnostic, and therapeutic medicine. Food production, which has benefited for over a century from improvements in agricultural technology, now depends not only on chemical fertilizers and pesticides, but increasingly on the application of basic advances in the understanding of biological and chemical processes. The development of alternative forms of energy, to replace rapidly diminishing and finite supplies of fossil fuels, also depends on the work of basic

[8] See R.W. Reid, *Tongues of Conscience: Weapons Research and the Scientists' Dilemma*, New York, Walker and Company, 1969, and the discussion of this and other accounts in S. A. Lakoff, "Science and Conscience," *International Journal*, 24, No. 4, Autumn, 1970, pp. 754–765.

and experimental science. Research into the process of human conception has already resulted in such significant applications as the birth-control pill, and future research will undoubtedly have even more radical implications, not only for the possibility of regulating population increase, but for unprecedented intervention into the genetic process of reproduction.

In all these instances, scientific work raises problems for which the Weberian conception of the scientific vocation has no answers. The very advances in medicine that result in such welcome benefits as the conquest of disease and the prolongation of the life span also raise agonizing problems of medical ethics, when choices must be made in allocating scarce medical resources, or when a decision must be made whether to sustain other vital functions after brain death has occurred. The use of chemicals as fertilizers and pesticides, although obviously beneficial in improving yields, has proved ecologically harmful. The effort to use atomic energy has led to a social dilemma over whether the safety hazards and the associated danger of nuclear proliferation are acceptable risks. Research into the nature of human reproduction, and more fundamentally into genetic replication, has already forced consideration of the extent to which such intervention is consistent with traditional notions of human dignity and with reasonable expectations of the human capacity to manage wisely such unprecedented powers of genetic control. More such problems are certain to arise in the future.

As their work touches these problems, scientists will inevitably be faced with the need to define their ethical responsibilities. If scientific work could somehow clearly reveal the meaning of life, it might be possible to infer this responsibility in readily applicable and unambiguous terms. For the reasons Max Weber noted, however, that is most unlikely. Instead, it will be necessary to define the nature and limits of this responsibility by a process of reflection, which can well begin in a recognition of the growing integration of science and society.

Science and Society: The Implications of Their Integration

Ethical responsibility is always and inescapably a problem for the individual; but the starting point for any examination of the meaning of the ethical responsibility of scientists must be the recognition that especially under modern conditions science is very much a social enterprise. The individual scientist is part of a larger community of researchers, and also of one or more sets of sub-communities. These communities could not function without the institutional infrastructure that sustains them. The "whole" of science, moreover, is in a sense more than the sum of its parts: Information produced by any one scientist or group of scientists, once it is made public, cannot usually be contained or monopolized for very long. It enters into the stream of knowledge and is available to all who are equipped to understand and apply it—and to extend it, for almost always it is one more link in an endless chain of understanding, whose full and final implications cannot easily be predicted in advance. Like the alienated proletarian depicted by Marx, the basic researcher usually finds that the products of his labor do not remain his own property. Even less than the proletarian, however, is the scientist apt to find a remedy in an attempt to gain power and establish the rule of his own class. Technocracy is no solution, because the would-be technocrats would have neither a common interest nor a common conception of what should be done with their knowledge. In view of the actual social nature of the scientific enterprise, the ethical responsibility of the scientist must involve him in concerted action with others—other scientists, other citizens, other human beings throughout the world.

The social nature of ethical responsibility is in a way the obverse of a practical dependence of science on society. Scientific activity now requires substantial public subvention. "Big science" can be accomplished only with elaborate apparatus, for which public patronage is essential. To some extent, this support is provided on the ground that science and higher education are worthwhile as ends in themselves and deserve support on the same basis as other cultural activities. The generally large scale of the support—larger in any event than the support provided the arts and humane letters—indicates that society expects tangible benefits from science, and that it is this expectation that accounts for the difference in level of support.

Sometimes, the political decision to support certain areas of research and application more lavishly than others will determine the direction of scientific activity, regardless of the priorities the scientists themselves might wish to see established. The United States space program of the 1960s has been criticized for distorting the priorities that would have been set by the scientific community. The same criticism has been made of programs aimed at inducing research with military applications. The system of support enables the political authorities to influence the course of scientific discovery, and it encourages scientists to become promoters of their own research, even if it means putting their economic and professional self-interest before their social responsibility. If the creation of the "scientific establishment" had indeed insulated researchers from the swings of political judgment and fancies, there would be no ground for misgivings. In fact, however, it is in the nature of the partnership that political direction and practical self-interest will lead to the emphasis on certain types of effort and the relative neglect of others. The decision to embark on a "war against cancer," before the scientific grounds for a campaign to eliminate the disease were well established, would probably not have been made had the decision been left up to the *general* scientific community, rather than to the combined force of the public representatives in the legislature and the subcommunities of committed researchers whose work would benefit from special subvention.

Because of the integration of science and society, it becomes almost axiomatic that research should serve social objectives. In this respect, it is true that an important distinction remains between basic and applied research. Whereas basic researchers must, to a considerable extent, determine their own agenda, applied research may be more readily influenced to flow into certain socially desirable channels. The basic researcher may properly consider it no particular business of his what other researchers choose to do in applying new knowledge, but even the basic researcher remains a part of a social system for the acquisition and application of knowledge. The support of his work is predicated on the belief that ultimately it may become useful, and his willingness to take part in the enterprise implies at least that he is aware that practical consequences may flow from the application of his work.

This is not to suggest that any one researcher can or should bear the full burden of ethical responsibility. Precisely because the scientific enterprise is a social enterprise, that responsibility is inevitably diffuse. The great danger, in fact, to the possibility of an operational sense of responsibility is that each individual may plausibly claim that his personal share is so small as to be of infinitesimal weight and of little or no practical consequence. There are critical points, however, when major decisions must be made, and it is at such points that a denial of all responsibility is unwarranted. The decisions sometimes involve the risks posed by research or experimentation; more often they arise out of a recognition of the potentially dangerous or unsettling consequences that can reasonably be expected to flow from a line of work. Those involved in this work can, especially if they act in concert, significantly affect the pace and direction of the work, and the social appreciation of its implications.

Knowledge and Its Applications

The main reason ethical responsibility attaches to the scientific vocation is that the knowledge provided by research often has important implications and applications. Scientists and technologists therefore come to stand with respect to the general public to some extent in a relation analogous to that of the physician toward his patients. The physician's moral responsibility is better recognized, better defined, and more institutionalized. One who violates the prescribed or implicit standards is accountable to his professional peers, who may bar him from the considerable benefits and privileges they control, and he may be held legally responsible by those who suffer the effects of malpractice. In recent years, scientific and engineering associations have also drawn up codes of ethical conduct, but these tend to be more vague and less meaningful than medical codes. It is not necessarily clear, for example, whether an engineer's responsibility to the public necessarily supersedes his responsibilities to his client, in case of a conflict, or under what circumstances he may have a positive obligation to become a "whistle blower," even if it adversely affects his client's interests. Since scientists and engineers do not depend as much as physicians do on the certification of their professional peers, the sanctions attached to such codes are likely to be considerably less effective. Without a law protecting the dissenter from reprisal, he is apt to put himself in a vulnerable position if he takes his obligation seriously, and his professional association, however sympathetic it may be, cannot be of practical help to him.

The analogy between medical and scientific ethics also breaks down on other scores. Basic research is aimed simply at the discovery of scientific truth, not at therapeutic intervention. The consequences of the application of what is discovered come at the end of a process linking the work of discovery and application, but the links are very tenuous; no one researcher can easily be held responsible for the outcome, since at most he contributes one small piece of information, which is combined with other information and then made use of in circumstances about which he may have no foreknowledge and over which he can have no control. Nor is there likely to be a single, clearly identifiable victim of scientific "malpractice." The harmful consequences of certain applications are sometimes felt only as statistical modifications to dangers affecting large populations. If "the public" is the patient, so to speak, protection is more likely to be achieved by the regulation of experimentation and the adoption of practical applications than by holding scientists, especially basic researchers, responsible.

This line of analysis has been challenged by those who contend that, because modern scientists must anticipate that their discoveries will probably find some practical application, they can no longer take refuge in the "old alibi" that they are not responsible for what is done with their findings.[9] One trouble with this objection is that it entails a wholly untenable conclusion. Since all new knowledge may conceivably be put to harmful use, and since it is impossible to foretell what these uses will be, the only logical conclusion a conscientious basic researcher could reach from this premise would be to refrain from all scientific work. If he were to refrain, however, he would be just as liable to the accusation that, by failing to use his talents to discover new truths about nature, he shows himself to be indifferent to the needs and sufferings of humanity, since his discoveries are also likely to be used for human benefit.

The fact is that the applications of knowledge often have both beneficial and potentially dangerous consequences. The laser has been used as a surgical device and as the guidance system for the "smart bomb." Atomic energy provides a remarkably efficient source of energy and at the same time poses safety hazards. In itself, knowledge is morally neutral. Moral issues are raised when applications are in question, and the mere existence of knowledge does not predetermine both beneficial and harmful uses, so the case for desisting from all research is by no means clear or convincing. To ask a basic researcher to weigh the ultimate mix of good and bad consequences that will follow from his discovery of a law of nature is to ask a question that no one, no matter how conscientious or perspicacious, could possibly answer. Nor would it be evasive for anyone asked

[9]See in particular Hans Jonas, "Straddling the Boundaries of Theory and Practice: Recombinant DNA Research as a Case of Action in the Process of Inquiry," *Recombinant DNA: Science, Ethics, and Politics*, edited by John Richards, New York, Academic Press, 1978, pp. 253–272; Sissela Bok, "Freedom and Risk," *Limits of Scientific Inquiry*, edited by Gerald Holton and Robert S. Morison, New York, Norton, 1979; and the discussion of the problems raised in these and other recent writings in S.A. Lakoff, "The 'Galilean Imperative' and the Scientific Vocation," *Ethics*, forthcoming, 1980.

such a question to answer that the decision with respect to the consequences is more properly one that must be made by society rather than by scientists.

There is, however, a middle ground, in which ethical constraints are both reasonable and meaningful, and not altogether foreign to scientific practice. Scientists adhere to certain norms in their work as scientists, such as the injunction against falsifying data. Although the opportunity to replicate procedures is a practical check against unintentional error, there is also a moral standard against deliberate falsification, which is enforced by fear of social stigma and institutional discipline. Some scientists have also taken the position that they must consider the social consequences of their work, at least in cases where the consequences are reasonably direct and important. Thus, the physicists who imposed censorship on their publications, lest the results aid the Nazi war effort, acted out of a recognition of the ethical responsibility their knowledge imposed. Similarly, the biologists who met at Asilomar to impose a limited moratorium on certain types of research whose consequences could not be predicted with a high degree of assurance, did so because they too recognized that the right to experiment carries an obligation not to endanger others. It may also be the case that the biologists acted in order to forestall efforts of social regulation that might have been even more restrictive, but their initative was essential if the public regulators were even to be alerted to the need for regulation. It is therefore doubtful that their action can be explained simply as a case of professional self-interest and not to a greater degree as an act of ethical responsibility.

In the case of applied science and engineering, the proximity of the work to actual end uses may well impose even clearer ethical obligations. Those who design nuclear reactor safety systems or aircraft and automobiles are helping to produce technologies that inevitably involve a degree of risk, not only to those who use them, but also to those who may be affected by their use. The designers' responsibility to those whose lives may be affected by their work is closer to that of the physician than that of the basic researchers. Even so, of course, it is by no means identical with that of the physician, for reasons that have already been indicated and for others as well. Engineering decisions are apt to be collective rather than individual. Engineers, moreover, always work under constraints imposed by market forces and considerations of consumer demand and taste. They are subject to pressure to produce systems that can be profitably marketed and, in the case of automobiles, that will suit prevailing styling expectations. Trade-offs need to be made between marginal increments of safety, cost increases, and styling appeal in which they cannot exercise control. In all such instances, however, the engineer is in the best position to know what the risks are of a given trade-off. To the extent that a product imposes significant risks, he is under a moral obligation to inform those who might be adversely affected. In recent years, a number of engineers and applied scientists have in fact taken courageous action, sometimes at considerable personal cost, to exercise precisely this sense of responsibility.

The most difficult area of moral responsibility is the zone of the greatest uncertainty. Traditionally, scientists have taken the view that there must be no limits placed on the process of inquiry. They have done so partly on the ground that freedom of inquiry deserves to be considered a human right, along with other forms of the exercise of free will. Many scientists also recognize, however, that scientific inquiry is at a far more advanced stage than the social capacity to assimilate new knowledge. It is undeniable that astonishingly sophisticated scientific inquiries take place in a world that is also marked by deplorable reversions to primitive hatreds, by reckless irresponsibility on the part of state systems, and by widespread popular ignorance and superstition. Under these circumstances, is it ethically responsible to confer new powers that may be profoundly unsettling to traditional notions of morality, and that may be used to condition or alter human behavior to such an extent that modern societies will be able to dominate not only physical nature, but human nature as well? The creation of formal and informal groups to study the ethical implications of certain forms of research and experimentation in biology and medicine indicates that this question is not easily answered. Nor is its consideration solely a matter that must concern the researchers themselves. In virtually all cases, but especially in those in which there is great uncertainty, ethical responsibility requires that scientists discuss the implications of their work with others who may have an equally good or better appreciation of the moral and behavioral complications that may ensue.

The Exercise of Ethical Responsibility

The ways in which ethical responsibility can be exercised by scientists will vary not only with their modes of work and the likely consequences of their researches, but also with the social environment in which the work is conducted.

The individual scientist must often make the lonely decision as to whether his most responsible course is to continue to work within a given system or to go outside it and even to work against it. He must decide among the choices vividly described in the title of Albert Hirschman's well-known study of such options: *Exit, Voice, and Loyalty*.[10] If he chooses to work within the system, he must accept its constraints and take responsibility for the outcome, to the extent that he contributes to them. Institutional loyalty, however, does not require or justify acquiescence in all respects. "Voice," or in other words critical concern for consequences, is often essential to the exercise of institutional loyalty, especially in institutions embedded in pluralistic and free societies that are more hospitable to dissent and vigorous argument than are institutions influenced by autocratic norms.

A scientist who decides that he must quit his institution or organization in order to obey his conscience, or that he must criticize a given decision in public,

---

[10]Albert O. Hirschman, *Exit, Voice, and Loyalty: Responses to Decline in Firms, Organizations and States*, Cambridge, Harvard University Press, 1970,

may well find that he is exposed to the hostility of his scientific peers (as Rachel Carson was) and deprived of access to research support. He may be stigmatized as a troublemaker simply because he does not raise his voice quietly and within the walls of the club. Scientists do not always take kindly to colleagues who pursue the glare of publicity and occupy themselves too much with social crusades, even those directly related to their scientific work. In some notorious cases, government agencies have penalized scientists who have taken unpopular political positions on subjects such as arms control, by blacklisting them and making them ineligible for service on advisory committees and as directors of research support agencies.

There are also instances, however, in which scientists have gained considerable stature, and even political influence, by speaking out in criticism of strongly entrenched policies or by lending the prestige of their reputations to some social cause. Scientists whose home bases are universities with faculties and student bodies well above the social norm in liberal political views may find the environment encouraging to political dissent rather than discouraging. On the campus, it is rather those scientists who wish to voice support of governmental or conservative positions who may find themselves stigmatized for their unpopular views.

It can be plausibly argued, however, that the most effective way for scientists to exercise responsibility is to do so in the course of pursuing an active career in research—a career that in many cases will inevitably put them into a position where they can contribute significantly to the discussion and consideration of sensitive policy questions. It is too simple to say, as many radicals do, that such people are simply co-opted by the system, or made the instruments of others who make the policy decisions. A more balanced view must recognize that, by providing their services, they gain access to the decision-making process, and that they may use this access either to influence policy decisions directly or to exert indirect influence by criticizing policies in other forums from a position of authority they would not have if they were not active participants in the research process.

The case of the nuclear scientists who blazed the trail by becoming involved in the Manhattan Project is instructive in this regard. Their work put them in a position where they could at least raise questions about the advisability of a military use of the bomb. Many of them signed a petition arguing against such use, but at the time there was no precedent for providing them with any formal consultative channels for expressing their views. A representative group was invited to suggest nonmilitary options. When this group was unable to suggest any satisfactory alternatives to military use, the policy makers were free to proceed as planned, but a precedent for consultation on policy issues had been established. It is arguable that if the scientists had been asked to take other factors into account, aside from the narrow question of whether a nonmilitary demonstration would have the desired effect, they might have been able to put a stronger case against military use (by recommending, for example, that the terms of the demand for unconditional surrender be spelled out to reassure the Japanese that they could retain their emperor). The lesson of this experience was not lost on the scientists who became involved in military and defense policy after the war, as the well-known case of the General Advisory Committee to the Atomic Energy Commission illustrates.

When, a few years after the end of the Second World War, this committee of scientific advisors was asked to study the advisability of a "crash program" to develop a thermonuclear weapon, the scientists believed that they had a broad mandate to consider not only the technical feasibility of the weapon but its political and moral desirability.[11] Although this attitude got the chairman of the committee, J. Robert Oppenheimer, into considerable difficulty, when he was accused of being an agent of the Soviet Union because of his "lack of enthusiasm" for a crash program, it proved to be a step toward the legitimation of a broader role for scientific advisors. In the debates over the antiballistic missile and over arms control, for example, scientific experts have entered into the decision-making process, and into the implementation process, in a variety of ways—as consultants, negotiators, diplomats, and shapers of public opinion.

Some scientists are suspicious of all entanglements with government agencies, on the ground that these agencies tend to be committed to the promotion of new technologies, regardless of social consequences. This proposition is open to challenge, in view of the institutionalization of efforts to control technology that has taken place in recent years, with the establishment of such bodies as the Arms Control and Disarmament Agency, the Environmental Protection Agency, and the Nuclear Regulatory Commission. In any case, scientists also have open to them the option of allying themselves with "public interest" science organizations or with political organizations of scientists, such as the Federation of American Scientists and the Union of Concerned Scientists.

Whichever route they choose, the scientists who participate in public debate and the decision-making process are performing a function that is vital to the effectiveness of democracy at a time when issues with significant scientific and technological content crowd the public agenda.[12] Unless they draw attention to the larger questions at stake, these issues are apt to be resolved by the more conventional force of interest group interaction. A controversy over nuclear power will be resolved by the interaction of the economic interests of construction workers and utilities with the countervailing pressures of environmentalists, as mediated by government agencies buffeted by domestic and international pressures. Intervention by scientists knowledgeable about the hazards and the benefits of nuclear power does not always make it easier for

[11]See Herbert F. York, *The Advisors: Oppenheimer, Teller, and the Superbomb*, San Francisco, W.H. Freeman and Co., 1975.

[12]See S.A. Lakoff, "Scientists, Technologists, and Political Power," *Science, Technology, and Society: A Cross-Disciplinary Perspective*, edited by I. Spiegel-Rosing and D.J. de Solla Price, Beverly Hills, Sage, 1977.

members of the general public to come to a decision. Often laymen find themselves with the impossible task of deciding which Nobel laureate to believe. Their intervention does, however, lift the debate to a proper plane, one on which the general public interest is as much in question as are the special interests and values of pressure groups. The creation of local committees of scientists and laymen to consider the safety and advisability of research on recombinant DNA is another example of the approach that needs to be cultivated, if scientists are to pursue the path of ethical responsibility in a reasonable and effective way.

In the decade of the 1960s, a certain fascination developed with a more radical possibility, reflected in such notions as John Kenneth Galbraith's identification of the "technostructure"[13] as the new source of decision-making in government and industry. Confidence in such mechanisms as well as fear of them rested on the belief that under modern conditions it might be possible to remove government from politics, to put decision-making on a purely rational, objective basis, and therefore in the hands of the competent skill-group elites. Experience has already shown that there is no satisfactory substitute for representative democracy, if the objective is to manage the process of technological innovation efficiently and at the same time preserve the right of the people to choose the risks and benefits they are to bear.

In nondemocratic systems, of course, the path of responsibility is much more arduous. There the choice of whether to work within the system or whether to become a dissenter may be harder to make with a knowledge of the consequences. One wonders whether, if Andrei Sakharov had his life to live over, he would again work on the thermonuclear weapon for a regime he has come to denounce as an enemy of human rights. Perhaps he would still find it necessary to do so, in order to gain the protected status that enabled him to become a voice for conscience. To this extent, there may well be a parallel between his behavior and that of American scientists who have emerged as champions of arms control after a career as defense researchers. There is, however, also a profound difference. The openness of liberal political systems to differences of opinion, whether expressed by adherents of opposing parties, branches of government, or segments of society, makes the expression of different points of view a legitimate activity. The active, independent citizen earns social respect, even if sometimes it comes grudgingly from those who disagree with him. In societies that are either fully totalitarian or are verging on totalitarianism, the dissenter is far more readily perceived as a disloyal "enemy of the people."

In some ways, it is probably true, however, that scientists are not particularly well fitted, as a general rule, to play an active political role. Preoccupied with matters that can be examined quantitatively and experimentally, they are apt to find the more qualitative, argumentative, interpretive, rhetorical, and uncertain quality of politics exasperating and confusing. Scientists qua scientists are reluctant to come to judgment until all the relevant evidence is in and checked. In

[13]J.K. Galbraith, *The New Industrial State*, 3rd ed., Boston, Houghton Mifflin Co., 1978.

politics it is frequently necessary to come to provisional judgments well before the evidence in favor of a policy is overwhelmingly clear. Nor are scientists apt to develop the skills of exposition and persuasion that come more naturally to politicians, and by academic training to lawyers and journalists, and that are vital to successful campaigning on behalf of even the most compelling causes. Still, there is ample basis for expecting scientists to play an active role in the discussion of the moral issues surrounding their work and in the debate over the formation of public policy with critical scientifc and technical components. They are in a position to understand the implications of their work earlier and in some cases more certainly than laymen. Their information and their reflections are badly needed. For their part, they depend upon public confidence and public support. When scientists disagree in public, some people may become disenchanted with science, but this is not reason enough for them to hold themselves aloof from important public controversies to which they can contribute by virtue of their special knowledge and insight. As Max Weber understood, science is neither a magical route to certainty about nature nor a revelation of ultimate values. The scientific vocation, however, remains a potential bulwark not only of individaul freedom, as he implied, but of the universal freedom that is the essential principle of all self-governing societies. It is by the vital contribution they can make to the process of self-government that scientists can best discharge the ethical responsibility that attaches to their vocation.

II. Science, Technology, and Arms Control

The first three selections in this section are drawn from a conference debate on the SALT II treaty. Kistiakowsky's reflections were written afterward and are designed to review the modern history of efforts to achieve arms control. Loehmannsroeben examines the impact of arms control negotiations on European security. Homer-Dixon and Glidden consider ways in which science and technology can be used to make arms control more feasible. York and Greb examine the possibility of a comprehensive nuclear test ban treaty.

4. The Case for SALT II

Herbert Scoville, Jr.

Now that a SALT II treaty has been achieved, the next problem is for the public to understand it and for the Senate to decide whether it is going to ratify it. I should like to stress first that it is very important to understand it. Everybody should make an effort to read the full treaty. It is an amazing document—extremely detailed technically, but understandable. You may not be able to figure out all the reasons why this or that clause was put in, but reading it and seeing firsthand the nature of the provisions in it is well worthwhile.

The treaty should be looked at from three main points of view. The first is: How will the treaty affect the balance of strategic weapons between the United States and the Soviet Union? The second is: What contribution will this treaty make toward controlling the very dangerous strategic nuclear arms race? And the third is: Can the treaty be adequately verified so that we can be confident that possible Soviet cheating will be detected before it causes any risk to our national security.

First, let us consider the military aspects of the treaty. I think that anyone who reads the treaty, even just the basic provisions without trying to understand all the fine print, must come to the conclusion that it improves the relative position of the United States vis-à-vis the Soviet Union as far as strategic weapons are concerned. The Soviet Union has to scrap and dismantle more than 250 delivery systems. These are not obsolete weapons; they are relatively modern missile launchers built since 1970. The Soviet Union also has to constrain many of its other programs.

On the other hand, the United States—and in fact this is one of the criticisms of the treaty—hardly has to constrain any of its programs. The United States does not have to destroy any existing weapons, at least not any currently operational weapons. It is true that thirty-three B-52s that were worn out in Vietnam and are mothballed out in Arizona will have to be scrapped by December 31, 1981. The United States arsenal of operational strategic delivery vehicles is well below the ceilings that are established, so the United States does not have to dismantle anything. The ceilings will not directly affect the United States until six or seven Trident submarines become operational in 1983 or 1984. In addition, all our existing programs, planned programs, or even seriously contemplated programs for new weapons are permitted to go forward under the treaty. From the point of view of arms control, this part of the treaty can be criticized because it probably does not clamp down as hard as it should on either country, particularly on the United States. Nobody can legitimately state that the treaty favors the Soviet Union. Virtually all the provisions in it are more favorable to the United States than to the Soviet Union.

One of the criticisms of the treaty by those who claim that it gives the Soviet Union an advantage concerns the provision that freezes the number of heavy ICBMs of both countries at their present level. It so happens that the Soviet Union now has 308, and the United States has none. But the reason that the United States has none is that we have neither had a program nor had even a desire for a program for heavy ICBMs. The Joint Chiefs of Staff have not requested any programs for these heavy missiles. So, although it is true that this is one area where the Soviets are allowed to have something the United States does not have, these missiles are something the United States does not want, has not wanted, and has no plans for in the future.

This is the only area in which the Soviets are allowed something that the United States is not permitted to have. In order to win that concession, the Soviets had to agree—and this provision was negotiated by President Ford and Secretary Kissinger at the time of Vladivostok—not to include in the ceilings the United States Forward Based Systems. These are the planes that the United States has in Europe and on aircraft carriers that are capable of striking the Soviet Union. These planes are not counted under the treaty—a unilateral advantage to the United States.

But the treaty does have one provision with respect to these heavy missiles. The main advantage they have is that they can carry more warheads than the smaller missiles. The treaty does set an upper limit on the total number of warheads the Soviets may have—specifically on the number of warheads that each type of missile may have. The Soviet heavy SS-18 ICBMs are to be limited to ten warheads each, which is the same number the United States would be allowed to have if it decided to procure the new MX missile. The MX, which can be developed under the provisions of the treaty, would be allowed to have ten warheads, the same number as the Soviet heavy SS-18. Thus, one of the main advantages that the Soviets have with their heavy missiles is taken away because they are limited to only ten warheads.

This overall constraint on warheads is one of the good features of the treaty as far as both countries are concerned, because in recent years we have seen a continual increase in the number of warheads that both sides have. The United States now has about 9500 and the Soviets have slightly more than 5000. The treaty will not stop this growth, but it does put an upper cap on the number of warheads that each country may have. The number of missiles that may have multiple warheads has an upper limit, and the number of warheads that each missile may have has an upper limit, so we can see a finite end to the race.

In my view, it would have been much better to have set lower limits, but that was not possible in SALT II. It would have been preferable if we could have kept the entire multiple warhead or MIRV missile problem so limited that there would not be a threat, or even a potential threat, to our ICBMs and Soviet ICBMs in the 1980s. Unfortunately, that opportunity was lost in SALT I. The United States government decided that it would not seriously try to negotiate a limit on MIRV missiles in SALT I, and the decision was made that instead, because the United States was five years ahead at that time, we would race the Russians in obtaining MIRV missiles. The decision was to follow the arms race road to security rather than the arms control path. Now we are paying the price for having chosen that route in 1972. As everyone predicted, the Soviets tagged along five years behind us; but now they have MIRV missiles, and they are getting the kind of accuracy that can threaten our ICBMs. And we are proposing a program costing $30 billion—or maybe $60 to $100 billion—to replace our ICBMs in an effort (rather hopeless) to make them invulnerable to this growing Soviet threat. That problem possibly could have been dealt with by arms control. There is no guarantee that SALT I could have succeeded in controlling the MIRV threat, but we should have made a greater effort.

Unfortunately, it was not possible to solve this problem in SALT II, but at least SALT II puts an upper bound on the total number of warheads allowed, so there is now an upper bound to the threat to our land-based ICBMs. It is incomprehensible to me how anybody can defend going ahead with the MX program to seek an invulnerable ICBM force, and at the same time oppose the SALT II treaty. Unless you have a SALT agreement with this upper cap on warheads, which we have achieved through negotiation, the MX system will not be worth a penny, and we shall be spending $30 billion or more on it. So SALT is an essential element to contain, even though it does not stop, the threat to the ICBMs. To be in favor of the MX and against SALT is a demonstration that one does not understand the issues.

Now let us turn to the arms control provisions of the treaty. First, it must be recognized that this is the first treaty between the United States and the Soviet Union that has required either side to scrap or dismantle existing, effective, operational strategic delivery vehicles. The Soviets will have to scrap 250 within the next two years. This is the first time that the race has been turned around. Until now the ceilings had always been established at higher than existing levels. The Interim Agreement of 1972 froze programs at existing or in some cases higher levels, and the Vladivostok Accords of 1974 set ceilings that were higher than either side had. As a consequence, both countries had incentives to build up to these ceilings. Then what happened? Of course, since the Vladivostok Accords were never put into effect, the Soviets went even beyond existing levels.

Now, however, we have a treaty that actually calls for lowering existing levels. Although this will be only a 10% reduction in the Soviet force, it is 10% in the right direction.

Second, I have mentioned that the treaty deals with warheads, not completely satisfactorily by any means, but at least it puts some upper limits on the number of MIRV missiles and warheads that each side may have. Further, for the first time the treaty puts some restrictions on qualitative improvements to weapons. Modernization of strategic weapons is probably the biggest driving force in the arms race today. The treaty does not deal with this problem perfectly, but it does include some very important limitations. For example, each country will now be limited to only one new ICBM between now and 1985, and that ICBM cannot be a heavy ICBM like the Soviet SS-18. It will have to be a light ICBM, and it will be limited to ten warheads, which is the maximum that either side will be allowed to have. In addition, there will be restrictions on the throw weight, launch weight, and types of fuel for missiles. Unfortunately, the treaty does not set any limitations on guidance systems, because these cannot be verified. It was decided that the requirement for verification was more important than the gain from a poorly verified limitation on the accuracy of a missile. At present nobody knows any reliable way to verify controls on improvements in guidance systems.

Having noted these features about qualitative limitations in the treaty, I must add that this permission for even one new ICBM on each side is, in my view, the biggest loophole in the treaty. It is one that I regret most strongly that we did not succeed in closing. And I regret to say also that the failure to close this loophole is primarily the responsibility of the United States government, which insisted on keeping open the option to go ahead with its MX missile. In the long run this may well be the decision that will light the time fuse and explode our long-term capability to control strategic arms. The United States insisted on keeping that option and is exercising it by the President's decision to go ahead with the MX missile. This, in my opinion, is a tragedy. It is my understanding that the Soviets agreed to a complete ban on new ICBMs at one point, but the United States refused.

On the other hand, I do not wish to label the Soviets as always being the "good guys" and the United States always the "bad guys." The Soviets may have been hiding behind the definition of what is a "new" missile. There is some evidence that they thought they could get their new missile by modernizing the ones they now have under development and are beginning to deploy. This righteous offer to ban all new ICBMs may have been a way of gaining an advantage for themselves. Both sides probably deserve some of the blame for not closing this loophole. This loophole is a serious one, which could lead to a very dangerous arms race in the 1980s and perhaps to the eventual disruption of the entire SALT process, because we may have no ability to verify and control these weapons in the post-1980 period.

Finally, let me discuss verification briefly, since it is going to be one of the key issues in our debate. Obviously, in a treaty of this importance we cannot rely on trusting the Russians. There may not be many incentives for the Russians to cheat under this treaty, because if they really wanted to do something forbidden by the treaty they would probably abrogate the treaty rather than cheat. Nevertheless, we cannot rely on that response. We must have independent means of verifying that any Soviet cheating that could significantly affect our national security could be detected in time. I have carefully reviewed the provisions of the treaty with the specific aim of checking verifiability. There are many useful provisions in the treaty that have not had much publicity until now. I am convinced that the treaty can be adequately verified and that we do not need to worry about Soviet cheating.

I say this also with the knowledge that we have lost two important intelligence bases in Iran. There is no question that the loss of those bases in Iran was a serious blow to our intelligence operations. They provided very specific, highly useful information on how Soviet missiles are constructed, particularly the first stage of Soviet missiles, since those sites were the closest to the Soviet ICBM test launch site at Tyura Tam. From those sites we could observe closer to the ground, and see earlier in the flight, than we can from any other ground station. Those sites are now lost, but that does not necessarily mean that we cannot verify the agreement by other means. I have gone through all the basic provisions of the treaty, and I am confident that we can verify the treaty so that we do not need to worry about Soviet cheating.

As far as the deployment provisions are concerned, we have satellites that can photograph the Soviet Union—the entire Soviet Union in a day if we want to—with moderate resolution so that targets of interest can be spotted. Then when something interesting is observed, or something new that we want more details on, we can zoom in on the target, in the same way a television camera does on a football field, and get high-resolution pictures. We can then make very precise measurements on all the details of Soviet missiles, launchers, aircraft, submarine pens, submarines when they are above the surface, and many other military items. The deployment restrictions are monitored primarily through satellites.

As far as MIRVs and multiple warheads are concerned, the sites in Iran were useful but by no means vital in order to verify the provisions that deal with multiple warheads, the most important qualitative provisions in the treaty. We have extensive means of intelligence collection at the re-entry end of their test ranges—ships, radars on land, airplanes equipped with all kinds of instruments, infrared devices, etc. Cameras observe the re-entry, so that we can not only count every re-entry vehicle that comes in but even weigh them if we want to. From these observations we can get a rough idea of what the useful payload of the test missile is. Even data from the earlier portions of the MIRV flight path, such as information on the dispensing of the MIRVs, can be collected without the Iranian stations, because these flights occur at such high altitudes that we do not have to be as close as Iran in order to make the observation. There is no problem with verifying any of the MIRV provisions.

The one area where there is a legitimate question lies in those provisions that deal with limits on testing of any new Soviet missile. The desired data on the first stage of the missile are harder to come by. Nevertheless, there are many alternative ways to get sufficient information to be confident that no militarily significant new missile could be launched by the Soviet Union without our detecting it. We have a multiplicity of ways of getting at this problem. For the Russians to build secretly a totally new missile that would have significantly different characteristics from those they already have would be impossible with the intelligence capabilities we have, even without our sites in Iran.

There is one provision in the treaty that was new to me until recently. When the Soviets decide that they are going to employ the option of testing their one permitted new ICBM, they must announce the date and time of that launching in advance. This gives the United States the chance to have all its intelligence capabilities geared at that time. It also provides the opportunity to acquire baseline information on the specific properties of the new missile. If we wish to rely on U-2 aircraft instead of the Iranian sites, in order to look closer to the ground, we can have the aircraft operational at the time of the launch. This notification can be a tremendous boon to our verification. In addition, if the Soviets fail to notify us when they test their new missile, they do so at their peril because there are so many different ways in which they can be caught. I believe that this provision requiring the test to be announced is a very important one.

In conclusion, there is no question that our security is far better off with the treaty than without it. I believe that the treaty takes some important steps toward getting the nuclear arms race under control. Unfortunately it has some loopholes. The loophole of allowing a new ICBM for each side is one that should be closed immediately. We should not depend on SALT III for this; we must close it now by unilateral actions on both sides. This is why I find the President's decision to continue with the MX particularly damaging; he may have set in operation some actions that will be very hard to reverse. It is important for our security that that loophole be closed at the earliest possible date. Finally, we can support this agreement, which is in the interest of our security, without having to worry about trusting the Soviet Union. We can independently verify its provisions with our technical intelligence assets even though we no longer have the Iranian bases.

5. The Case Against SALT II

William Van Cleave

I shall try to cover the three points that Dr. Scoville addressed in order to present an explicitly different view. These are the strategic balance and how SALT influences it, the contribution of SALT II to controlling the so-called but nonexistent strategic arms race, and the question of verification. I shall also address the major administration arguments for SALT II.

We all know what has occurred in the strategic balance, and what has caused it. There is no need for me to present data you are all acquainted with—data that have become so abundantly clear that there is no longer any major debate concerning the facts. Whatever specific numbers, comparisons, or analyses are used, the trends in the strategic balance during the SALT decade have been seriously adverse to the United States and to international stability. This has been due to the Soviet drive and willingness to expend enormous sums of money to pursue highly ambitious strategic force goals. During the ten years that we have been engaged in SALT, during the three or four years prior to that, in which we were preparing for SALT, and during the next six years in which we would be governed by SALT II, the pace of the United States has been slow and modest, while the Soviet effort has steadily intensified. No one here should entertain seriously the rhetoric of an "arms race," much less the notion that the United States has somehow been responsible for driving that "arms race." What we have witnessed has not been action–reaction, much less inaction–inaction, but inaction–reaction. The United States has presented opportunities, and the USSR has seized them.

The Chairman of the Joint Chiefs, whose job depends upon understating, not overstating, the threat, testified earlier this year that the greatest Soviet military gains have come since the signing of SALT I, and went on to say further that these trends will continue through the middle 1980s, with or without SALT II. He said that we are facing an imminent and acutely dangerous imbalance.

In this year's FY 1980 Department of Defense Report, Harold Brown warns against underestimating the Soviet military buildup, and states: "Our most serious concerns, which we need to act now to meet, are about the period of the

41

early-to-mid 1980s." Make no mistake, this is by no means an exaggeration. To the contrary, the FY 1980 DOD Report, overall, is carefully tailored to soften the situation in an attempt to justify the continued exceedingly low rate of strategic force expenditure, which has fallen below the amount budgeted for retirement pay. This period, of course, is precisely that which would be covered by the proposed SALT II treaty.

It is no wonder that this situation has come about. The Soviet Union has been outspending the United States, by CIA estimates, or underestimates, by a factor of 2½ to 3 on strategic offensive arms, and by a factor of 7 on strategic defensive arms. If we just took the amount of money that has gone into investment for arms—nothing else but investment for arms—for the five years after the SALT I agreement, the difference spent by the Soviet Union over that spent by the United States would have paid for the entire planned B-1 program; 500 MX in a survival basing mode; the entire cruise missile program, including not only the conversion of the B-52Gs but a new cruise missile carrier aircraft; the Trident I and the Trident II programs; and, just for good measure, 7000 MX-1 tanks for the Army, 7000 combat infantry vehicles, and all the fighter aircraft now planned for the Navy and the Air Force.

You may recall that when the SALT I agreements were submitted to Congress in 1972, the administration treated the particular limitations with restrained enthusiasm. They were, even at that time, heavily one-sided in favor of the Soviet Union. Indeed, the administration even appended a formal declaration to the agreements stating that, if they were not superseded within five years by an agreement providing for more complete and equitable strategic arms limitations, the United States' supreme interest could be jeopardized. The principal argument for the agreements was that they were a necessary first step in the SALT process, which would lead to those more satisfactory limitations. And in the meantime they would at least slow the momentum of Soviet strategic arms programs so that we would be relatively better off with the agreements than without them.

I should like you to note these arguments well, for two reasons: First, they are a far cry from the expectations with which the United States entered SALT. They mark a major diminution of our SALT objectives. Second, they are essentially the major arguments now being used by this administration for SALT II.

I do not have the time to review the course of SALT I, but to set the record clear let me respond to something that Dr. Scoville asserted. Dr. Scoville contended that the refusal of the United States to attempt to control MIRV during SALT I is responsible for the predicament that we find ourselves in today. At least he acknowledges a predicament. But the MIRV question is much too complicated to be reduced to such a simple debating proposition, and it was especially so during 1969 and 1970. Several points should be kept in mind. First, the United States did offer the Soviet Union a MIRV ban option in the Spring of 1970, and the Soviets rejected it. Now I am aware that it may be argued that the offer was not sincere, inasmuch as the proposal required on-site inspection, but it was formally offered, and if the Soviets had been seriously interested, it could have been negotiated.

Given our record of backing down from verification requirements in SALT, the Soviets might well have had a MIRV ban without on-site inspection had they pressed negotiations for it or for other negotiable MIRV limitations. After all, the SALT delegation at that time was heavily weighted in favor of a MIRV ban. ACDA, the Department of State, and the chairman of the delegation all favored a MIRV ban. To that point I might mention that in the early summer of 1970 the Secretary of Defense complained to the President of the United States that the SALT delegation was pushing too hard for the MIRV ban option at the expense of what was to have been given equal weight—a proposal for reductions of offensive forces. The point is that the Soviets were never seriously interested in a MIRV ban, or in stringent MIRV limitations. They have not been interested at any time since, and at Soviet insistence there are no effective limitations in SALT II on the Soviet MIRV buildup. Soviet MIRVs, I might add, are clearly designed for counterforce rather than "assured destruction" purposes.

Now, what about the SALT process? The administration's case for the proposed limitations is so weak that it tells us not to become preoccupied with the details but to look at SALT as a valuable process. Let us do that. It is a process that is not working. It is producing a situation and a dynamic never intended when we entered it.

In reality the SALT process has been less an exercise in arms control or a means to enhance American security than a charter for continued Soviet buildup and continued self-restraint on the part of the United States. SALT II would continue this. It is contended that SALT II would limit and even reduce Soviet strategic weapons. But Soviet strategic weapons—missiles and warheads—are not limited. SALT II applies to *launchers* for those weapons, not to the number of weapons. We know that the USSR, for example, has produced a larger number of ICBMs than it has SALT-counted launchers for those ICBMs. Estimates range in excess of 1000. If they wish to produce an even larger number of missiles, SALT II does not prohibit it. And the USSR currently has four very active ICBM production lines, whereas the United States has none. Such additional missiles could be used to reload Soviet silo launchers, or they could be launched by entirely different means. Even though SALT II applies to launchers, it does not contain any satisfactory definition of what constitutes a launcher. ICBM launcher limits are intended to apply to silos, or to identifiable tractor launchers for mobile missiles. It is not necessary that ICBMs be launched from either. With the aid possibly of an electrical extension cord, Soviet canisterized ICBMs could be launched from parking lots, or street corners. Incidentally, the reason that launchers rather than missiles are the currency of SALT is that the decision was made in 1969 that we cannot reliably count or verify the number of missiles, but that we can verify the numbers of launchers for those missiles. The assumption is clearly false.

How strategically significant are additional ICBMs? Not 1000, but only 100 additional ICBMs, if a mix of the SS-18 and the SS-19, would add to the Soviet arsenal more MIRVed payload than exists in the entire United States ICBM force.

We have long recognized that limitations on launchers alone are relatively meaningless; that enormous leaps in capabilities can be made while "launcher" numbers remain stable. Counting *only* those ICBMs on SALT-counted silo-launchers, during SALT II Soviet warheads of a hard target nature will approximately double; Soviet MIRVed ICBM throw weight will triple; and hard target capability will increase, perhaps tenfold. Are these "reductions"? These figures do not count the Backfire; they do not count the SS-20; they do not count missile reloads; nor do they count the new SLBM programs. Even launcher numbers are limited or reduced only if one ignores the launcher definition problem, and ignores as well, as SALT does, the Backfire bomber and the SS-20 missile launcher. These are only a few examples of defects in the SALT II treaty. They should suffice to show that the agreement does not restrain the Soviet buildup, and certainly does not in any way amount to reductions in that buildup.

There are many loopholes and ambiguities in this agreement, as well. It is even less tight and precise than the SALT I agreements, which allowed differing interpretations and led to trouble concerning compliance and verification. These loopholes and ambiguities are in the agreement precisely to disguise differences that could not be resolved, and because the Soviets insisted upon them. We can expect the Soviets to exploit them. Let me give just one example.

Dr. Scoville says that each side is limited to one new ICBM to 1985. This is untrue. They are limited to one "new *type*" ICBM. What does this mean? The agreement defines a "new type" as one that differs by more than 5% from a previously tested ICBM in four general parameters: length, diameter, launch weight, or throw weight, or that differs in number of stages or type of propellant (meaning only solid or liquid, not advanced new propellants of the same solid or liquid type). No data base is given from which to determine the 5%. The Soviets have tested many ICBMs, with many differences. Which of these is to be the standard for the 5%? The 5% rule allows new ICBMs with new propulsion, new post-boost vehicles, new guidance systems, more warheads, and 5% larger in both launch and throw weight, and they still will not be a new type. In short, it allows Soviet testing and deployment of its fifth generation of ICBMs—all of them—without violation of the one "new type" rule.

The situation is even worse because it is clearly so easy to circumvent this putative limitation by other means. For example, no range is specified for the throw weight. A missile with a range of 5500 kilometers is an ICBM, but one could have a "light" ICBM at 10,000 kilometers that would be a throw-weight "heavy" at, say, 6000. Or one could develop a 5000-kilometer "IRBM" with no restrictions on its throw weight. It would be possible to build a missile of any size and load it with enough re-entry vehicles and ballast (and the SS-20 has been tested with ballast) so that the range is under 5500 kilometers and it can be called an IRBM. Keep in mind that the SS-20 is not a small missile. It is nearly equal to Minuteman in lift and in throw weight, and it carries three one-half megaton re-entry vehicles to 4000 or 5000 kilometers. It can reach 8000 kilometers with a reduced payload. Another loophole, of course, is the SLBM loophole. One can develop, test, and deploy any new types of SLBMs that one wishes. The SSN-18, for example, carries seven re-entry vehicles to 5000 nautical miles, and a new

SLBM, yet to be tested, will have a 5000-pound throw weight at 6000 nautical miles. Such missiles can be developed, tested as "SLBMs," and later deployed as ICBMs. After all, we considered for a long time developing the D-4 into a land-based as well as a sea-based missile.

The story of how these loopholes came to appear in the agreement is one of the Soviets' negotiating, again, so as not to restrict any of their planned programs. The effect of the agreement, predictably, would be that the United States will be restricted to developing only one new ICBM during the agreement (presumably the MX), while the Soviet Union would be free to develop and field as many as they wish. This story would not be complete if I did not mention that the United States sought to include five more specific and precise "new type" parameters, in addition to the general ones noted. These would have been more helpful in establishing the "new type" rule, and they are essential for verifying compliance with it. The Soviets refused. This refusal must be seen in the light of the verification provision. The Soviets successfully had this provision worded so that encryption of missile telemetry is expressly permitted, not prohibited, as long as it does not *deliberately* deny information necessary to verifying compliance with the agreement. The specific parameters rejected by the Soviets are therefore not cited in the "new type" rule—by agreement, then, not information relevant to compliance—and the Soviets may encrypt the information, with our agreement.

Such things make a sham of SALT II.

At the outset I said that I would comment on the subject of verification, which was raised by Dr. Scoville. I do not want to take the time to point to specifics that are unverifiable, or inadequately verifiable, even though there are some important ones. That the administration speaks not of verifiability but of *adequate* verifiability, and then defines it in the subjective terms of what is "strategically significant," should tip off anyone. The definitional ambiguities in the text, the lack of adequate data bases, and other loopholes magnify the problem. Verification is important but in my view it is of secondary, if not tertiary, importance. Why? Simply because even a thoroughly verifiable bad agreement remains a bad agreement, and SALT II is a bad agreement. As Charles Burton Marshall is fond of pointing out, a good contract can be debased by cheating, but a bad one cannot be redeemed by faithful performance. The best way to ensure compliance is to permit the other side to do whatever he wants without "cheating," which seems to me, really, to be this administration's approach to securing Soviet compliance. Our strategic situation will not be improved even by faithful Soviet compliance with the agreement; it will be worsened. That strikes me as the most important factor. I side with Dr. Scoville entirely in urging you to read the agreement itself with great care, to see for yourself what loopholes exist, and to determine whether or not you think that real limits exist.

It is probably in the light of such defects in the agreement that the "SALT-sellers'" case has so radically shifted in recent months. Earlier, SALT-sellers such as Mr. Warnke argued that the treaty would provide a "measurable advance" in American security and arms control. As such arguments crumbled in the debate, the enthusiastic SALT-sellers were replaced by new strategists, who then switched the argument to warnings of dire consequences should SALT II be rejected. As these were parried, the two SALT I arguments re-emerged: (1) This agreement will make SALT III possible, and SALT III will be more helpful; and (2) the Soviets will do more in the absence of the agreement—that is, do not be concerned with the inequities or with the superiority the Soviet Union will gain with the agreement, as it would be even worse without it. To these stale arguments, a new one was added: The agreements leave the United States free to do what it wants, so how are they *harmful?* At least this argument acknowledges implicitly that the debate is over whether the agreement is harmful, or merely worthless. Let me address these three points briefly.

That good agreements will follow bad ones and that follow-on arms agreements will be unburdened by the principles and precedents of earlier steps in the process cannot be seriously entertained by anyone. SALT II did not correct the defects of SALT I, it built upon them; and the same would be the case for a SALT III agreement based on this SALT II treaty. Even the limitations contained in the Protocol, due to expire in two years, are precedential, which is why the Soviets insisted on keeping them. This is also why the Soviets insisted on including in the Basic Guidelines for SALT III negotiation of the issues raised in the Protocol. Defects should be corrected *before* an agreement is reached, or certainly before it is approved by the Senate and ratified by the United States Government, rather than accepted on the naïve proposition that we can correct later what we cannot correct now.

The argument that the Soviets would do more in the absence of an agreement is a poor argument for bad agreements. It is an argument of resignation. It also ignores two central facts. A clear lesson from the decade of SALT is that the Kremlin will not agree in the first place to any limits that would prohibit or change what it plans to do. The truth about SALT agreements is that they reflect what the Soviets want to do. Secondly, the Soviets are already pretty close to the margin of what is practically feasible in their military effort. Given their effort, and the percentage of gross national product being spent on the military, it is difficult to imagine what more could be done. The real threat is what the Soviets are allowed to do by the agreements, not what is hypothetically possible in the absence of the agreement. The agreement, anyway, does not restrict them from producing and stockpiling more missiles, warheads, and bombers, if they wish.

To argue that we are free to do what we wish under the agreement is a curious rationale for arms control agreements; it certainly obscures the purposes for which we sought arms limitation, and underscores the questionable value of the agreement. It is also a specious agreement for three reasons: The administration that has accepted the agreement also defines what we want to do (in accordance with the agreement, naturally). We may be free, but we are not doing much, and there are no plans at this time for doing more. Many of the programs that are planned, particularly the MX program, are post-SALT II deployments; in their case, SALT II is irrelevant in what it frees us to do.

But SALT II would not leave us free to address our most urgent problems—ICBM survivability, for example. SALT I and II say that we cannot defend the ICBMs, we cannot conceal them, we cannot deploy them in a multiple aim point mode, we cannot proliferate them, and we can only make them quasi-mobile, if the Protocol does in fact expire and the Soviets do not object. We are, in fact, reduced to two options: doing away with land-based ICBMs, or resorting to launch on warning, not attractive options to say the least. SALT II contains other direct constraints as well—for example, those involving cruise missiles and cruise missile carriers. But SALT has always constrained us indirectly, beyond the terms of the limitations, and there is little reason to believe that this would not continue under SALT II. The process does have an inhibiting effect surpassing the precise terms of the limitations. We are reluctant to "jeopardize" on-going negotiations by program actions on systems under negotiation. The truth is that the United States does have actions it could take that would significantly reduce growing vulnerabilities, enhance stability, and correct some of the dangerous deficiencies in our strategic forces, and would do so before 1985. But are we really likely to be taking such steps if SALT II, as it stands, has been approved by the Senate, and SALT III negotiations are underway? Are such actions really compatible with the assurances about the worth of the agreement now being given by the administration? There is no doubt in my mind that ratification of this agreement would forestall our ability to avert the threat we face in the immediate future.

Our attention now should be focused not on SALT, but on what we must do to prevent what the Chairman of the Joint Chiefs terms "an acutely dangerous imbalance" in the early to middle 1980s. This imbalance would mean not only unprecedented American insecurity, but the likely insecurity of other countries as well, since the Soviets are most likely to move to take advantage of the imbalance. The matter is urgent, and major actions are now required. In my view, these actions, necessary to security and strategic stability, are not incompatible with arms control in the future. Arms control agreements tend to reflect the reality of the situation at the time of their negotiation and for the period they would cover. To obtain agreements that would contribute to American security, we must first improve that reality; otherwise, agreements will tend to build on our weaknesses.

Arms control is supposed to contribute to United States security. It fails if it does not do so. Soviet strategic and SALT objectives have not been compatible with successful arms control. United States restraint, encouraged by the SALT process, has been exploited by the Soviets for their unilateral advantage. Senate rejection, or substantial amendment of SALT, would not cause a failure of arms control. It would reflect it. But this would certainly not end the chances for arms control. On the contrary, it could possibly rescue the SALT process and improve the longer-range prospects for meaningful arms limitations. If accompanied by the right programs, such rejection or amendment would announce to the USSR: "No more unilateral advantage from SALT; the United States does not intend to accept worthless agreements or the continued shift in the strategic balance. We have been put in a position where we must act decisively to prevent an unacceptable situation, and we are capable of doing so." Perhaps then the Soviet leaders would see that they have more to gain from the mutual benefits of real arms control and a moderation of their own strategic programs and objectives. Perhaps not. But I find nothing in SALT II that would lead in that particular direction.

6. SALT II and Beyond: Toward a Common Platform for Hawks and Doves

Jeremy J. Stone

I am not so concerned about the specific terms of the SALT II agreement as I am about its prospects for leading to disarmament and strategic stability—in other words, where is all this going?

SALT is not like the Panama Treaty; it is part of a process. And a very important question is, "How is this piece of the process going to be linked to the next stage in the process? One of the reasons it is so difficult for me to describe my position is that much of what I should like to do depends on what other people are going to do, especially in the forthcoming Senate debate.

I think that if the treaty is approved, it should be accompanied with a resolution giving strict instructions to the negotiators that they should not come back again with an agreement having as little substance as this one. In other words, as the treaty is passed, one could, with a resolution of instruction, increase the chances that the next round will produce a better treaty. If SALT II is going to pass, I think it should include that kind of resolution. Indeed, I believe that such a resolution would significantly help the passage of the treaty, because all those who are dissatisfied with it can agree that, if the next round addresses the issues they are disturbed about, this treaty would be more acceptable.

If the treaty is not likely to be passed, if in fact it is likely to be defeated, I for one do not want to keep silent about its arm control inadequacies and let the floor be held only by those who look at the treaty from a perspective very different from mine.

The main point is that one has to worry about the future of arms control negotiations. When all the shouting is over, if we are not careful we are going to find that SALT III has already been fixed in concrete and is beyond our control. Just as happened with MIRV, we may find that the horses we wanted to control have bolted the barn.

In some regards, I am the bearer of bad tidings. It is like the story of the emperor's new clothes. The things I want to say about the SALT agreement are only too clear.

Everyone will ask you to read the treaty closely, and I cannot disagree with that. But what you should do, in my opinion, is step *back* from the treaty and take a good look at what is really going on. Consider the best arguments of both sides, but remember that you represent the next generation of thinking about this issue. And if you do not approach this agreement from a different perspective, any different perspective, I shall be very disappointed indeed, because along with fixed positions come fixed presumptions.

To help you to think with a fresh outlook, I shall try to explain to those of you who are not graduate students in arms control more or less where the arms race is. Otherwise, I am afraid that your mind will be boggled by all the specifics. Consider these five quick ways of deciding who is ahead in the arms race.

1. The first thing you must remember is that both superpowers are really behind in the arms race for obvious reasons. If these weapons go off, we are going to be blown to smithereens. It is not necessarily true that the whole world will be blown up—the Bolivians, the Chileans, the Australians, the New Zealanders, the Africans, the Ceylonese, and the Indians may not. A lot of people are not in the line of fire. But the superpowers have put themselves into a peculiar fix, a unique fix. If all this weaponry goes off, through a war that no one wants, they are both going to be destroyed.

2. At the next level, there is what Leo Szilard called "saturation parity." In other words, there are so many warheads on both sides that it does not make much sense to say who is ahead. We have 10,000 warheads at the ready; the Russians have 5000, and under this treaty they may increase this number to 10,000. Dr. Van Cleave may be able to give a more precise estimate, but they are permitted to have several thousand more warheads, beyond any question. There are only 100 major cities on each side—so this is what is meant by saturation parity.

3. At the next level of analysis, there is incomparability. The Russians are ahead in some ways, and we are ahead in others. This is because we have not purchased exactly the same strategic weapon systems. If a Martian compared these systems, he would be amazed at how similar they are—but they are not exactly alike. There are differences, and the Russians have been catching up, as you might expect—the power that is behind usually does. But we have not been overlooking such possibilities as the cruise missile, which is a great leap forward from the point of view of the Defense Department. And our development of the cruise missile would leave the Russians again, as they often are, back in mass production of whatever it was we had before. As usual, we are working on the next dimension and the next round.

4. At the next level of analysis, I believe that most people would agree, and perhaps Dr. Van Cleave also, that we in this country would not want to trade our whole strategic weapons system and posture for that of the Russians. I think, by

and large, that if you want to say who is ahead in the sense of which of these strategic forces you would rather own, the Joint Chiefs and others would buy our system. (Of course, they might say that if these trends go on indefinitely they would not.)

5. Finally, one can ask whether there is anything that one side can do that the other cannot. Here we face the problem that is disturbing the hawks. In the early 1980s, the Russians will have the ability to knock out, on paper at least, one arm of our strategic deterrent, our land-based missiles. This does not mean that we can be disarmed. We have 5000 warheads on the submarines. But it means that something new is happening to the land-based missile arm of our deterrent, which in some ways we replaced years ago with the submarines. For our part, we shall not be able to do the same thing to the Russians' land-based missile force. By that time, I think that we shall be able to destroy perhaps 50% of it. And if the MX missile, which has been proposed, is actually built, we would destabilize the whole Soviet land-based missile force too, a few years after they did it to us. Now that is more or less where the "who-is-ahead" analysis brings us.

The next thing you have to understand is that since World War II there have been *two* debates in the United States on strategic policy. The reason we have this dialogue of the deaf between doves and hawks is that the doves and the hawks in this country look at the danger in completely different ways. Since the end of World War II there has been one school that claims that the problem is the "Russians." One must deter them, stay ahead of them, and prevent them from getting the wrong idea. The Russians are the ones who may attack us in Europe and attack us at home or elsewhere; we have to maintain our defenses against the danger.

There is another school that holds that, since about 1950 if not before, the Russians have been thoroughly deterred from nuclear attack, but that there is great danger in the arms race—that the arms race itself *is* the danger; that the war is likely to be a war nobody wants, not a deliberate breakdown of deterrence, and that we therefore have to control the arms race.

These two different views of where the danger lies have brought about two quite different debates. When the hawks look at the SALT treaty, they say, "Is it better for us or for the Russians? Do the terms favor us or the Russians? Who is going to get a military advantage out of it?" The doves ask, "Is it controlling the arms race? What is it doing vis-à-vis the arms race?"

Thus, the reason you have such a confused debate over this treaty is that neither side is really satisfied. The hawks look at the treaty and say, "This is not stopping the Russians from continuing their upward momentum," but they cannot blame the treaty for that. After all, this treaty offers, if anything, something for nothing. It is not preventing us from doing anything we had planned to do. Whether it is stopping the Russians from doing anything they plan to do depends on what their plans are. But one can make a case that it is constricting them in certain dimensions. And one can argue that they accepted the constriction because they had enough momentum going for them so that they do not mind—

and because they very much want this treaty for political reasons, even as our leaders want it for political reasons. But I do not think one can argue that this treaty is strategically to our disadvantage.

The hawks' main concern is with strategic balance. The Committee on the Present Danger claims that "probably no SALT treaty is safe at this time." They would rather have us go forward with other weapons projects, among which they have in mind the MX, in particular, to reestablish the strategic balance—after which SALT treaties *may* be a sensible thing. In the hearts of the hawks, there is the feeling that the treaty is going to be a soporific. It is going to persuade the United States not to build those weapons that the hawks want us to build in order to reestablish the balance they feel is lost. And in the heart of the doves is the feeling that if the treaty is defeated it will be a victory for the hawks, who will get all those weapons that, in fact, the doves do not want.

The doves do not worry about these imbalances, because they have considered the Russians deterred for a long time. Their approach is to ask what the treaty will do for the arms race. There is thus a big difference in how worried they are about Soviet throw weight, and the other dimensions in which the Russians may be ahead of us.

Now what I am worried about is the end result. I compare the United States and the Soviet Union in the SALT process to two alcoholics who agree that they will not go on the kind of binge that neither one of them had in mind anyway, but who feel that another drink or two on either side will not hurt, and that by a proper definition of alcoholic content one could probably have his wine while the other has his beer.

In 1974, when the limits of this agreement were first set at 2400 vehicles on each side and 1320 MIRVs, many people, Dr. Scoville among them, thought that these limits were much too high and that the treaty was probably worse than nothing. I remember very well arguing with him that we ought to support the treaty, when he felt that we should not. Since then, five years later, the big victory in the negotiations has been our right to go forward with the cruise missile—a Pyrrhic victory in terms of impact on arms control because, eventually, the Russians will have it too, and because cruise missiles are so small and so cheap that they are very hard to verify. This is going to make arms control a lot harder to achieve. Another victory, a minor one to the arms controller, has been fraction-ation limits—limiting the number of warheads on each missile to existing numbers.

According to President Carter, "The agreement constrains none of the rea-sonable programs we have planned to improve our defenses." That is true. As I said before, it is not stopping the United States from doing anything it was planning. The President added, "Moreover, it helps us respond much more effectively to our most pressing strategic problem, the prospective vulnerability in the 1980s of our land-based misiles. Without the SALT II limits, the Soviet Union could build so many warheads that any land-based missile system, fixed or mobile, would be jeopardized." In other words, the President bought an argument that would make even a dove blush. He bought the argument that these limits, although high, will prevent the Russians from getting so many warheads that they could blow to bits the MX missile, which he had called for. But the MX missile is not going to be deployed until the late 1980s and the early 1990s. This treaty runs out in 1986. What the President is telling us is that we should base a $30 billion strategic-weapons program on a piece of paper—a piece of paper that expires before the MX missile is deployed, a piece of paper we are going to have to renegotiate with the Russians.

How would you like to go to the Russians in 1986 and say, "Fellows, you know we would like you to continue this treaty because we have a $30 billion investment in this weapon system that you otherwise would be able to blow to smithereens." I do not want to have to deal with the Russians on that basis.

So what is happening is that SALT is being turned on its head. The MX missile system is going to be expensive, and vulnerable to Soviet attack. It raises serious questions about the value of the SALT treaty if one is going to have to give away MX missiles for it. And it is quite possible that we are taking giant steps backward with SALT purchased in this way.

Let us not forget that there are partial alternatives to the SALT process. I am a strong supporter of the SALT process, but I recognize that there is an alternative, which is to "buy only what we need." If the situation gets bad enough, we can break off negotiations and take the position that we do not have to keep up with the nuclear Joneses over there; we should buy neither less nor more than we need.

We do not have to give away weapons systems periodically in order to get two-thirds of the Senate to agree that negotiations with the other side are desirable. There is an alternative. I do not suggest that it be tried now. But I do propose that, if the SALT process, in the next round, does not make progress commensurate with its costs, we should keep in mind that there is another track. We do not have to accept any agreement that is presented as a fait accompli.

So in considering this agreement, you have to ask yourself the following question. Is this arms control? To be arms control, it has to deal with the probability of war, or the cost of war, or the cost of preparing for war. It cannot just be what an Assistant Secretary of the State Department referred to when he said, "Well, arms control is basically just an exercise in mutual self-confidence." That is not what I signed up for seventeen years ago, in working for arms control—"an exercise of mutual confidence." People who are going to use these treaties as an exercise of mutual confidence are fighting the last war over again, the war for détente. There was a big struggle in the sixties for détente. Treaties in that more hostile period were very important to maintain a modicum of good relations, but we are beyond that point now. We have to make real progress, substantive progress.

So, here is my proposal for the SALT II agreement. It will help its prospects for ratification, and it will help the SALT process. First, in conjunction with this treaty there should be a resolution that says, "Do not come back without having made important progress." We must see if we can get reductions in MIRVed land-based missiles on both sides. We have to start decelerating the arms race. A resolution of instruction relating to this treaty does not require amending the treaty and does not require the Russians to agree to it. It does involve the Senate's stating that it will not settle for another agreement that is "cosmetic only" in its rough outlines.

I am not saying that there is nothing positive in this treaty. But there is not enough to justify seven years of negotiation, and six more years under such a treaty. After all, the situation in the arms race is going to be worse when this treaty is over—just as it is worse now than it was when the negotiations started. We cannot continue to say that things are getting better, when in fact they are getting worse.

I propose that we try to construct a new consensus. We must reach some agreement between the hawks and the doves as to exactly what we can accomplish in this process. Now I grant, as I have tried to explain, that the hawks and doves look at this from very different points of view. And although, in logic, disarmament that solves strategic disabilities is a common ground between the two of them, logic is not what governs the political process; "psychologic" is a closer description. So it may not be possible to reach some common ground on what should be done next. And if it is not possible to reach a decision, there is not going to be any arms control, because the doves cannot do it by themselves. It may be that my proposal is apolitical, but it is also apolitical to think that we can establish a policy of arms control with just the doves alone. If the Russians are not willing to engage in serious disarmament, then we are going to have to decide what to do about it. I for one would go to a policy of "buy only what we need."

That policy does not mean that all arms control would be set aside. The ABM treaty is very important. The Comprehensive Test Ban is very important. And we could go forward on those fronts, even if we failed with offensive weapons control. But offensive arms control is looking a little bit like a tar baby today. We have to make a major effort to achieve real progress out of SALT. And now is the time to press for it. We do not have to reject the SALT treaty; we can press for progress with SALT. But do not let the treaty be defeated only by hawks. If it is going to be defeated, be sure that you express your discontent about it, because people in favor of arms control have as much right to complain about this treaty as anybody else.

7.   Reflections on Arms Control

George B. Kistiakowsky

At the time the Student Pugwash Conference was held, only weeks after President Carter and President Brezhnev solemnly signed the SALT II treaty on the limitation of strategic nuclear arms, expectations were high that this event signified a turning point in the nuclear arms race. Even then, however, some of us were troubled, because the Carter administration, apparently for domestic reasons, had delayed the finishing touches on the treaty for well over a year; because in the meantime the Soviet Union was engaged abroad in near-military activities inimical to American interests; and because, after signing the treaty, Mr. Carter announced that our nuclear arms needed modernization and that the treaty did not interfere with any of our plans for new strategic nuclear weapon systems.

Half a year later, as I write this, President Carter, in response to the military invasion of Afghanistan by the Soviets, has postponed indefinitely consideration of the treaty by the Senate, to which it was submitted last summer for ratification. Before the Soviet invasion, he had committed himself, in order to propitiate the hard-liners in the Senate, to large increases in military expenditures, so that in fiscal year 1981 the military budget may be nearly twice as large as the one he inherited in 1977 from President Ford. The Carter administration has also pressured our NATO allies to agree to the deployment in Western Europe of the new "theater nuclear weapons" targeted on the Soviet Union and thus conflicting with the spirit of the SALT II treaty limiting the number of such weapons. The result of the Soviet invasion and of President Carter's actions is that the SALT II treaty is probably dead, our relations with the Soviet Union resemble those of the cold war, and efforts to bring a halt or a slowdown to the nuclear arms race are being suspended indefinitely.

To gain better insight into what the future now holds for us, one needs to survey not only the extraordinary vagaries of the Carter administration policies but also the record of earlier negotiations and agreements for nuclear arms control.

1946–1960: Limited Progress

The beginning took place only a year after the bombing of Hiroshima and Nagasaki. A Washington panel, which included Robert Oppenheimer and was chaired by David Lilienthal, then the head of the Tennessee Valley Authority and later the first chairman of the Atomic Energy Commission, generated the concept for the internationalizing of atomic energy facilities and technology—for peaceful as well as military uses—under the auspices and control of the United Nations.

President Harry Truman chose Bernard Baruch to present and negotiate a detailed plan based on the Lilienthal report in the United Nations that fall. The proposal put forward by Mr. Baruch, a tough and pragmatic politician, stipulated that the United States would keep its nuclear weapons and would not disclose its nuclear technology secrets until all the nonnuclear states, including the Soviet Union, became totally open to international inspection.

The Soviet Union rejected this proposal. Knowing now that at that time the Soviet Union was feverishly engaged in developing its own atom bombs, one can surmise that no plan for internationalization of atomic energy would have been acceptable to Stalin. However, a more balanced proposal from the United States might have prevented a long hiatus in serious attempts to put the brakes on the nuclear arms race that followed the Baruch proposal.

In the 1950s, the frequency of nuclear weapons tests kept increasing and so did their explosive force, especially after the invention of the fusion or hydrogen weapons. As these tests were held mostly in the atmosphere, near or on the ground, they created massive quantities of radioactive products that were lifted into the stratosphere in mushroom clouds, then carried by high-altitude air currents and deposited in varying amounts over the entire northern hemisphere.

Soon, radioactivity was being detected in agricultural products, notably in milk, and the fear of radiation exposure generated a public movement to stop nuclear testing. This movement was geographically broader and included more varied elements of the population than the more recent movement, which opposes nuclear electric power plants because of concern about inadequate safety and radiation exposure in the event of catastrophic failure.

In the 1950s, Jawaharlal Nehru, the first Prime Minister of India, was especially forceful in his opposition to the testing of nuclear weapons, as expressed in his appearances before the United Nations and in other international forums. There was under way also a mass movement involving many countries and spearheaded by thousands of scientists. It was skillfully organized, led by such leading scientists as Linus Pauling, and endorsed by Albert Einstein, Bertrand Russell, and some other world notables.

The pressure of public opposition was strong enough to induce Nikita Khrushchev, who by then had succeeded Stalin, to propose in the spring of 1958 a moratorium on nuclear testing. President Eisenhower agreed, provided the moratorium would be succeeded by a verifiable treaty. After some false starts the moratorium went into effect in the fall of 1958 and lasted nearly three years, when it was abruptly terminated by the Soviet Union with a large series of tests in the summer of 1961.

Eisenhower's acceptance of the moratorium was influenced by the conclusion of a high-level technical panel in Washington that clandestine nuclear explosions can be detected by a variety of technical means, so that a treaty prohibiting tests could be monitored and compliance could be verified.

Armed with these conclusions, a delegation of American experts met in extended sessions with their Soviet opposite numbers in Geneva in the summer of 1958 and after hard bargaining hammered out a fairly detailed joint plan for a worldwide monitoring system to detect nuclear explosions in the atmosphere, on the ground, in the oceans, and underground. This meeting of experts was followed by formal negotiations between the United States, the United Kingdom, and the Soviet Union for a nuclear test ban treaty, which went on into 1960.

Within a short time it became clear that the meeting of the minds was far from complete despite the experts' plan. The central conflicts involved Soviet insistence that they retain a veto power over key decisions of the monitoring organization. They also wanted to circumscribe narrowly the freedom of action of the international teams that would be sent to inspect a locale suspected to have been the site of a clandestine nuclear test because of signals from the monitoring instruments. To this day the issue of "on-site inspections" has played a major role in arms control negotiations between the United States and the Soviet Union. We consider them to be important for verification, but the Soviets look upon them as a cloak for general intelligence gathering.

As 1958 turned into 1959, new obstacles arose. The analysis of seismic signals from the American underground explosions in September 1958 led to the conclusion in Washington that the seismic monitoring system agreed to by the experts earlier that year would be inadequate and that a far more elaborate one was needed for effective monitoring. The Soviets rejected this as an American subterfuge intended to wreck treaty negotiations.

By this time active opposition to the nuclear test ban treaty had hardened in the United States, the foci being the Air Force officials in the Pentagon, the Atomic Energy Commission, especially its chairman, and a group of civilian scientists led by Edward Teller. Before I was sworn in as Special Assistant for Science and Technology in July 1959, President Eisenhower told me that he wanted me to work hard for a nuclear test ban "consistent with our national security." That stipulation became the problem, because by then a number of plans, some involving technically exotic schemes for circumventing the treaty (such as testing in outer space *behind* the sun), had been invented by its opponents as "proof" that the treaty would be harmful to our security. In hindsight, it is ironic that the most vehement opponents of the treaty were those who believe in preserving our security through military superiority maintained by continuing the arms race, for had the test ban treaty come into force at that time it would have almost frozen the state of nuclear weapons technology and thus preserved our qualitative superiority. We enjoyed that position then, but the subsequent years of development by both sides have matured the technology and equalized it on both sides.

By the spring of 1960, my friends and I had some success in developing a treaty framework that was acceptable to President Eisenhower, Prime Minister Macmillan, and even Chairman Khrushchev. But then our U-2 plane piloted by Gary Powers was shot down by Soviet missiles over the Urals. The subsequent bumbling and mishandling of press releases by Washington ended when President Eisenhower personally assumed responsibility for the U-2 photographic intelligence flights. This incident invited the political disaster of the Paris Summit Conference—wrecked by Khrushchev, who refused to deal with Eisenhower, ostensibly because the latter was "engaged in spying."

A short time after returning from Paris, President Eisenhower said to me sadly that in his later years he had focused all his efforts on ending the cold war, that he was heartbroken by the U-2 affair and its aftermath, and that he now saw nothing positive to accomplish in his remaining months in office. Indeed, the Soviets gradually withdrew from all the tentative agreements on the test ban, and the negotiations were terminated.

Perhaps the brightest spot on the gloomy panorama of the nuclear arms race is the treaty on Antarctica signed in Washington in 1959. It is a multinational agreement that completely demilitarizes the Antarctic continent and provides for cooperation in its scientific exploration, which has been quite successful. The thorny problem of the conflicting territorial claims by several states largely of the southern hemisphere could not be resolved, but it was at least postponed. Unfortunately, the likelihood that oil may be present under the off-shore waters of that continent, coupled with the worldwide oil shortage, is activating and intensifying these old territorial conflicts and may yet lead to the abrogation or termination of the Antarctica treaty.

1960–1969: Treaties and New Developments

Trilateral negotiations for the comprehensive nuclear test ban were resumed by President John F. Kennedy, but neither he nor Khrushchev showed great eagerness to make concessions to achieve the treaty. In Washington, political forces demanding continuing development of nuclear warheads were in the ascendancy, and Kennedy, with his anticommunist stance, did not oppose them forcefully enough, apparently because he was less concerned than was Eisenhower about the probable doomsday-like end result of a continuing nuclear arms race.

Strong public opposition was generated again by the resumption of nuclear weapons testing on a grand scale by the Soviets in the summer of 1961 and some months later by the United States, creating political problems both for Washington and for Moscow. This difficulty, coupled with the reluctance of the superpowers to accept a comprehensive test ban, was resolved when they agreed in the summer of 1963 to a partial test ban treaty that allowed underground tests if they were so carried out that no radioactive products would vent into the atmosphere and be carried beyond national borders.

The agreement to continue the tests underground meant that the treaty would interfere relatively little with the plans of nuclear weaponeers, except in the already well-developed and therefore less important domain of multimegaton warheads. When President Kennedy stated further that the development of warheads would be speeded up if the treaty came into force, only those on the extreme wing of the antitreaty forces, such as Dr. Teller, continued to oppose treaty ratification in the Senate hearings.

Compliance of the superpowers and Great Britain with the treaty drastically reduced the worldwide dissemination of radioactive fallout, and this brought to a virtual end the public test ban campaigns. Only in Australia and New Zealand did such opposition continue and eventually win out over France, which had refused to join the treaty and for more than a decade carried out atmospheric nuclear tests on its colonial possessions in the South Pacific. The People's Republic of China, also a nonsignatory, to this day carries out infrequent atmospheric tests, and only in Japan (which occasionally gets some fallout) are public voices still heard protesting these activities.

New negotiations on a comprehensive test ban were started by the Carter administration, but opposition from the nuclear weapons laboratories and their allies has been so strong that these negotiations seem to be in a state of deep coma. One must conclude sadly that the nuclear test ban conceived under strong public pressure for a major arms control measure has failed to play that role in its present form. What we have is highly beneficial as an environmental antipollution measure, but it is a measure that is costly to the progress of disarmament forces, because it has led to the demise of the organized and vigorous public opposition to nuclear weapons technology that was quite effective in earlier years.

In the decade following the establishment of the partial nuclear test ban, several multinational treaties became part of the record. None of them curtailed on-going military activities, but like the Antarctica treaty they imposed some bounds to the diversification of the nuclear arms race.

A nonnuclear zone comprising much of Latin America (but not the largest countries, Argentina and Brazil) was agreed to, thus denying that part of the world as a place for the storage and transit of nuclear weapons. It was also agreed not to place "weapons of mass destruction"—that is, nuclear warheads in the first instance—on artificial earth satellites and deeper space probes. Similarly prohibited was the deployment of nuclear weapons on or under the floor of the oceans.

There are strong military reasons against engaging in either of these activities. The satellites are too easy a target for incapacitation by enemy missiles to rely on them in peace time as the carriers of nuclear arms. Besides, one must take as part of the bargain public alarm and indignation accompanying the inevitable re-entry into the atmosphere of some of these nuclear-armed satellites. Some indication of that response may be gauged from public reaction to the break-up over Northern Canada of a Russian satellite carrying only a small nuclear reactor, and to the break-up over Australia of our Spacelab, which had no radioactive components.

As regards the placement of nuclear mines and similar weapons on ocean floors, the likelihood of the other superpower or perhaps some third party fishing out these devices to learn their supposedly well-guarded technical secrets—or to arm itself—is much too real to be disregarded. To appreciate the perceived importance of such a loss, one needs only to recall the extended efforts and the millions of dollars spent by the United States Air Force to recover an H-bomb that was jettisoned some years ago into the Mediterranean deep by a B-52 bomber in trouble.

Distinctly different in its objectives is the Nonproliferation Treaty of 1970. As originally worked out by the superpowers and submitted in 1967 to the United Nations Committee on Disarmament, it called for no commitments on the part of the superpowers except the promise of assistance with peaceful nuclear technology to the nonnuclear weapons states that would adhere to the treaty. In return for this vague promise the nonnuclear states were to commit themselves not to acquire nuclear weapons and to open their nuclear facilities to international inspection.

The reaction of the nonnuclear states to this proposal was quite negative. They saw it as leading to the division of the United Nations members into two permanent classes with different degrees of sovereignty, and they objected to the substantial surrender of their own sovereign rights in return for so little.

The main result of an extended debate was the addition of another article to the treaty in which the nuclear powers committed themselves to work without delay toward their own nuclear disarmament. A large number of states have by now adhered to the treaty, but the ones that have declined or have postponed doing so, including Argentina, Brazil, India, Israel, Pakistan, and South Africa, are those that have the technical and intellectual resources needed to acquire nuclear weapons technology soonest.

Now, nearly ten years later, disarmament among the nuclear weapons states has not even begun; instead, the superpowers have more than doubled the number of their strategic nuclear warheads and have added thousands of other warheads to their arsenals. Moreover, the evidence is decidedly scarce that adherence to the treaty by the nonnuclear states netted them preferential progress in peaceful nuclear technology. Consequently ferment is ripe among these states, and the next quinquennial review of the treaty is likely to see a strong challenge to its continuation.

During the early and mid-sixties the United States was in a position of undisputed superiority in strategic nuclear arms—in ICBMs, heavy bombers, and submarine-launched ballistic missiles. The prevalent doctrine of assured destruction called only for the possession of strategic forces so strong and invulnerable that even after a successful surprise attack on our forces by the Soviet Union (or by China) the surviving strategic arms would be more than adequate to inflict unacceptable damage on the Soviet Union (and/or China). "Unacceptable" was arbitrarily defined as such damage to its industrial structure and such loss of the skilled work force that the Soviet Union would cease to exist as a modern industrial state.

The outcome of the Cuban missile crisis in 1962 was a severe political setback to the Soviet Union, and it occurred at a time when President Kennedy was engaged in a further expansion of our offensive strategic forces beyond the level approved by President Eisenhower. These factors were probably central in the decision of the Soviets to undertake a massive buildup of their own strategic nuclear forces. As a result, by the early seventies they had deployed in hardened silos and in submarines quantities of offensive missiles that matched our own, and more were under construction. The Soviets were also engaged in initial deployments of antimissile defense systems (the ABMs), which were later discontinued as ineffective. They were also involved in a country-wide deployment of an antiaircraft missile system capable of bringing down even supersonic high-altitude bomber. These several defensive activities of the Soviets caused much concern in Washington, especially among the hawks and their perennial leader, Paul Nitze, who argued that Soviet defenses may mean the failure of our retaliatory strikes in Soviet eyes and therefore the loss of our deterrent to their surprise attack.

The first response to this hypothetical threat was the deployment of multiple nuclear warhead re-entry vehicles on Polaris missiles, which increased the number of targets with which the Soviet ABM forces would have to contend. A more advanced project of multiple independently targetable re-entry vehicles (MIRVs) was also begun, initially in great secrecy.

The development of an American ABM, which began in the late 1950s, was pushed, and in 1968 began the deployment of a system of "light" defense of several cities, supposedly against a weak attack, such as by Communist China. In the Pentagon, however, the decision by the Johnson administration to deploy this Sentinel system was welcomed as the start on a massive defense system to neutralize the threat of a Soviet nuclear attack.

In the meantime, strong representations were being made to the Soviets, and personally to Prime Minister Aleksei Kosygin, that deployments of ABM would be responded to on both sides by an endless expansion of offensive missile forces in order to preserve adequate retaliatory strike capabilities. Whether for this or for other reasons, in 1968 the Soviets accepted a proposal from the United States to start SALT, the Strategic Arms Limitations Talks. President Lyndon Johnson was preparing to depart for a visit to Moscow, there to open these talks, when the Soviet Union invaded Czechoslovakia and the United States postponed the talks indefinitely.

As the work on the Sentinel system progressed, the prospect of having in their suburban backyards these nuclear antimissile missiles, an attractive target for Soviet attack, upset our citizenry, and in response to widespread protests the Sentinel project was liquidated early in 1969 by the incoming Nixon administration. In place of Sentinel, the Nixon administration, overcoming substantial opposition in Congress, began the deployment of Safeguard, a missile defense installation aimed at protecting Minuteman ICBM silos. Although the name given to this system was new, actually it employed almost the same technology as the Sentinel system.

1969 to the Present: The SALT Process

In the fall of 1969 the Nixon administration began the SALT negotiations, which, after a relatively short time, led to agreement on an ABM treaty and an interim agreement on offensive missiles; these were signed in the summer of 1972. This SALT I treaty so severely limited the deployment of existing hardward and so restricted the development of new forms of ABM that since that day the defense against missile attacks has ceased to be a major policy issue and appears to be dormant for the time being.

SALT I also included an interim agreement on offensive strategic missile launchers, to last five years. This agreement froze the American deployment at the existing level of 1000 Minuteman silos, 54 Titan silos, and 656 launchers on Polaris-type submarines. The Soviets were allowed to complete many ICBM silos and quite a few missile submarines then under construction, but to make no new starts. The agreement specified numbers of launchers rather than numbers of missiles because of the far greater difficulties in monitoring the missiles. When Soviet construction was completed, they had deployed about 1400 land-based ICBMs and about 950 submarine launchers.

SALT I thus allowed the Soviets to deploy considerably larger numbers of missiles. Some of them, moreover, such as the SS-9, SS-18, and SS-19, could lift much heavier payloads of warheads than those in the American strategic force. The defense of SALT I by the Nixon administration against its critics stressed (a) that the United States possessed a far stronger fleet of strategic bombers than the Soviet Union, and (b) that the Soviet submarines were handicapped by the limited choice and geographic location of their home ports. The most potent defense was that, whereas the Soviets may have more missiles, the American force would have more independently targetable warheads. This is because in 1972 MIRVs began to be deployed on the Minuteman III ICBMs and on Poseidon missiles, which were carried by the modernized Polaris submarines called Poseidon.

Although the original impetus to the development of MIRVs was provided by concern about the growth of Soviet ABM defenses, the MIRV project was not canceled when these defenses were gradually recognized as existing largely in the minds of our paranoic hawks. We all must gravely regret the decision of the Nixon administration not to propose a ban on testing and deployment of MIRVs early in the SALT sessions, when it might have been still acceptable to the Soviets. Opposed to this proposal was a powerful group of Washington hawks, who chose to disregard the obvious consequence of deploying our MIRVs—a similar deployment by the Soviets, which indeed began about five years later. Both sides are now acquiring the means to knock out several hardened ICBM silos by one missile equipped with accurate guidance and several independently targetable warheads. This possibility weakens the effectiveness of a strategy based on ability to inflict unacceptable damage in a retaliatory strike. On the contrary, it creates at least a theoretical possibility of disarming the opponent by using but a fraction of one's MIRV-equipped missile force. Hence it increases the temptation to stage a surprise attack against the adversary's ICBM silos in times of international crisis and so to ensure against a similar attack on one's own forces. The stability of nuclear deterrence is being undermined, and the probability of the escalation of a conflict into an exchange of strategic nuclear forces is being enhanced.

Other politically important consequences of SALT I were (a) the commitment not to interfere with the "national technical means"—that is, with spy satellites, radars, etc.—needed to monitor compliance with the agreements, and (b) the setting up of the standing joint consultative committee to discuss and, to the extent possible, resolve challenges of noncompliance. Except for the deployment of Soviet SS-19 missiles in the silos of the older SS-11, which complied with the letter but not the spirit of the interim agreement (because the SS-19 missiles were considerably larger than the SS-11, although fitting into the same silos), the consultative committee functioned well, and the monitoring has not been interfered with by the Soviets since 1972. The deployment of the SS-19 was apparently remonstrated against but was probably expected and was not deemed important enough by the United States to abrogate SALT I.

In 1974 President Ford and President Brezhnev signed the so-called Vladivostok agreement, which set an upper limit of 2400 on the total number of strategic launch systems, including heavy bombers. As this agreement was never ratified, the Soviet Union has kept its 2500 strategic launchers to this day, of which 156 are heavy bombers, so that it complies with the SALT I limits on its missile launchers.

Negotiations on the SALT II treaty began soon after the ratification of SALT I. About that time the United States authorized the development of long-range precision guided cruise missiles, supposedly as a bargaining chip for SALT II. The Trident submarine project was also started about that time.

Jimmy Carter campaigned for the presidency on promises to cut military budgets drastically and to end the nuclear arms race. Shortly after assuming the presidency, he bypassed the SALT II negotiations and in March 1977 sent Secretary of State Cyrus Vance to Moscow with a proposal to reduce strategic forces fairly drastically. Unfortunately for this venture, the reductions emphasized heavy ICBMs, the mainstay of the Soviet strategic arsenal, and affected American forces only slightly. Rumors have it that certain dedicated hawks in Washington had a heavy hand in shaping this proposal and that the significance of their activity was lost on the new administration. Be that as it may, the Soviets flatly rejected the proposal instead of accepting it as a basis for negotiations; Secretary Vance returned empty-handed.

George F. Kennan, the veteran Soviet expert and former ambassador to Moscow, said about the venture: "The new administration has made just about every mistake it could in these Moscow talks and has defied all the lessons we have learned with the Soviets since the last World War." Regrettably, that was apparently not the inference drawn by President Carter, who instead blamed the Soviets for the failure.

The Geneva negotiations on SALT II were resumed, and by the beginning of 1978 only details of the treaty were incomplete. Then followed a year and a half of what can only be called foot-dragging by the Carter administration, evidently unable to reach internal consensus on the treaty and the accompanying protocol. Meanwhile, some portions of the still-confidential treaty were leaked to unfriendly senators and the media by enemies of the treaty within the administration, with the intent of showing that the treaty was harmful to our security. President Carter's defense of the treaty was less than effective, partly because, instead of emphasizing our strength, he began conceding to the hard-liners that the United States was falling behind the Soviet Union in strategic arms.

Several geopolitical events were also taking place, which were linked by the opponents of the treaty to the question of its desirability. It was seen as an essential part of détente with the Soviet Union, although in the meantime the Soviet Union provided military aid to Hanoi, essential for its invasion of Cambodia, and revolution took place in Iran, led by elements unfriendly to the United States and resulting in the loss of our intelligence installations, which had gathered electronic information on Soviet missile tests. The Soviet Union assisted, if it did not engineer, the wars initiated by the Marxist governments of Ethiopia and South Yemen. A coup supported by the Soviet Union took place in Afghanistan. A Soviet combat brigade was supposedly discovered in Cuba, and President Carter's inept handling of the incident added to anti-Soviet feelings in Congress.

Thus, notwithstanding President Carter's increasing promises of money and new weapons to the Pentagon, the SALT II treaty was losing ground in the Senate. By the fall of 1979, substantially less than two-thirds of the senators were either in favor of or leaning toward the treaty. Then President Carter allowed the Shah of Iran to enter the United States, the United States embassy in Tehran was attacked in response, and over fifty hostages were taken. The Soviet Union took a negative stance toward President Carter's Iranian policies, and that of course increased the opposition to the SALT II treaty in the Senate.

The death knell to hopes of ratification was dealt by the ruthless invasion, overthrow of the government, and occupation of Afghanistan by Soviet forces. President Carter responded with several "punitive" actions, among which was a request to the Senate to postpone indefinitely the debate on treaty ratification. He noted, however, that the treaty was advantageous to our national security and should eventually be ratified.

About the desirability of ratification there can be no rational doubt, even though as an arms control measure it leaves much to be desired. It provides a limit to strategic launchers that is lower than that of the Vladivostok agreement or present Soviet deployment; it limits MIRVs numerically—not enough, to be sure, but better than would be the case in the absence of any limitation. And it spells out clearly that "national technical" means of intelligence gathering must not be interfered with. All these steps are advantageous to arms control and to our national security. Without them, the Soviet Union, which is in the midst of a large "modernization" program of its strategic missiles, could rapidly expand its ICBM forces and deny to us knowledge of the location of its new missile silos.

One should therefore welcome President Carter's proposal that the United

States will continue to abide by the limitations imposed by past agreements, as we and the Soviets have done since the SALT I interim agreement expired in the fall of 1977. It is to be hoped that the Soviet Union will respond favorably to this proposal, and thus a foundation will be preserved on which to reconstitute the treaty when·and if the political climate clears enough to resume negotiations.

To expect this to happen soon, however, would be showing totally unrealistic optimism. The actions taken or promised by President Carter in response to Soviet aggression in Afghanistan augur the beginning of another cold war with the Soviet Union, which could become a hot war if the Soviets make further military moves toward the oil sources of the Middle East. Moreover, the commitment to abide by earlier agreements does not mean the cessation of the strategic arms race. The MX mobile ICBM, the Trident II missile, which is also designed for explicit counterforce use, and a new strategic bomber can not only be developed but also deployed, replacing some obsolescent weapon systems; there is nothing to stop deployment of air- and ground-launched cruise missiles by the thousands or to prohibit deploying intermediate-range missiles in western Europe. In plain words, President Carter's proposal allows even a costlier and more dangerous nuclear arms race than anything engaged in heretofore. A truly enlightened leadership is needed to prevent these developments from taking place. Although we have not yet reached the state of the 1962 Cuban missile crisis, we have traveled a frightfully long distance toward that state from the situation that existed when Jimmy Carter became our President.

Has this grave worsening of relations with the Soviet Union been caused by or at least stimulated by the arms control negotiations, as has been asserted by some analysts? Or has the SALT II treaty been the victim of both the Soviet's aggressive expansionism and Carter's foreign policy gyrations? In damning the arms control negotiations as counterproductive, their opponents point to their failure to stop the nuclear arms race and, worse still, to the deployment of new weapons systems that were begun as bargaining chips in anticipation of new treaties.

But consider the alternative to the existing agreements. A small clue is offered by conventional arms, about which only token efforts for international controls have been made. Annual world military expenditures have exceeded $300 billion; about 25 million men and women are under arms; the annual weapons trade exceeds $20 billion, and much of it involves poor, underdeveloped nations. Surely all this is not evidence that absence of arms control agreements is benevolent. In comparison, in fact, the proliferation of nuclear arms might be almost called restrained, thanks to arms control agreements. Without them, the Antarctic continent would certainly not be demilitarized any longer; nuclear warhead deployments might be proliferating on ocean floors and in space. Without the partial test ban treaty, atmospheric and exo-atmospheric testing would undoubtedly be continued—for instance, to develop novel ABM warheads, or to test effects of nearby explosions on missile silos.

The absence of the ABM treaty would invite the deployment of more advanced missile defense systems than those developed in the 1960s. These deployments would be in full swing and would stimulate limitless expansion of offensive strategic forces to preserve the capabilities for counterforce and retaliatory strikes. Furthermore, in the absence of the SALT I treaty, the policy of noninterference with our intelligence satellites, which the Soviets adhered to in the sixties, could easily come to an end. Satellite intercepts, electronic jamming, etc., would cause the loss of much of our knowledge about Soviet military capabilities and hence would contribute to the proliferation of wildly exaggerated estimates of them. This was the situation in the 1950s, and it caused much harm to our policies.

On balance, then, one is led to conclude that nuclear arms control negotiations and agreements have a beneficial effect, but their effectiveness is unquestionably very low. Why? And why the failure of the last three years? It is my conviction that the answer lies partially in the failure of our citizenry to question and reject as false the propaganda of the hard-liners, which has lately blanketed the country. Another important factor is that in the last few years some other concerns, such as civil rights, protection of the environment, and the hazards of nuclear reactors, have taken precedence among activist citizens, and the "peace lobby" has nearly vanished. To change this trend one must somehow convey the brutal fact that none of the above concerns will be of the slightest relevance if nuclear war is upon us. Therefore our first priority must be the prevention of nuclear war—which cannot be assured or even made probable in the climate accompanying the nuclear arms race.

8.　SALT II and European Security*

Hans-Gerd Loehmannsroeben

Introduction

In the post-World War II era, Europeans have learned, consciously or unconsciously, that they have to live with ever-growing nuclear forces in the two adversary blocs, the North Atlantic Treaty Organization (NATO) and the Warsaw Treaty Organization (WTO). With the frontier of those blocs dividing the continent, the people in Central Europe have had to accept the fact that in case of a nuclear exchange between the superpowers their home countries are likely to be the battlefields. They have learned this through numerous "war games" in NATO maneuvers and through various books by generals painting a frightening picture of a nuclear war in Europe. Elaborate studies have analyzed the consequences of a nuclear nightmare in detail,[1] without significantly influencing the European perception of the dangers of a nuclear confrontation and the necessity of arms control and disarmament.

Nevertheless, the governments in NATO have taken an active part in the Strategic Arms Limitation Talks (SALT) from the very beginning. There was controversy about a few details in SALT I, but, seeing that their vital interests were not endangered by that agreement, they overwhelmingly greeted it with satisfaction and relief. There was, however, some anger about the fact that the consultations of the West European governments by the United States did not extend much beyond a willingness to inform them of what had already been decided.[2]

---

*In writing this paper I have benefited from discussions with Dr. D. Schroeer, Associate Professor of Physics at the University of North Carolina, Chapel Hill.

[1]A detailed study for West Germany can be found in Carl Friedrich von Weizsaecker, ed., *Kriegsfolgen und Kriegsvorbeugung* (*Consequences of War and the Prevention of War*), Muenchen, Carl Hanser Verlag, 1971. See also Herbert F. York, "The Nuclear 'Balance of Terror' in Europe," *Ambio*, 4 No. 5/6, 1975, pp. 203–208.

[2]A good examination of the relationship between the United States, SALT, and Western Europe is provided by Ian Smart, "Perspectives from Europe," in *SALT: The Moscow Agreement and Beyond*, Mason Willrich and John Rhinelander, eds., London, The Free Press, 1974.

In the course of the discussion of the proposed SALT II agreement, however, it has become more and more evident that the security of Western Europe is going to be the crux of a very controversial debate. The various positions include severe warnings about "the threat from the gray zone," warnings that appear now even in official declarations by the governments,[3] and in statements of high-ranking NATO officers.[4] Politicians question the sufficiency of the NATO strategy of flexible response, claiming that Russian superiority in conventional and tactical nuclear weapons (TNW) makes the principle of limited retaliation by the Western side in case of a limited aggression inapplicable.[5] The proposed SALT II agreement is claimed to be part of the reason for an allegedly rapid buildup of Russian forces in Europe.

In return, advocates of the new treaty deny the existence of a Russian superiority and declare that the United States still can guarantee West European security. In their opinion, the principle of mutual assured deterrence is still valid and will protect Western Europe from Russian aggression.[6]

It seems to me that these diverging positions result from a confusion about fundamental aspects of the present political and military situation in Europe; a confusion both in military/technical and political/ideological terms.

As a result of these misunderstandings and uncertainties, Central Europe appears presently at the threshold of a nuclear arms race. To overcome this dangerous new facet of the worldwide arms race, we have to eliminate the previously mentioned confusion by exposing improper and irrelevant considerations.

A European Nuclear Arms Race?

Tactical Nuclear Weapons in Europe: Since the early 1950s the NATO nations have had to tolerate a certain numerical advantage of the WTO in conventional forces in Europe; a "calculated conventional inferiority" was accepted. To minimize this imbalance, however, the United States started in 1953 to bring the then so-called tactical, or battlefield, or theater nuclear weapons to Europe. Although their deployment in Europe was not negotiated with NATO member nations, they were tacitly accepted, particularly by the major recipient, West Germany, at that time not yet a NATO member. The Soviets responded in 1957 by deploying

[3]See, for instance, "Security and Detente," Declaration by the Minister for Foreign Affairs of the FRG, Hans-Dietrich Genscher (February 16, 1979), recorded in *Statements and Speeches*, German Information Center, March 2, 1979.

[4]A good example is the statement of General Alexander Haig, "What NATO Faces Today," *The Atlantic Community Quarterly*, 16, No. 4, 1978, pp. 425–439.

[5]Conrad Ahlers, "NATO's Flexible Response Strategy No Longer Sufficient," *Deutsche Zeitung*, February 23, 1979, reprinted in *The German Tribune*, March 4, 1979.

[6]See, for example, Theo Sommer, "Weltuntergang als Notloesung" (World Collapse as the Ultimate Solution), *Die Zeit*, March 16, 1979.

TNW themselves in Eastern Europe. The number of these new weapons increased steadily until the end of the 1960s, when the policy of détente started to reduce the tensions between East and West. The political détente culminated with the signing of the Final Act of the Conference on Security and Co-operation in Europe (CSCE) in Helsinki on August 1, 1975. In order to continue military détente also, the Mutual Force Reduction (MFR)* negotiations were initiated in Vienna on October 31, 1975. These negotiations have not yet reached a stage of significant progress and furthermore have been focused on general-purpose ground forces and tanks, rather than on nuclear forces. Tactical nuclear weapons have been excluded for a variety of reasons, one of them being the great difficulty of evaluating adequately the entire package of nuclear arms. Within NATO a certain consensus exists today on dividing nuclear weapons into three categories:

1. Battlefield nuclear weapons, including mostly artillery and short-range battlefield missiles. The number of these weapons is very hard to estimate, both because of problems of numerical verification and because of their continuous modernization. For locations in Europe the most frequently mentioned numbers are 7000 for the United States and 3500 for the USSR. The latter number has acquired unwarranted validity mainly through constant repetition in the press and in the academic literature. There is, however, no publicly available evidence on their actual presence among Soviet forces in Europe.[7] It was finally decided in Vienna to negotiate the reduction of these weapons as part of the MFR discussions.

2. Medium-range weapons, so-called theater nuclear weapons. Until recently these consisted mainly of submarine missiles from the United States and the USSR and of the French and English nuclear forces. These "gray" weapons are not included in the MFR debate, nor in the SALT discussions.

3. Long-range strategic delivery systems negotiated in the SALT talks.

Threat from the "Gray Zone"? As we have seen in the preliminary analysis, medium-range weapons are not covered by any limitation talks—that is, neither by SALT nor by MFR. On these grounds this area is called the "gray zone." It is in this intermediate area where most of the military and defense analysts in the NATO nations see a new Russian threat coming.

Two Soviet weapons systems are considered to be the particular cause of what is called a critical situation for the security of Western Europe:

1. Soviet medium-range offensive weapons, in particular the most modern missile in this category, the SS-20. With a range of more than 4000 km (2500

---

*The negotiations were first proposed under the name Mutual Balanced Force Reductions (MBFR), but at the exploratory talks in 1975 the Russians insisted upon deleting the term "balanced."

[7]A fundamental treatment of the problems of TNW in Europe is provided by "Tactical Nuclear Weapons. The European Predicament," contributions by several authors, *Bulletin of Peace Proposals,* **9** No. 4, 1978, pp. 378–386.

miles) this missile can be equipped with multiple warheads. The latest numbers indicate that around 100 SS-20s are available for the Soviet forces in Europe.[8]

2. Soviet medium-range TU-126 jets, known under the NATO codeword Backfire, which can fly roughly 6000 km (3750 miles) without refueling.

Defense experts argue that the increasing numbers of these weapons in Soviet arsenals are leading to an advantage for the Soviets much higher than is needed for purely defensive tasks. The rate of growth of the Russian nuclear armament is said to be inexplicable from a defensive policy perspective. If we assume that Soviet intentions are offensive, the conclusion of most military experts is the same in the United States and Western Europe: in order to maintain a credible deterrence, the modernization of NATO's tactical nuclear weapons is inevitable, for a weak link in the chain of escalation would make a strategy of flexible response impossible. The Atlantic Alliance, they claim, cannot afford to lag behind the Soviets in nuclear capabilities. As result, the deployment of cruise missiles and the "neutron bomb" is advised to keep pace with the Russians. These new weapons systems are assumed to be necessary as a counterweight to the "gray-zone gap."[9]

West Europe and SALT II: At this stage of our considerations the question arises: How did we get into the present situation? Why is the "threat from the gray zone" suddenly in vogue among military and defense experts, among politicians and journalists?

As early as 1974, after the Vladivostok agreement, the Strategic Arms Limitation Talks were said to be focusing disproportionate attention on intercontinental capabilities, lacking sensitivity for the medium-range potentials of the superpowers. Some authors even introduced the terms "first front" and "second front" to describe the competition for intercontinental and medium-range delivery systems, respectively.[10] This concern obviously stems from and was enhanced by the Russian strategy in the SALT negotiations. From the beginning, the Soviets tried to get the American Forward-Based Systems (FBS) and the French and English nuclear potentials involved in the negotiations.* On the other hand, they refused to accept restraints on their Backfire bomber and on the

---

[8]A detailed analysis of Soviet and NATO theater nuclear weapons can be found in Air Vice-Marshal Stewart Maraul, "The Shifting Theatre Nuclear Balance in Europe," *Strategic Review*, 6, No. 4, 1978, pp. 24–35.

[9]See, for instance, Hans Ruehle, "Cruise Missiles, NATO and the 'European Option'," *Strategic Review*, 6, No. 4, 1978, pp. 46–52.

[10]Fred Charles Iklé, "SALT and Nuclear Balance in Europe," *Strategic Review*, 6, No. 2, 1978, pp. 18–23.

*The FBS are mainly American aircraft based in Europe and on carriers. These aircraft are capable of delivering nuclear warheads to Soviet territory. The French and British nuclear forces consist of aircraft and submarines from which approximately 120 warheads could hit targets in Russia.

SS-20, claiming that these medium-range systems were not able to reach the United States and therefore were not in the realm of SALT.

The SALT I agreement resolved the problems of the FBS and the third-country nuclear potentials by ignoring them, whereas the Backfire bomber was one of the main obstacles in reaching a new SALT II agreement. Neither the Backfire bomber nor the SS-20 is covered by the SALT II treaty, although the Soviets have given assurances that the rate of production of the bomber, now about 30 per year, will not be increased. In the light of these agreements, the military and political leaders of Western Europe are skeptical as to whether Western European interests have been considered and whether there still is a guarantee of Western Europe's Security. Besides being disturbed about President Carter's wavering decisions in the cases of the neutron bomb and the B-1 bomber, they are afraid that the United States, in the wake of Vietnam and Watergate, no longer is ready to take enough global initiative to make Western protection convincing.

But at the same time no European member of NATO has been willing to bear responsibility for constructive arms control proposals. Opportunities afforded by the Alliance's Nuclear Planing Group (NPG) have been little exploited. Separated by national interests and unable to state their views clearly, they rather happily accepted a postponement of European participation in strategic arms control. Since only a future treaty including medium-range potentials (particularly cruise missiles) would be of any significance, and since these weapons systems will be based mainly in Europe, a multinational approach to SALT III is absolutely necessary.[11] On starting to negotiate the avenues to a third SALT agreement, the NPG already faces initial problems, and it is safe to forecast more major difficulties within the Atlantic Alliance.

Inadequacies in the Discussion

Up to this point the views of NATO's most prominent military leaders, politicians, and experts have been represented. They express themselves as a kind of intellectual monopoly, occupying almost the whole "public" discussion. Their basic assumptions, which vary only in details, are rarely challenged.

But there is a difference between published and public opinion. This was learned by the surprised West German and American governments during the massive public uproar against the deployment of the neutron bomb on German soil. In addition, there have been numerous attempts by scientists and politicians to find alternatives to the prevailing methods of dealing with the arms race, especially in Sweden, Norway, the Netherlands, and Great Britain.

In this section we shall follow some of these alternative arguments and examinations of the present political and military situation. It seems all too obvious that "the superpowers have indulged in subterfuges and half-truths, with their closest and usually most dependent allies following suit or keeping silent.

---

[11] For proposals of joint SALT negotiations, see Lothar Ruehl, "NATO Europeans Call for a Say in the Drafting of SALT III," *The Atlantic Community Quarterly*, 16, No. 1, 1978, pp. 46–50.

On balance, there has been no real advance toward limitation of armaments. The competitive race between the two superpowers has steadily escalated, and the militarization of the economy and national life of almost all countries has intensified."[12]

Numerical Comparison of Forces: A starting point of almost any analysis of the "balance of power" is the comparison of the potentials of NATO and the WTO. We are told almost daily about some kind of "superiority" of the Russians. Often, however, the analyses are not very careful, as they do not take into consideration the immense complexity of weapons systems and their supporting subsystems. For an effective assessment of two similar systems a set of weighting parameters is necessary; that is, it is necessary to understand what are the most important criteria of the system.

Consider, for instance, the differences between Russian and American ICBMs. The former have an outstanding payload, whereas the latter are smaller and of significantly higher accuracy. Which are the better missiles? The often-used formula for "circular error probability," $CEP \propto Y^{1/3}$ where CEP measures the accuracy, and Y is the yield, is rather arbitrary, although developed from the energy distribution in space. It does indicate that accuracy is far more important than yield. But in this case we are considering only two parameters. Hot or cold launch, solid or liquid fuel, single or multiple warheads, range and liability—all these characteristics play an important role in the final judgment of the "quality" of a missile.

A direct comparison, of course, often becomes impossible because of the great asymmetries in the development and deployment of systems in the Soviet Union and the United States, as in the case of the different emphasis put on bomber forces by the two superpowers. These asymmetries may have historical or technological reasons, or they may result from rivalries within the military; in any case, they make a comparison of potentials very difficult and are a major obstacle in arms control efforts.

Moreover, modern weapons systems are highly dependent on a series of subsystems, as well as on operational characteristics, mobilization, logistics, etc. In order to make a judgment of two competing weapons systems, it is necessary to make a complete systems analysis regarding not only all technical aspects of each system and its subcomponents but also these characteristics for supporting and implementing the system. Because of the fundamental differences in the structure and weapons of the NATO and WTO forces, it is evident that even an elaborate systems analysis is only of limited value. An incomplete consideration of just one or of a few of the numerous components would be inadequate and dangerous!

Worse problems are yet to come. Weapons at least have technical values that can be compared. But military forces have to be regarded in the light of history and geography also. The long border between the United States and Canada, for

[12]Alva Myrdal, *The Game of Disarmament, How the United States and Russia Run the Arms Race,* New York, Pantheon Books, 1976, p. XV.

instance, does not have the same significance as the one the Soviet Union shares with China. To assess accurately the manpower actually available for war-fighting purposes, one has to look at the several functions of the forces.[13] It can be seen, for example, that the Russians employ uniformed personnel in some support functions for which the United States uses civilians. After comparison of all different functions and, most important, after reducing the Russian potential tied down at the Chinese border, the Soviets end up with an advantage in Europe of less than 10%. This is far less than the figures usually given, and it does not even take into consideration the politically rather unstable situation in some Eastern European states, where Soviet forces may still fulfill some of the tasks of an occupational power, so that their value and readiness for a confrontation with NATO is further reduced. And this assessment does not consider the physical and psychological training, or the motivation and discipline of the troops.

These are only a few of the factors influencing the military potential of a nation or an alliance. Only if all factors are adequately considered can we get a realistic impression of the military capabilities of a country's forces. Otherwise we are in danger of falling into new hysterias like the old ones over supposed "missile gaps" or "bomber gaps." Those two historical examples of wrong assessment of Soviet potentials could not be corrected early enough. The American reaction was to go one step further along the spiral of the arms race.

The immense problems encountered in analyzing the weapons systems and manpower of the two hostile superpowers can quickly lead to the point of incomparability. As the International Institute for Strategic Studies (IISS) noted: "The military balance inherently resists analysis. Not only is the balance affected by so many qualitative factors, to which it is difficult to give proper weight, but also any single, static and compressed comparison of opposing forces can only give limited insight into what might happen under the dynamic conditions of conflict."[14]

Evaluation of Central Concepts: In the preceding section our critique was limited to rather technical considerations in order to reveal basic flaws in the present assessment of military potentials. Now we turn to the underlying principles of the military and political perceptions by the policy makers in the Soviet Union and in NATO. And we shall find, as before, significant variations in the understanding of fundamental concepts. We have discussed the problems of classifying tactical nuclear weapons. The reason for these difficulties is that there is no consistent definition of "tactical" nuclear weapons shared by all the countries of the confronting alliances. They disagree on what the difference is between tactical and strategic nuclear weapons, what their main applications are, and, finally, whether the term "tactical" is even applicable to nuclear weapons.

Evidently, in order to decide whether a weapon is a tactical or a strategic

[13]Philip Morrison and Paul F. Walker, "A New Strategy for Military Spending," *Scientific American,* **239,** No. 4, 1978, pp. 48–61.

[14]International Institute for Strategic Studies, "The European Military Balance," *Strategic Survey,* 1975, pp. 62–67.

weapon, one must consider its characteristics, such as yield, range, location of delivery system, and target, the main purpose of its use, and the consequences of such use. In the past, the distinction between tactical and strategic was at least partly based on the result of the implementation, particularly on consequences in a context broader than just military—for example, the involvement of the civilian population. Consider, for example, the "strategic" bombing of Hamburg and Dresden as opposed to the "tactical" raid against Schweinfurt in World War II. That context was diluted further by the horrible destructive potential of the atomic bomb. Today, the main criteria in the American definition are the range of the delivery system and the operational purpose. Any system able to deliver nuclear devices over intercontinental distances is considered to be strategic. Anything else is a tactical weapon, so that the latter is defined by exclusion. In particular, any weapon used in the zone of hostilities is tactical. This implies that the very same explosive device may be deployed with strategic bomber aircraft and with tactical aircraft. With regard to tactical nuclear weapons, the American belief is that they do not involve global delivery, but they may well include medium-range missiles based in Western Europe, capable of reaching the territory of the Soviet Union.

On the other hand, the Soviets define anything that is able to reach targets in Russia from a NATO base as strategic. Therefore one and the same missile in Europe is called "strategic" by the Russians and "tactical" by the Americans. The two rounds of SALT presented a whole series of problems and misunderstandings resulting from these different definitions.

However, it is most important to note that both the Americans and the Soviets define strategic and tactical in terms of their home countries. The terms for nuclear weapons do not have any significance for nations in Europe. In Europe, all nuclear weapons are automatically strategic if one takes into account their disastrous consequences. One has the suspicion that it was in order to conceal this fundamental truth that the term "theater nuclear weapon" was introduced.

In conclusion, the inspection of the distinctions between tactical and strategic weapons provides us with two important findings: First, divergent interpretations are made by the Soviets and the Americans, and second, these terms are not applicable to the European situation. It is to be expected that similar discrepancies will appear in other concepts and models. For a serious and profound consideration of the problems of arms control and disarmament, a deep reexamination of those concepts is necessary. Expressions like "parity," "balance," "equivalence," and "stability" are vague concepts derived from the physical sciences and of only limited value for describing highly complex processes. Such reexaminations cannot be done in this brief paper. Instead, in the following sections the Western European reaction to SALT II will be considered, with particular attention to the difference in European and American perceptions and interests.

The West European Reaction to SALT II

Arming in Order to Disarm?: Observing that the completion of the new SALT II agreement is accompanied by a strong demand for the deployment of new weapons systems in Europe, we must conclude that the new treaty does not cope with the arms race at all. In fact, SALT II not only includes quantitative constraints but also attempts to control the quality of new weapons development. This holds, however, for only a few types of new technologies. Limitations are put on the numbers of multiple warheads permitted per rocket, and the testing and deployment of ground- and sea-launched long-range cruise missiles are prohibited. The experience of SALT I shows that arms control agreements limit only what is specifically covered by the agreement. In SALT II, the two sides are agreeing on ceilings, but in the meantime new developments make even lower numbers of weapons more efficient and destructive, as in the case of the dramatically increasing accuracy of missiles, which is not taken into account at all.

Once again, technology is outstripping disarmament, and the arms race seems bound to continue. Arguing that the new treaty will simply redirect the arms race, Senators George McGovern, Mark Hatfield, and William Proxmire, all strong supporters of arms control, hesitate to support it.[15] Deployment of cruise missiles in Europe would mean an intensified expansion of the nuclear arms race into new geographic and technological realms. The military advantage of cruise missiles is at the same time their disadvantage for arms control: they are very small and do not need large launching systems. Thus, their detection is nearly impossible. Because of the qualities of high accuracy, low cost, large numbers, and high penetration ability, the Soviets will almost be forced to install cruise missiles in Eastern Europe themselves. Thus, Europe appears right now on the threshold of a new nuclear arms race. Cruise missiles could end arms-control efforts forever, because any agreement concerning them could hardly be verifiable.

Even a decision not to deploy cruise missiles but to counterbalance "Russian superiority" by building up a medium-range missile force in connection with enhanced radiation weapons (ERW), like the neutron bomb, would result in a massive increase in armament in the European threater. Our century has seen several times a rush for armament in limited geographic areas. The rationale for such a military buildup has frequently been that it creates a position from which agreements on disarmament can follow. This argument is derived from the assumption that only competitors in an equal position can reach disarmament agreements. But this "peace through strength" approach has never in the past allowed an actual reduction of armaments. On the contrary, as the relaxation of tensions is said to be a result of military balance, continued vigilance is then seen as a prerequisite for reducing tensions even further. Whatever the situation, arming is said to be necessary. The insanity of arming in order to disarm is summarized in Willy Brandt's words: "He who conducts his foreign policy as a function of military strategy will remain a prisoner of the vicious circle of atomic armaments."

Objections to the Deployment of TNW in Europe: Seeing Europe at the edge of an abyss, opponents of an arms buildup are trying to initiate a new discussion of the military situation and of the political prospects in Europe.[16] Their critique of the premature cry for more nuclear weapons in Europe is based on two arguments:

1. The fact that a large Russian missile force is directed against European targets is not new. It has been the case for more than two decades. As early as 1959 the present Chancellor of Federal Republic of Germany pointed out this imbalance and discussed the consequences for the Atlantic Alliance.[17] During the subsequent time the Russian advantage never played a decisive role in European or world politics. If the USSR ever intended to use TNW as an instrument of coercion, it was successfully negated by joint NATO efforts. Moreover, this imbalance has been reduced over the last few years, as the Soviets started to transfer some of their military arsenal to the Chinese border in the Far East.

2. To provide a military potential for European security does not necessarily mean that it has to be based in Europe. "There can only be an overall balance in Europe's Central Sector if strategic nuclear weapons are drawn into the equation";[18] this has always been a principle of the Atlantic Alliance. And there is no reason suddenly to change this principle.

Since considerations like these are finding response and support in the general public, the West German government was forced to modify its former willingness to accept TNW for basing on German soil. Under considerable pressure, Chancellor Schmidt in 1978 finally declared that West Germany could not carry the burden of nuclear defense alone. He demanded that any new generation of nuclear weapons would have to be stationed in at least one additional nonnuclear NATO nation in Europe, raising a contentions issue that has since been resolved.

Schmidt's decision was reportedly not well received in Washington, and subsequently "signals of irritation" were exchanged between the two nations. It became even worse when a few left-of-center politicians in Bonn publicly thought about alternative ways to defend Europe. The expression "self-Finlandization" emerged. In the parlance of NATO's hard-liners, that is the ultimate form of capitulation and surrender to Communism. Even the *Washington Post* did not hesitate to publish suspicions about secret messages between Bonn and Moscow. According to these messages, if West Germany were to gradually back out of NATO, "Moscow will help take care of Bonn's requirement for oil, when, as seems certain, the West's energy supplies become unbearably tight."[19]

[16]Theo Sommer, *op. cit.*, gives a good introduction to alternative arguments and considerations.

[17]Helmut Schmidt, *The Balance of Power*, London, Kimber, 1971. See especially Chapter 7, "The Position of the West," and Chapter 3, "The Implications of Modern Weapons."

[18]*Ibid.*, p. 113.

[19]Rowland Evans and Robert Novak, "The Kremlin's Bait for Europe: Oil," *Washington Post*, March 21, 1979. For a rebuttal, see the front-page article: "Kartenspieler sind wir nicht mehr" (We Are No Longer Gamblers), *Die Zeit*, April 6, 1979.

Divergent Interests of the United States and European NATO Members: Although nobody should pay serious attention to unconfirmed reports, disagreements within the Atlantic Alliance have become more evident in the last two or three years. They are largely the result of contrasting concepts of deterrence in the United States and Europe. Because of the nuclear parity of the superpowers on the strategic level, military planners in the United States developed the concept of containing a war in Europe as long as possible in order to solve the conflict politically before it becomes global. Thus, defense in Europe has to be sufficient at least to delay a quick conquest of Western Europe by the Russians. If we assume this to be a possible intention of the enemy, the American understanding of deterrence is to deny the enemy the ability to attain his objectives ("deterrence by denial"). Europeans, however, realize that fighting a modern war in Central Europe would necessarily mean its complete devastation. To them "defense through destruction"—that is, through the nuclear extinction of whole nations in a future war—cannot be a promising strategy. Therefore European strategists have for a long time remained committed to the notion of deterrence through nuclear retaliation ("deterrence by punishment"), hoping that an automatic escalation to an all-out nuclear exchange would transfer the deterrence to the European level.[20]

As a result of these conflicting conceptions of deterrence, the doctrine of "flexible response" is formulated very vaguely. At the time this doctrine was introduced (1967), it was realized that "the least doctrine is the best doctrine." But even this ambiguity could not impede the gradual erosion of the common standpoints of the NATO nations. This erosion occurred chiefly at two points:

1. From the beginning there was scepticism in Europe about the willingness of the United States to unleash a global war in order to defend the allies. This anxiety was expressed by Henry Kissinger as early as 1959: "The defense of Europe cannot be conducted solely from North America, because however firm allied unity may be, a nation cannot be counted on to commit suicide in defense of a foreign territory."[21] Since then the fear of a decoupling of American and European defense has been steadily increasing and has even been incorporated in various studies and "war games."[22]

2. On the other hand, the demand for a defense short of major destruction is heard more and more loudly. Europeans realize that turning the area between the Elbe and the Seine into a radioactive wasteland is no defense. Again it was Kissinger who stated this most clearly: "There is a contradiction in conducting a war fought presumably to maintain the historical experience and tradition of a

---

[20]A good discussion of these disagreements can be found in Manfred Woerner, "NATO Defenses and Tactical Nuclear Weapons," *The Atlantic Community Quarterly*, 16, No. 1, 1978, pp. 22–32.

[21]Henry Kissinger, "The Search for Stability," *Foreign Affairs*, 37, No. 4, 1959, p. 548.

[22]One of the recent examples is given by General R. Close, *L'Europe sans Défense* (*Europe without Defense*), Brussels, Lucien DeMeyer, 1976. The author claims that a conventional attack on West Germany by the WTO nations could bring their armies to the Rhine within 48 hours. He believes that the conflict will not escalate to a nuclear war and that the Americans will not interfere, for they are not able to make a decision to do so in such a short time.

people with a strategy that is almost certain to destroy its national substance."[23]

Although these discrepancies have existed for more than two decades, considerations of a realistic and "constructive" defense were hardly ever discussed publicly in Europe. In a mixture of confidence and fatalism, West Europeans were generally apathetic about strategic policy.[24] In recent years, however, the destructive potentials of the two superpowers have grown to a level where using it against each other's homelands has become completely unthinkable. As a result there has emerged the strategy of a "limited" nuclear war in a theater outside the superpowers, rather than a general confrontation. As Alva Myrdal puts it: "The homelands of the superpowers become 'sanctuaries' while wars, if and when they would occur, are to be fought in the territories of lesser powers."[25]

The protection of Western Europe by the American strategic forces now goes hand in hand with the risks of being dragged into a nuclear confrontation between the superpowers. Evidently this situation is producing different views in Europe and in the United States. But before regarding the consequences, let us look briefly at how most Western Europeans perceive Soviet intentions. Since they are more directly exposed to Soviet policy than the Americans, the fear of West Europeans of the "Russian threat" has been reduced since the cold war. It is believed by many that today the USSR is as interested in peaceful coexistence as is the West. Often the belief is expressed that détente is more than a "working relationship," that it is a necessary basis for solving the world's problems. The theory of an ultimate convergence of the two adversarial systems and, to a certain extent, the phenomenon of Eurocommunism stem from these views.

Consequently, in spite of often repeated warnings from the military and defense establishment, West Europeans find it, to quote Thomas Schelling, "more and more difficult to remember why we consider the Soviet Union our chief enemy." The result is a wide divergence in the assumptions about the "threat" from the WTO nations. The likelihood of a deliberate attack on Western Europe is questioned more and more. Since such an attack is the basis for the "classic" approach to European defense and arms control, this "classic" approach itself is also being questioned.

New Concepts of Defense and Disarmament in Europe

The "Lesser Powers" and Disarmament: In our preceding analysis, the different aspects of defense and military policy for the United States and the European

---

[23]Henry Kissinger, *Nuclear Weapons and Foreign Policy*, New York, Harper & Brothers, 1957, p. 244.

[24]Among analysts and scholars who studied questions of strategy and deterrence, there was obviously the fear "that talking about these questions might actually hasten the decoupling of American commitment of strategic forces to the defense of Europe." Lynn Davis, "Limited Nuclear Options: Deterrence and the New American Doctrine," *Adelphi Papers*, No. 121, London, The International Institute for Strategic Studies, 1976, p. 11.

[25]Myrdal, *op. cit.*, p. 30.

NATO members have become evident. It would therefore seem only logical if Europeans would focus on different concepts of defense and disarmament, or would emphasize other aspects than the Americans do. This, however, is only partly the case. Particularly in West Germany the military establishment has always closely followed American military policy, although often disagreeing in details.* But one should expect that especially West Germany, in its geostrategic position where East and West confront each other, might search for alternatives in coping with the arms race. With a future war most certain to destroy the country, with the exposed situation of West Berlin, with the "enemy brother" East Germany, and with its historical links to the small Eastern European countries, West Germany is privileged and at the same time obligated to take a lead in East–West communications. This is even more the case because the superpowers have failed in their attempts to curb the arms race, and because they do not show any ability to find constructive approaches to disarmament. It would seem to be the responsibility of the so-called "lesser powers," such as West Germany, to provide new models of limiting the arms race and easing worldwide tensions.

Because of its strategic importance, and because of its fairly stable conditions resulting from the absence of a direct East–West confrontation in the last decades, Central Europe can play an essential and constructive role in the search for new political strategies.

Alternative Defense and Disarmament Strategies: When approaching the problems of defense and disarmament, it is important to remember the magnitude of the challenge. The following proposals are not to be considered as complete. Rather, their intention is to provide a conceptual basis for new strategies. To achieve this, I shall focus on discussing intersystemic communications (ISC) and confidence-building measures (CBM) in Europe, consider a nuclear-free zone between the blocs, and look at the meaning of the term "national security" for smaller European nations.

Intersystemic communication should be seen as a process invoking the flow of information, opinions, and ideas between countries belonging to different social and political systems in East and West.[26] It consists not only in official contacts between governments but also in communication between private and semi-private institutions, and between groups and individuals. ISC was for the first time explicitly recognized in the Final Act of CSCE (Helsinki). The Helsinki conference established a certain communication between East and West, but it could not really change the patterns of secrecy in the Eastern European nations,

---

[26]See for example, Karl Birnbaum, "Détente and East–West Communication," *Bulletin of Peace Proposals,* **8**, No. 3, 1977, pp. 216–218.

*The situation is quite different in other European NATO nations. French defense policy is very peculiar, and in Holland, Great Britain, and Norway scepticism about NATO doctrine is a tradition. Why this is not the case in West Germany is a very interesting and important topic, which cannot be pursued here.

particularly the Soviet Union. Now that at least a first step has been taken at CSCE, it is necessary to extend ISC. Only if applied in a free and frequent manner can ISC reduce tensions of mutual threats as perceived by competing systems. Communication is the first step to cooperation.

ISC is also the first step to the creation and implementation of new confidence-building measures, a concept again introduced by the CSCE. "With respect to the military situation in Europe confidence-building involves the communication of credible evidence of the absence of feared threats."[27] The practice of prior notification of maneuvers and the invitation of observers has in the meanwhile been well established in Europe. Now it is important to extend the list of CBM. Notification of military movements and publication of data about military expenditures have been proposed. Confidence-building measures establish the linkage between political and military communication that precedes arms control and disarmament. They reduce the uncertainties generated by inadequate information. Out of them could grow a constructive atmosphere in the MFR talks in Vienna, an atmosphere desperately needed after five years of failure.

A New Agenda for the MFR Talks: So far, the Western posture in these negotiations has been predicated on a defense view in order to cope with a deliberate WTO attack. The suggested ISC and other CBM should ensure that the Soviets' interest in the stability of Europe dominates any interest they might have in conquest. And, conversely, the WTO nations should be assured of NATO's peaceful intentions. Then, with a direct outbreak of violence in Central Europe considered as unlikely, the MFR agenda could put more emphasis on impeding the escalation of local conflicts arising, for example, in southern Europe. Today, the establishment of command, control, and consultation arrangements to cope with a conflict that neither side wants seems at least as important as balanced force reductions.

With the help of these communication facilities it may well be learned that West Europe does not need more tactical nuclear weapons, that on the contrary the deployment of those weapons will make the situation in Central Europe more unstable and more dangerous.

One of the long-term goals of disarmament in Europe is the creation of a nuclear-free zone to reduce the complexity of the military balance gradually. After World War II, a lively discussion of such zones went on (for example, the Eden, Rapacki plans); the obvious advantages of such zones have been pointed out on several occasions.[28] A nuclear-free belt between East and West from Scandanavia through Germany and Austria to Italy could not be realized in the immediate postwar era, mainly because of the resistance of the superpowers, notwithstanding their verbal support for the idea. As was pointed out above,

[27]Johan Holst and Karen Melander, "European Security and Confidence Building Measures," *Survival*, 19, No. 4, 1977, pp. 146–154.

[28]See, for instance, Myrdal, *op. cit.*, pp. 195–207, and Helmut Schmidt, *Defense or Retaliation*, New York, Praeger, 1962, pp. 125–161.

today's defense strategy of the superpowers makes a denuclearized zone in Europe even less feasible. But there is no reason that the "lesser powers" and the nonaligned countries could not make proposals themselves, even if contemporary plans would have to be more modest in scope. A beginning was made with the discussion of a Nordic nuclear-free zone a few years ago.*,[29] Realizing their own vital interests, and insisting on their independence from the superpowers, the European nations must make attempts like this. Realistic proposals for a nuclear-free zone in Europe could also revitalize the MFR talks or, rather, be a starting point for true disarmament negotiations in Europe.

The Absurdity of Weapons Security: A nation that is, partly or completely, in a nuclear-free zone, or even in a zone totally free of weapons, will of course have to change its perception of national security. Today, it is usually argued that military power is the prerequisite for the maintenance of peace and security, or that the absence of military power will lead to the loss of national independence. This concept of security might be called "weapons security."[30] It is, however, evident that, except for the United States and the Soviet Union, all-out weapons security is impossible for any nation, and it is in some ways undesirable. Consequently all lesser nations must use other means to ensure their national security. This is done mostly by establishing military alliances; in both the NATO and the WTO, weapons security is provided by one of the superpowers. With the requirements for obtaining weapons security becoming higher and higher, as the military capabilities on both sides become more and more sophisticated, a point of "counterproductivity" has been reached. In spite of intensified armament, worldwide security is diminishing, as is forcefully shown by Herbert F. York.[31] Considering nations in Central Europe, we see that the striving for weapons security has led to the prospect of complete nuclear devastation—indeed an ultimate absurdity!

This dilemma can be avoided only by converting the now prevailing concept of defense by weapons security into another form of defense, as long as a defense is needed at all. One alternative may be the model of "social" or "civilian" defense. This concept, developed by Gene Sharp,[32] is based on the consideration that in

---

*Some progress in establishing a nuclear-free zone has been made in Latin America. Twenty nations signed the Tlatelolco Treaty, which was concluded for the prohibition of nuclear weapons in Latin America. It is necessary to analyze how far this pioneer attempt can serve as a model for other parts of the world.

[29] Keijo Tero Korhonen, "Regional Arms Control in Europe: A Nordic Nuclear-Free Zone?", paper presented at the 24th Pugwash Conference, Baden, Austria, August 1974, mimeo, pp. 1–7. For a review of two responses, see Johan Holst, "The Nordic Nuclear-Free Zone" and Pertti Joenniemi, "Why a Nordic Nuclear-Free Zone," *Bulletin of Peace Proposals*, 6, No. 2, 1975, pp. 148–151.

[30] For a discussion of fundamental concepts of security, see Bert Roeling, "Feasibility of Inoffensive Deterrence," *Bulletin of Peace Proposals*, 9, No. 4, 1978, pp. 339–347.

[31] Herbert F. York, *Race to Oblivion*, New York, Simon and Schuster, 1970, pp. 228–239.

[32] Gene Sharp, "The Political Equivalent of War—Civil Defense," *International Conciliation*, 65, 1965, pp. 208–251.

highly industrialized nations any occupation must depend on the cooperation of the occupied country. The objectives of social defense are described as: "(1) retain the life-style of the population, (2) if possible make converts among the enemy and (3) save the territory for future uses."[33] To elaborate on these ideas, much work needs to be done. Today not only is the theory incomplete, but the lack of any practical initiatives is overwhelming. There are no successful examples of social defense in history; attempts in Czechoslovakia during the Russian occupation of 1968 were followed by total defeat and unequivocal retreat from these objectives.

But this does not necessarily mean that there are no alternatives to a defense by weapons security. Since the latter is leading to more and more nuclear proliferation and to higher and higher overkill capabilities—that is, to a more and more unstable and dangerous situation—today's societies have to find effective ways to ensure national security. There must be viable alternatives.

Concluding Remarks

In the course of our considerations two major findings have been derived:

1. Arms control and disarmament cannot rely on technological solutions alone, but require basic political responses. It is hoped that with the intersystemic communications and confidence-building it will be possible to get closer to such a political responsiveness. Moreover, it has become evident that partial and limited arms control efforts, such as SALT II, cannot halt the arms race any more. On the contrary, they lead to an ever-intensified rush for armament among the superpowers and furthermore transfer the arms race to other theaters such as Europe. Therefore, fragmentary arms control efforts must be replaced by comprehensive disarmament moves. As one example, the reexamination of the concept of a nuclear-free zone in Europe is proposed.

2. A wide divergence of political and strategic interests between the United States and Western European NATO nations, or more generally between superpowers and "lesser powers," has been found. Particularly the hegemonial definition of the use of nuclear weapons and the resulting strategy of flexible response cannot be in the interests of the European nations. Therefore, the "lesser powers" and the nonaligned countries have to develop their own approaches toward disarmament; they can no longer leave all the initiative to the United States and the USSR.

The arms race continues. Many years of efforts at arms control have not changed this situation. As a result, apathy and resignation are spreading, and nobody is talking about disarmament anymore. Making proposals for disarmament in this climate seems to be very idealistic or even naïve. Does not the long history of defeat of disarmament attempts show that to reach effective agreements for disarmament is impossible?

[33]Johan Galtung, "Two Concepts of Deterrence," *Bulletin of Peace Proposals*, 9, No. 4, 1978, pp. 329–334.

Mankind cannot accept a defeatist view when millions of lives and the existence of whole countries and cultures are at stake. The arms race is irrational and must be curbed by concerted actions. Alternative conceptions and models of defense and disarmament must be developed and discussed and applied to change the course of the present *Realpolitik*. Strategies for change and the will to realize them are what mankind needs most today.

9. The Qualitative Arms Race and the Role of the Scientist

Fraser Homer-Dixon

> Signs are accumulating that efforts at arms control are in deep crisis.
>
> —M. THEE (Norway)

> Despite numerous proposals and initiatives aimed at reaching agreements on disarmament and military disengagement, which have sometimes led to international agreements, the arms race has not been slowed down; it has continued to speed up, recording during the last few years an unprecedented rate of growth.
>
> —A. CORIANU (Romania)

> It cannot be concluded that negotiation has totally failed, but it clearly has not resulted in substantial progress. While it should not be abandoned as a mechanism for reducing and eliminating weapons, it might be assisted if other methods could be pursued simultaneously.
>
> BETTY G. LALL (United States)

> (Quotations selected from papers submitted to the 26th Pugwash Conference held in Mühlhausen, German Democratic Republic, August 1976.)

The Momentum of the Arms Race

Although this conference has the theme Science and Ethical Responsibility, this paper is not directly concerned with the moral issues surrounding the scientist and arms control. The question, "Does the scientist have the ethical responsibility to involve himself in arms control issues?" is assumed to have been answered "Yes!" Rather, this paper will outline how a scientist might effectively use his talents to aid the arms control lobbies. The talents of the scientist are vital to the comprehensive arms control strategy that will be set forth here.

Since the Second World War, we have seen the most formidable arms race in history. Never before in a period of general peace has such a quantity of resources been diverted to military uses. We once imagined that an arms race took place between two discrete states or alliances, and that it could be easily identified in both space and time. Now, in addition to the familiar two-power arms races, we have regional and continental arms races. We have qualitative and quantitative arms races. We have distinctions between conventional, nuclear, strategic, and tactical arms races. On the global scale, each type of arms race overlaps and cuts through other types in a multidimensional confusion of military spending.

The statistics are staggering. A recent study by the United States Arms Control and Disarmament Agency entitled *World Military Expenditures and Arms Transfers 1967–1976* reports that in 1976 world military outlays were $380 billion.[1] This was 11% more than world spending on public education, and 242% more than world spending on public health. Extrapolating this $380 billion figure in constant dollars, using a conservative 2% annual growth rate, we find that total military expenditures may already exceed $400 billion per year. A series of graphs published by the Stockholm International Peace Research Institute (SIPRI) is reproduced in the Appendix to this paper; these graphs indicate the momentum of the arms race.

It is not only a momentum of quantity; it is also a momentum of quality. The size of the military budgets can give no idea of the destructive power and efficiency of the weapons now produced. In his book *Destin de la Paix*, Jules Moch calculated that the world's nuclear stockpile in 1969 was sufficient to destroy the human population 690 times.[2] One of the new American Trident submarines will be capable of leveling 408 separate cities, and the United States Navy is planning to build thirteen of these submarines. The cruise missile, the product of advances in electronic miniaturization and jet engine science, will be able to fly up to 5000 km and strike within 10 meters of its target.[3] And the U.S. Director of Defense Research and Engineering, William J. Perry, has claimed that the American forces will soon be able to "see all high-value targets on the battlefield at any time; to be able to make a direct hit on any target we can see, and to be able to destroy any target we can hit."[4]

The arms race has become institutionalized:

The arms race has become politically connected with the vested interests that President Eisenhower termed the "military–industrial complex." In military matters, no limit is set by market forces, by competitive demand, or by prices. Every new plant for military production, every new production contract, increases the weight of these vested interests. In democratic countries these interests, both

[1]"Expenditures and Transfers," *Scientific American*, **239**, October 1978, p. 85.

[2]Jules Moch, *Destin de la Paix*, Paris, Mercure de France, 1969, p. 211.

[3]Kosta Tsipis, "Cruise Missiles," *Scientific American*, **236**, February 1977, pp. 20–29.

[4]Philip Morrison and Paul F. Walker, "A New Strategy for Military Spending," *Scientific American*, **239**, October 1978, p. 58.

labor and business, often become rooted in the parliaments and the provincial assemblies, whose representatives are expected to defend local interests. In authoritarian countries, these vested interests should be easier for a government to control, but apparently they are not.[5]

Bureaucratic and scientific interests can be added to this list. Concerning bureaucracies, Charles Ostrom has proposed that the complexity of the decision-making process within military bureaucracies leads to incremental budgeting, with the previous level of funding as a base.[6] As for the scientists, nearly half the world's scientists and engineers who are employed in research and development are engaged in military research and development.[7] In the constant struggle for resources and recognition, each of these vested interests contributes to a general increase in military spending, and tends to exacerbate the rift and animosity between the powers involved in the arms race. As the superpowers strive for preeminence, the arms race compounds itself in an action–reaction cycle based on a fear and a systematic "mirror image" misconception of the other side.[8] Various arms control agreements, such as the Strategic Arms Limitation Treaties (SALT I and II), have only codified this process.

The conjunction of immense military budgets with the unprecedented capabilities of modern weapons has produced the first credible threat to the survival of humanity. Global problems such as population growth, resource depletion, and the increasing economic disparity between the rich and poor nations will soon create severe international pressures. In November 1975, a group of nuclear and political scientists from Harvard and The Massachusetts Institute of Technology stated that these pressures would almost certainly result in an atomic war before the year 2000.[9] Even if the atomic war were begun by Third World nations, it might be impossible to prevent it from escalating into a superpower conflict. As Henry Kissinger wrote in 1965: "No one knows how governments or people will react to a nuclear explosion under conditions where both sides possess vast arsenals."[10]

In the light of the danger, progress toward meaningful disarmament or arms control since World War II has been dismal. From the early 1960s until May 1979, seven multilateral and six bilateral agreements were concluded. All of these, except one, must be considered arms control rather than disarmament

[5]Alva Myrdal, *The Game of Disarmament, How the United States and Russia Run the Arms Race*, New York, Pantheon Books, 1976, p. 10.

[6]Charles W. Ostrom, Jr., "Evaluating Alternative Foreign Policy Decision-Making Models: An Empirical Test Between an Arms Race Model and an Organizational Politics Model," *Journal of Conflict Resolution*, 21, June 1977, pp. 235–263.

[7]Myrdal, *op. cit.*, p. 13.

[8]For information on the "mirror image" phenomenon, see Urie Bronfenbrenner, "The Mirror Image in Soviet–American Relations: A Social Psychologist's Report," *Journal of Social Issues*, vol. 17, No. 3 (1961), pp. 45–56.

[9]Jan Tinbergen, coordinator, *Reshaping the International Order, A Report to the Club of Rome*, New York, Dutton, 1976, p. 46.

[10]Cited by Sidney D. Drell and Frank von Hippel, "Limited Nuclear War," *Scientific American* **235**, November 1976, p. 37.

agreements. None directly impinged upon the power of the interests that help to perpetuate the arms race. For example, the SALT I agreement and the subsequent Vladivostok Protocol placed numerical limits on strategic delivery vehicles that were much higher than the number of vehicles possessed by either superpower. These agreements placed no limits on either the size of the warheads or the number of warheads per delivery vehicle. Yet, at the time, the major thrust of the strategic arms race was in the deployment of multiple independently targetable re-entry vehicles (MIRVs). By 1974 the United States had been deploying MIRVs for four years, and immediately after the Vladivostok Protocol the Soviet Union began its comparable program. It is this technological thrust that should have been the object of any arms control agreement, yet it was conveniently skirted. It was skirted because the military, industrial, and political forces on each side that supported the MIRV programs could not be challenged. Ironically, in terms of security it is now widely recognised that the high-accuracy MIRVs have reduced the credibility of both sides' second strike deterrent.

The other arms control agreements that have been achieved are in areas of no direct concern to the military interests. There are limitations on weapons deployment on the moon, on the seabed, and in Latin America (the latter treaty has not been signed by the USSR because of its interests in Cuba). Specific weapons technologies that have been banned include antiballistic missiles and bacteriological weapons. But neither of these technologies was regarded as really useful, and it is difficult for the military interests to advocate weapons systems that have no real merit. As Alva Myrdal points out: "Experience has taught us that governments are more prone to renounce weapons when they are understood to be of questionable military value, either because they are obsolescent or because they are dangerous in handling or have a boomerang effect."[11] Arms control agreements have been planted all around the purview of the military interests. None has impinged upon it.

The approaches to arms control and disarmament that have been tried to this point have failed. They have failed because they were not founded on a thorough understanding of the forces behind the arms race. They have failed because the arms control lobbies have tried to confront the military interests directly from a position of tremendous political and economic inferiority. The military interests have deflected the efforts of the arms control lobbies so that any agreements that have been concluded do not threaten their influence or security. We need a new approach to arms control—an approach that will exploit the weaknesses of this arms race juggernaut.

Technological Suffocation

In her book *The Game of Disarmament*, Alva Myrdal comments on the "technological imperative" in Western society.[11] The drive for modernization often seems to be its own impetus. Technological change is the hallmark of our time; it is the cement binding together our belief in progress.

[11]Myrdal, *op. cit.*, p. 269.

The technological thrust is one of the most pervasive and destabilizing elements of the arms race. It is most apparent in the strategic nuclear arms race, where the United States has maintained a lead of five to seven years over the Soviet Union. Since the Second World War, we have seen advances in this area along two separate lines: warhead technology and delivery system technology. The bomb dropped on Hiroshima on August 6, 1945, weighed 4 tons and had an explosive power of about 12 kilotons of TNT. This crude fission weapon was soon replaced by thermonuclear technology and coupled with developments in miniaturization. Today a warhead with the explosive equivalent of 200 kilotons weighs no more than 250 pounds.[12] There have been even more amazing advances in delivery systems, as measured by range, accuracy, flexibility, and survivability. We have seen a succession of weapon delivery systems, from the manned bomber through to the missile and the maneuverable re-entry vehicle (MARV). The accuracy of American ICBM warheads will soon be within 30 meters.[13]

The technological thrust undermines arms control agreements. The SALT negotiations were continually confounded by the introduction of new weapon systems on both sides. Any arms control agreements that are concluded, weak as they may be, will inevitably be circumvented by new technologies. Furthermore, the technological thrust compounds the fear that drives the action–reaction cycle. If it were a purely quantitative arms race, the present military capabilities of each side could be accurately estimated from the quantities of materiel deployed, while the future capabilities could be judged from the economic potential for each side. But technological change introduces the unknown—a psychological factor that makes the arms race more dynamic. If a quantitative arms race could be described as linear, then technological change is another dimension, and an arms race with this new dimension would be represented on a plane. The future capabilities of each side become difficult to predict, for the range of possibilities is increased enormously. And there is always the perceived chance of that freak discovery, that technological breakthrough by one side that will completely alter the nature of warfare, and will lift that power into unchallengeable predominance. This is another legacy of the atomic bomb. A single new weapon ended a war within days of its first use, and the United States transformed itself into a superpower. It was a final lesson that technology could be the deciding factor in war. Even in the West, there is a subtle fear that the tables could be turned, and that we could be on the receiving end of a radically new weapons technology.

The result is that the technological thrust is almost an autonomous force. Technological change is infrequently directed by specification founded on strategic policy; more often it becomes a mad race of possibility. Once the effectiveness and feasibility of a new weapons technology have been demonstrated, even if only in the crudest terms, it must be developed. A military planner might think in these terms: "If the weapon looks as if it might have potential, and

---

[12]Stockholm International Peace Research Institute, *World Armaments, The Nuclear Threat*, Stockholm, SIPRI, 1977, p. 18.

[13]*Ibid.*

we can possibly develop it, we had better start now, or else the other side might get to that new technology first."

This psychological dimension based on fear of the unknown is a fertile environment for the expansion and entrenchment of the vested interests mentioned previously. The uncertainty about the other side's intentions or capabilities gives some momentum to the cycle on its own, but it is also used to justify the ossification of arms production interests. In many cases, these interests may aggravate the uncertainty for their own ends.

If arms control is to succeed, the strength of the technological thrust must be lessened. If the present superpower arms race could be changed to a largely quantitative arms race, this would lead to a decrease in fear and to an increase in security on both sides. Military spending by the superpowers is now so great, and the weapons they produce are so terrifying, that subtracting the technological impetus from the arms race could result in a real desire to disarm. The fear and distrust produced by a quantitative arms race might not be enough to justify the current levels and capabilities of armaments. If the superpowers moved toward arms control, this would be a good example for other countries with large military budgets. This whole process could be called a deflation of the arms race.

But how can the technological force behind the arms race be enervated? Firstly, the disarmament and arms control lobbies must realize their relative weakness in any struggle with the interests behind the arms race. As long as the arms control lobbies tackle issues that the military interests consider vital to their spheres of influence, the arms control lobbies are bound to fail. For example, the cruise missile is recognized as an effective deterrent, and the military interests long ago included it within their purview. Any threat to the deployment of the cruise missile will be met with the full opposition of these interests, for it will be regarded as a general threat to their prestige and influence. If the United States Air Force loses the cruise missile, it will be a blow to the pride of the service. The loss of the political struggle for the cruise missile will probably decrease the prestige of the service's military leaders. Even if the arms control lobbies could mount a concerted attack on a number of new weapons systems, the threat would unite all the military interests behind the single goal of maintaining the existing order. In any direct struggle, the arms control lobbies cannot match the economic or political power of the military interests. In such a struggle, the arms control lobbies are wasting their time; the proof is the lack of effective arms control agreements despite thirty-four years of effort. These lobbies must design a strategy that can make effective use of their limited power.

The arms control lobbies are caught in a dilemma. They do not have the power to successfully tackle issues that the military interests consider important. Yet it is these issues that they must tackle if they wish to slow the arms race. The military interests are most committed to the most effective and powerful weapons: the Tridents, the cruise missiles, the SS-20 missiles, and the Backfire bombers. Yet it is precisely these weapons that must be restricted. To succeed, the arms control lobbies must impinge upon the purview of the military interests; yet this is beyond their power.

A more effective approach to arms control would be technological suffocation.

This would require considerable foresight and patience. The arms control lobbies could try to establish technological barriers fifteen or twenty years in the future. After carefully studying trends in the development of weapons technology, these lobbies would push for multiparty agreements restricting the deployment of specific forms of weaponry. By doing so, they would not be directly confronting the military interests, and the lobbies would have a higher chance of successfully concluding agreements.

For technological suffocation to be effective, two criteria must be met. Firstly, the technological barrier has to be placed far enough in the future that the military interests do not yet fully recognize the potential of that technology for weapons purposes. Once the knowledge of an effective new weapons system has filtered its way through the military–industrial–political complex, it is probably too late to stop its deployment. The technology must be blocked before the potential weapon system is included in the purview of the military interests. Secondly, the agreement limiting the new technology must be as specific as possible. Similar agreements in the past have been so general that their provisions have been easily avoided, and it has been difficult to determine if a violation has occurred. The agreement could specify the new technology to be limited (for example, enhanced radiation technology), or it could specify the form of the new weapon (for example, cruise missiles, which incorporate several broad technological innovations).

With luck, after a period of years all the specific agreements would form an interlocking comprehensive barrier to the technological thrust. The beauty of this approach is that the arms control lobbies could afford to be wrong with their technology predictions countless times, if only one agreement prevented a new form of weapon from being deployed. Even if the interlocking barrier were not comprehensive, such an approach could significantly slow the technological momentum of the arms race.

These agreements to limit weapons technology could be reached through the United Nations or through organs such as the Conference of the Committee on Disarmament in Geneva. Once the barriers are in place, it may be many years before the technological arms race encounters them. The political gains that could be derived by proving that an opponent country was violating an agreement would lead each party to search for and publicize such violations scrupulously. Violations would be controversial, and they would probably damage the guilty country's international credibility. Thus, the arms control agreement would develop its own constituency and could be considered self-enforcing.

This process would be buttressed by a new development in the international arena. Despite the claims of cynical "realists," it became possible in the 1970s to speak of a world public opinion, which is articulated through the United Nations. Mass communication has not only bound us closer together, it has increased global political awareness. Today, there is world concern about the arms race and violations of arms control agreements, and this environment of concern would strengthen any technological suffocation agreements.

Some might say that the technological suffocation approach wrongly emphasizes the law over the will to obey that law. Such critics might claim that a technological suffocation agreement would only be a "piece of paper," which would be disregarded by any country with a desire to develop that weapons technology. Of course no international regulation can be effective if there is no will to abide by it. But the self-enforcing nature of a technological suffocation agreement combined with the growing weight of world public opinion should produce a will to abide by the agreement. When the leading edge of the technological arms race encounters a limitation on weapons technology, the desire to violate the limitation will not be as strong as the desire to avoid the international opprobrium that would result if it were violated.

Figure 1 is a diagram of the stages leading from research to the deployment of a new weapon. The technological barriers proposed would probably interrupt the process between the research and the development stages. It would be impossible to stop basic research, since this is often abstract and seemingly unrelated to the end-product weapon. But translating that research into the development of a new weapon would require a conscious commitment by the organization sponsoring the research. Further, it would require a large capital investment, for today's weapons based on advanced technology are often very expensive. Instead of investing resources in the development of a weapon whose deployment has been specifically banned by prior international agreement, research will probably be directed into other areas. In some cases the basic research may be extended to the initial development of the weapon and its testing, to allow for the possibility that the agreement might be violated by the other side. But the weapon would probably not be manufactured, since this would imply a commitment to deploy it.

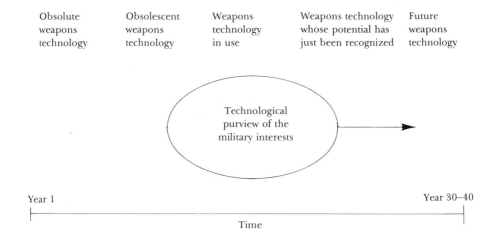

FIG. 2. The movement in time of the purview of military interests.

An analogy can be drawn to a young, fast-growing plant. The first tendril is flexible and easily trained, and it skirts obstacles in its path. But as this part of the plant grows and ages, it becomes hardened. That first tendril is similar to the leading edge of the technological thrust—it is flexible, and it can be deflected into other areas. But attempts to change the direction of the thrust further along the sequence, say at the manufacturing stage, will be difficult if not impossible.

By using the technological suffocation approach, the arms control lobbies will effectively deal with the problem that has always made their efforts futile—their inability to confront the military interests directly. As has already been noted, the arms control lobbies have never successfully impinged upon the purview of the military interests. The nature of this purview is always changing in a variety of dimensions. For example, the military interests may recognize the military potential of deploying their forces in a new area of the globe, and this might lead to an expansion of their purview in a geographic dimension. On the other hand, the purview of the military interests is continually expanding in a quantitative dimension as force levels increase. Arms control can involve blocking the avenues for change or expansion of this purview before the change takes place. Using the two examples just mentioned, instances of such arms control measures would be the Treaty of Tlatelolco and the quantitative restrictions of SALT II. But in both the geographic and quantitative dimensions, change does not have to take place; the momentum of the arms race might follow other paths. Thus, such arms control agreements might never effectively restrict the arms race.

But technological change is an intrinsic element of the arms race. Technological change occurs with time, and we can think of the technological purview of the military interests as moving through time, as pictured in Fig. 2. Any comprehensive barrier to the development of new forms of weapons technology is guaranteed to eventually impinge upon the purview of the military interests. In this way the arms control lobbies can establish agreements now that will impinge upon the purview of the military interests in the future, by taking advantage of the inevitable change in the weapons technology within the purview of the military interests. By establishing a series of technological barriers ten or fifteen years in the future, the arms control lobbies would eventually cut to the heart of the arms race, and stifle one of its most vital elements, technological change. This then is technological suffocation, a method noted by Hedley Bull:

[many] steps intervene between the conception of a weapons system, and its becoming operational, and impose a barrier of time, sometimes of a decade or more. A prohibition of military development, or of certain branches of military development, which involved controls at all these stages . . . might very considerably affect the direction in which innovation flowed. It could not prevent the conception of military innovation, but it might impose obstacles to the exploitation of this conception.[14]

[14]Hedley Bull, *The Control of the Arms Race, Disarmament and Arms Control in the Missile Age*, 2nd ed., New York, Praeger, 1965, p. 198.

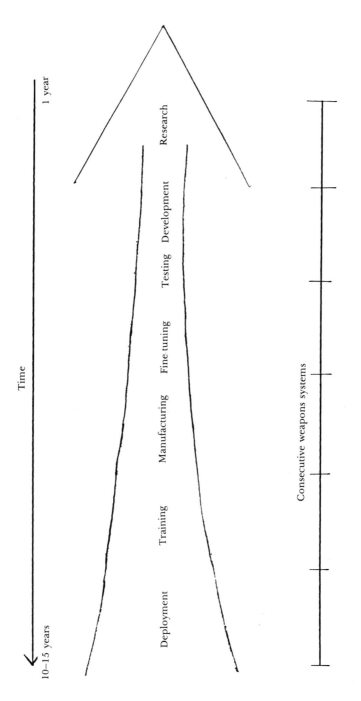

Fig. 1. Stages in weapons development.

Bull also observes that

The closing of particular channels of military innovation will divert innovation into different directions, but might be worthwhile none the less; and if there is no question of ending all military innovation, it is the most that can be undertaken.[15]

Bull believes that barriers to military innovation can be self-imposed by the major powers. But there has been no evidence that they are willing to do this.

Applicable Arms Control Agreements

There have been relatively few qualitative arms control agreements in history. The first and perhaps one of the most successful examples was the St. Petersburg Declaration on Exploding Bullets of December 11, 1868. Following the Crimean War, the Russian Army developed an exploding bullet for use against hard targets such as ammunition wagons. Recognizing the inhumanity of this weapon if it were to be used against people, the Czar had the Imperial Russian Cabinet convene an International Military Commission to "renounce the use of explosive projectiles under 400 grammes weight." The declaration was signed by most of the major powers, and it was eventually included in the Hague Regulations dealing with land warfare. This declaration banned a specific weapon before it had been included in any arsenals. It may have been instumental in preventing the weapon's widespread use in subsequent wars.

The First Hague Conference of 1899 banned the use of expanding bullets and asphyxiating gases. Neither Great Britain nor the United States supported these declarations, since both powers recognized the effectiveness of these weapons under certain circumstances. Great Britain's representatives felt that the extra stopping power of expanding bullets was needed to halt a rush of "uncivilized tribesmen."[16] Captain A. T. Mahan, chairman of the United States delegation, stated that asphyxiating gases were no more inhumane than many widely used weapons, and that gases might be needed in a future war.[17] The Second Hague Conference of 1907 adopted a declaration prohibiting the "discharge of projectiles and explosives from balloons" or any other form of air-to-surface bombardment. By this time there was a general recognition that this form of warfare could be very useful, and the declaration was ratified by only two of the major powers, Great Britain and the United States. This declaration was not binding on the signatories when they were at war with a nonsignatory power. Moreover, it was to remain in effect only until the end of the Third Hague Conference, which was to be held in seven years. The Third Hague Conference was never convened.

[15]*Ibid.*, p. 199.

[16]Richard A. Preston and Sydney F. Wise, *Men in Arms, A History of Warfare and Its Interrelationships with Western Society*, 2nd ed., New York, Praeger, 1970, p. 218.

[17]Trevor N. Dupuy and Cay M. Hammerman, eds., *A Documentary History of Arms Control and Disarmament*, Virginia, Dupuy Associates, 1973, p. 56.

The Washington Treaty of 1922 prohibited the use of submarines to destroy commerce and the use "noxious" gases. The signatory powers were limited to the United States, the British Empire, France, Italy, and Japan. The effectiveness of a submarine force in a blockade had been vividly demonstrated during World War I, and both the weapons mentioned in this treaty were already an integral part of many defense forces. In 1925, the Geneva Protocol on Chemical and Biological Warfare was concluded. It prohibited the use in war of most gases and all bacteriological agents. Although gases had been used in previous warfare, serious research into bacteriological weapons was not begun until the mid-1930s.[18] The Geneva Protocol did not ban the manufacture or deployment of these weapons, just their use in war. As a result, military research and development was not directed away from this area. Even though the protocol was not ratified by the United States Senate at the time, it is possible that this agreement has discouraged the widespread use of bacteriological weapons.

A series of agreements concluded in the late 1950s and during the 1960s established nuclear-free or demilitarized zones in the Antarctic, in Latin America, in outer space, on the moon, and on the sea floor. Although these agreements do not limit new military technology, they are related to technological suffocation, since they attempt piecemeal limitation beyond the purview of the military interests. Except for outer space, these regions are of little strategic importance to the major powers.

The outer space agreement is interesting. The original declaration renouncing the stationing of weapons of mass destruction in space was issued by the Soviet Union and the United States in 1963, when the military establishments in both countries had not yet committed themselves to positioning nuclear weapons in permanent orbit. The initiative to bring the two parties together on this point was taken by Canada. There was little opposition from the military interests in either country. Without this agreement, it is quite likely that within a decade the nuclear arms race would have expanded to outer space. The proliferation of spy satellites and the present development of "killer-satellites" by both superpowers is proof that neither country has any compunction about using outer space for military purposes. As an aside, the "killer-satellite" would have been a prime candidate for limitation under a technological suffocation approach.

As a final example of qualitative weapons limitation, there is the Treaty on the Limitation of Anti-ballistic Missile Systems of 1972 (SALT ABM Treaty). Here we have a restriction of a specific weapons technology, but it is a restriction dotted with loopholes, and it is a restriction of a technology that many experts regard as ineffective.

The dearth of qualitative arms control agreements makes falsification of the technological suffocation approach difficult. Except for the serendipitous and not entirely applicable outer space declaration, none of the previous agreements can be used as an example of this approach. In one way or another, they are all

[18]Bull, *op. cit.*, pp. 126–127.

missing an important aspect: Some are not specific enough, some try unsuccessfully to limit weapons whose effectiveness has already been recognized by the military, some impose restrictions in areas of little strategic interest to the military, some prohibit the weapon's use in war and not its deployment in peacetime, and some of the agreements have not been ratified by important powers.

One final note in this section about test bans. Many arms control specialists consider that the test ban is the most effective method of limiting the qualitative arms race. The Partial Test-Ban Treaty was signed in 1963 after five grueling years of negotiations. Nuclear testing was not stopped. The superpowers simply moved it underground. It is doubtful that this treaty slowed the improvement of warhead technology. At the 1978 Tenth Special Session on Disarmament of the United Nations General Assembly, Prime Minister Pierre Trudeau of Canada suggested that the technological arms race could be suffocated by halting both warhead tests and the flight tests of delivery vehicles. He is probably right—if these tests were stopped, technological improvement would be difficult. But there is no chance of halting these tests as long as they are of use to the military interests on either side. As long as the United States Air Force needs to test its cruise missile, it will move to block such limitations. As long as a branch of the Soviet military bureaucracy depends for its funding upon the approval of a new mobile missile, it will lobby the government to reject suggestions that testing should be curtailed.

A comprehensive warhead test ban has been under negotiation for decades. It seems that the Soviet Union, the United States, and the United Kingdom might finally be close to agreement on this issue. But testing is probably no longer needed. Most physicists believe that the highest level of warhead efficiency has nearly been reached.

The Role of the Scientist

Now we can tie the technological suffocation of the arms race to the theme of this conference. At the same time we can deal with the weakest stone in the theoretical foundation of this approach to arms control. Can new weapons technologies be successfully predicted? Certainly the arms control lobbies would need to study the trends of weapons development, and it is here that the scientist becomes vitally important. Scientists from the disciplines relevant to weapons technology should be encouraged to join with the arms control lobbies to research thoroughly and predict weapons developments as part of a broad technological suffocation strategy. The Pugwash movement can play a vital role in this effort. It is the creativity and ingenuity of the scientist that has produced today's weapons, and we should harness these same talents to advance beyond the horizons of current technology so as to predict future developments.

Granted, such predictions would not be foolproof. But we can afford to be frequently wrong, as long as we are occasionally right. Further, the chance of the unforeseen breakthrough occurring that will change the nature of warfare and that is beyond current trends in weapons development is actually quite low. As technology becomes more advanced, improvements tend to be incremental rather than dramatic. And all technological change, whether incremental or dramatic, is founded on the broad wealth of scientific theory that is available to all scientists, both inside and outside the military establishment.

In 1960, Hedley Bull listed the following new technologies that could be limited:

The development of cheap missiles, for example, that would place this weapon of instantaneous destruction in the reach of minor powers; the development of techniques that would cheapen and simplify the production of nuclear weapons, and thus exacerbate their spread among a widening circle of powers; the development of weapons of increasingly greater destructive power, increasingly indiscriminate in their effects; the development of missile accuracy to a point where it threatened to make possible a successful disarming attack.[19]

Bull did not mention specific weapon systems, but he did clearly outline some trends and areas for attention. The cheap missiles and high-accuracy missiles that he foresaw have been developed, and they have become a serious concern to arms controllers in the late 1970s. Bull is not a scientist, and if he could predict general weapons developments up to fifteen years ahead of deployment, then surely specialists should have been able to predict more specifically these new technologies some years before 1960.

In 1968, Nigel Calder edited a book entitled *Unless Peace Comes*, in which a number of specialists discussed possible weapons of the future. Many of their forecasts have already come true. Some of their long-range predictions included the massive use of drugs that will alter the psychological balance of the enemy, microbiological weapons for use against crops, the creation of earthquakes, robot devices for the battlefield, floating aircraft and submarine bases anchored at sea, destruction of sections of the earth's ozone layer, and automatic command and control procedures that will take crucial decisions out of the hands of political and military leaders.[20]

In a Pugwash symposium on The Impact of New Technology on the Arms Race held at Wingspread, Racine, Wisconsin, in June 1970, a number of new technologies were mentioned that could alter the future course of the arms race. These included high-altitude nuclear explosions, weapons incorporating super-heavy elements, the use of controlled fusion to produce large amounts of plutonium, pure fusion trigger devices for nuclear weapons, the use of laser and other electromagnetic technology to jam enemy surveillance efforts, and extremely sensitive sonar for antisubmarine warfare.[21]

---

[19] Bull, *op. cit.*, p. 199.

[20] Nigel Calder, ed., *Unless Peace Comes, A Scientific Forecast of New Weapons*, New York, Viking, 1968, pp. 231–243.

[21] Bruno Brunelli, "Report on the Pugwash Symposium on 'The Impact of New Technology on the Arms Race'," *Disarmament and Arms Control, Proceedings of the Third Course Given by the International Summer School on Disarmament and Arms Control of the Italian Pugwash Movement*, New York, Gordon and Breach, 1972, pp. 15–22.

The second edition of *Weapons Technology*, compiled and edited by the Royal United Services Institute for Defence Studies in London, contains a detailed analysis of current weapons technology along with many insights into future developments.[22] The prediction of future weapons technology is certainly not beyond the capability of arms control lobbies.

V. Conclusions

Today's global arms race cannot go on indefinitely. With each year the security of individual countries is decreasing. Even if the world's leaders are always sane and rational, the great weight of military establishments increases the chances that an error or systems breakdown will lead to a major war. The bigger and more numerous the armies, the faster and more powerful the weapons, the less room there is for mistakes. In human terms the waste of resources is a crime. When tens of millions of people are starving to death each year, when a very large fraction of the earth's population does not even have safe water to drink, we are channeling our money into capital-intensive, nonproductive, obsolescent armaments.

As has been pointed out in this paper, the arms control lobbies must find a new approach, since none of their previous efforts have had a significant effect. Certainly, belaboring the public with doomsday statistics is counterproductive. Such an approach drives people away from the issue, for it increases their feeling of impotence. Technological suffocation may be a worthwhile new approach to curbing the arms race. It is a limited and interim strategy, but it is realistic nonetheless.

Why has there been so little consideration of technological suffocation? Probably the main reason is that this approach demands planning many years in the future, and it offers no instant results. The arms control lobbies attack the immediate threats, the immediate problems, even though they have no hope for dealing with them. When there is doubt whether the world will survive the next decade without a nuclear holocaust, it seems foolish to ban weapons that may be developed fifteen years in the future. It is the existing weapons that pose the immediate threat to mankind, not the hypothetical ones of the future.

But the arms control lobbies were reasoning in the same way twenty years ago. Their approach of tackling immediate problems has achieved virtually nothing in those twenty years. Yet if they had attempted technological suffocation, we might not be burdened with cruise missiles, MARV warheads, and neutron bombs in 1979.

The arms control lobbies with the help of scientists have the potential to outwit the institutionalized arms race by using foresight. They can compensate for their weakness in the present by transferring their energies to the future. They can push through technology-limited agreements in areas where the military interests have not mustered their forces.

---

[22]Royal United Services Institute for Defence Studies, *Weapons Technology*, London, Brassey's, 1978.

Appendix

World and Regional Military Expenditures*

Military expenditure is a measure of the quantities of resources devoted to military use. SIPRI estimates show that total world military expenditure in 1976 was about $334,000 million, a 3000% increase since 1900 (in constant prices.)*

The major alliances—NATO and the Warsaw Treaty Organization—spent about 70% of the 1976 world total. The Third World (excluding China) spent $51,000 million, or 15% of the total. The Middle East spent about 53% of total Third World (excluding China) expenditures, the Far East (excluding China) and Africa (excluding Egypt) each 13%, South America 11%, South Asia 8%, and Central America 2%. China has given no budgetary data since 1960, but a rough estimate of its military expenditure indicates that in 1976 it spent about 10% of total world military expenditure.

The military expenditure of individual Third World countries vary considerably. In 1975, for example, the three top spenders out of 93 Third World countries were Iran, Egypt, and Saudi Arabia, which spent (in current prices) $7300 million, $5400 million, and $4400 million, respectively. These three countries accounted for 36% of Third World military expenditure. Israel spent $3600 million, and India spent $2800 million. These five countries together accounted for 49% of Third World expenditure. Each of another seven countries (Argentina, Brazil, Indonesia, Iraq, Libya, Nigeria, and South Africa) spent between $1000 and $2000 million. These top twelve countries—five of them in the Middle East—together accounted for 70% of Third World military expenditures.

An example of the burden that military spending can become is given by the fact that Israel's gross domestic product per capita (in current dollars) in 1975 was

*Source: Stockholm International Peace Research Institute, *World Armaments, The Nuclear Threat*, Stockholm, SIPRI, 1977, pp. 6–11.

*If the U.S. intelligence estimate of the dollar cost in the United States of 1976 Soviet military activities ($130,000 million) is used instead of the SIPRI estimate of Soviet Military expenditure ($61,100 million), then the total world military expenditure in 1976 exceeds $400,000 million. The SIPRI figure is calculated from the Soviet national budget, converted to U.S. dollars by using SIPRI estimates of the rouble/dollar purchasing power parity. The SIPRI calculation takes into account the fact that the costs of some military activities—such as military research and development, military aid and stockpiling, and the military elements of the nuclear energy program—are considered not to be fully covered by the Soviet budget. The U.S. intelligence figure is a calculation of how much it would cost in the United States to develop, procure, man, maintain, and operate a military force like that of the Soviet Union.

100

WORLD MILITARY EXPENDITURE, 1908–1976 (CONSTANT 1973 PRICES)

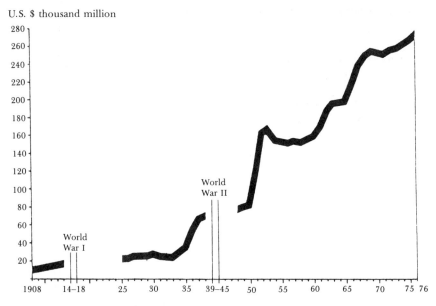

Source: SIPRI data

about $3600, out of which about $1050 was spent on the military; for Egypt the figures are $302 and $140, respectively. In comparison, the United States per capita military expenditure (the highest of all developed countries) in 1975 was about $425.

Total World Regional Military Expenditures, 1957–1976
(U.S. $ million at constant 1976 prices and 1976 exchange rates)

Region	Total expenditure (including 1976)	Percentage of world total	Percentage of third world total	Military expenditure in 1976
Middle East	145,200		34	27,300
Far East (excluding China)	86,800		20	6,420
South America	75,300		18	5,600
South Asia	53,200		13	4,000
Africa (excluding Egypt)	47,800		11	6,490
Central America	15,900		4	1,190
Total Third World	424,200	8	100	51,000
NATO and WTO	4,341,000	79		231,160
China[a]	530,000	10	36,000	
Total world	5,500,000	100		334,000

[a]Rough estimates
Source: SIPRI data

Wᴏʀʟᴅ Mɪʟɪᴛᴀʀʏ Exᴘᴇɴᴅɪᴛᴜʀᴇ, 1957–76

U.S. $ thousand million, at 1973 prices and 1973 exchange rates

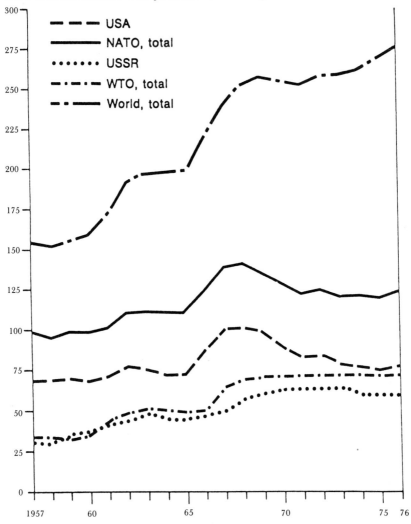

Source: SIPRI data

WORLD MILITARY EXPENDITURE AND WORLD AID TO THE THIRD WORLD IN 1976
(CURRENT PRICES)

U.S. $ thousand million

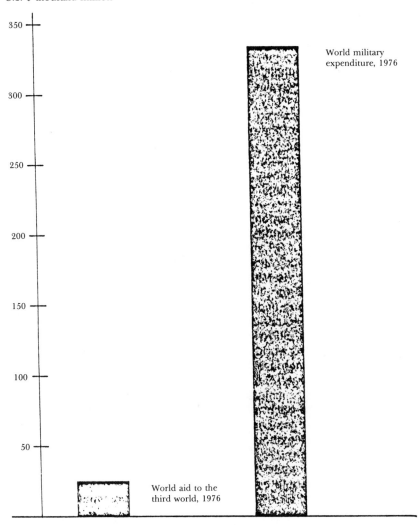

World military
expenditure, 1976

World aid to the
third world, 1976

Source: SIPRI data

World Military Expenditure, 1956–1976[a]
(U.S. $ million, at 1973 prices and 1973 exchange rates)

	1956	1960	1065	1970	1975	1976
United States	68,234	68,130	72,928	89,065	75,068	77,373
Total NATO	97,479	99,180	110,085	127,450	120,719	125,232
USSR	31,600	32,700	44,900	63,000	61,100	61,100
Total WTO	34,200	35,658	49,498	70,498	71,307	72,107
World total	152,292	158,086	198,845	254,130	268,220	276,031

[a]Source: SIPRI data.

Third World Military Expenditure, 1957–1976

U.S. $ million, at 1973 prices and 1973 exchange rates

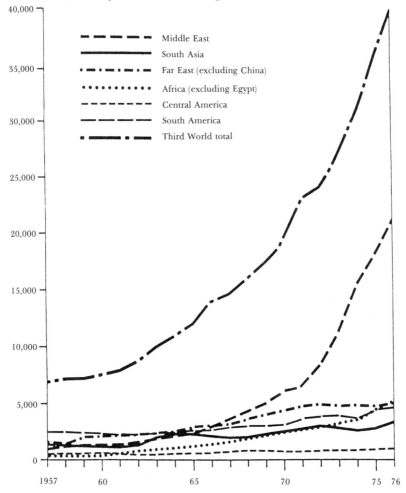

Source: SIPRI data

RATE OF THIRD WORLD MILITARY EXPENDITURE, 1957–1976 (CONSTANT 1973 PRICES)

Index: 1957=100

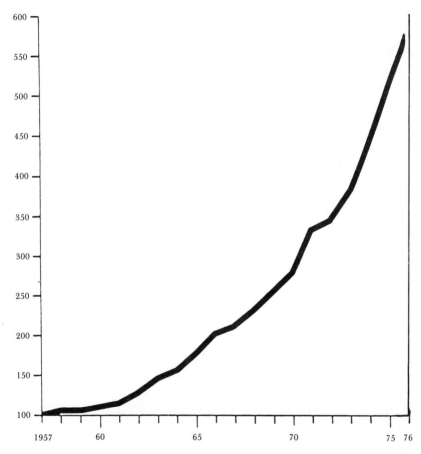

Source: SIPRI data

10. The Qualitative Arms Race and the Problems and Prospects of a Missile Test Flight Limitation

Craig B. Glidden

Despite peaceful coexistence, détente, and SALT, the arms race between the Soviet Union and the United States proceeds untrammeled. As Gerard Smith, Chief Delegate for the United States to SALT I, has said, "With or without SALT, we are in for a continuation of the fateful competition with the Soviets in the field of strategic weapons."[1] The inability of such arms agreements to halt weapons competition testifies to the complexity of the problem. It is tempting to rely on static indicators and technological comparisons to gauge the course of the race, but once begun, an arms race acquires a tempo and direction dictated by political rivalries. As long as foreign policy-makers are confronted with an adversary who appears to pose a political and military threat, and who cannot be thwarted by the acquisition of balance-turning allies, efforts to improve weapons and expand one's arsenal continue to seem a prudent response.[2] The modern technological arms race between the superpowers, however, imprudently threatens the foundation of stability.

A variety of sources have been held responsible for the modern arms race. Since Secretary of Defense Robert McNamara blamed the "action–reaction phenomenon," it has become customary to rely on systemic explanations for the spiral in military activity. Although this type of analysis is correct in some cases, it remains simplistic because it presumes a unitary actor making decisions without consideration of the internal struggles of military and bureaucratic power constellations.

Other theorists, such as Graham Allison, focus on the internal workings of domestic constituencies as a source for the arms competition. Thus, "American

---

[1] Gerard Smith, "SALT Perspectives," (from a speech given at Tulane University, April 1979), p. 1.

[2] Colin S. Gray, *The Soviet-American Arms Race*, Lexington, Massachusetts, Lexington Books, 1976, p. 16.

and Soviet actions or reactions are the negotiated or legislated resultants of domestic contention."[3] Organizational interests determine policy decisions and result in competing views.

A variant of the bureaucratic model centers on interservice rivalries as illustrated in recent deployment decisions stressing offensive rather than defensive technology. The decision to deploy MIRVed systems rather than the ABM carried with it a severe organizational loss for those who backed the defensive system.

Regardless of whether one places emphasis on internal or systemic sources of arms competition, history has shown that any arms agreement is actually three agreements—the first two among the relevant parties within each government, and the third between the respective governments.

Technology itself has also been suggested as a major generating element of the arms race. Research and development produce inevitable pressure for the production of new weapons, often regardless of the need or of the destabilizing effect the weapons may have upon the nuclear balance. Unfortunately, a technological gatekeeper does not exist to prevent destabilizing weapons from passing into nuclear arsenals. Recent technologies have been refined and advanced by incremental evolutionary changes that evade conscientous scrutiny. Even if there were such a gatekeeper, many of the technological innovations of the last decade would probably have passed unnoticed. Astute observers have dubbed this phenomenon "technology creep."

Contributions from the civilian sector and inexpensive innovation have transformed the capabilities of weapon systems, once designed to bolster stability, into systems that threaten the foundation of mutual deterrence. The NS-20 guidance system is a case in point. "When researchers from the Federation of American Scientists tried to find evidence in the federal budget for the device which provided such a tremendous reduction in CEP,* they could find no line item devoted to it."[4] No wonder these incremental advances escape the notice of Congress and the arms control community.

Yet not all technological breakthroughs pass unnoticed; some are sanctioned by decision-makers because they offer ready solutions to strategic problems. The decision to deploy MIRVs in the early 1970s has been pointed to as an example of such a "technologically sweet" solution: "Because these weapons work, and work well, it is taken for granted that they will be deployed, regardless of their actual contribution to national security."[5]

Herbert York holds that the "root of the problem is not maliciousness, but

---

*Circular error probability (CEP) is the figure used for the radius of a circle in which 50% of the munitions aimed will fall.

[3]*Ibid.*, p. 29.

[4]Deborah Shapley, "Technology Creep and the Arms Race: ICBM Problem a Sleeper," *Science,* 201, September 22, 1978, p. 1105.

[5]William C. Potter, "Coping with MIRV in a MAD World," *Journal of Conflict Resolution,* 22, No. 4, December 1978, p. 6.

rather technological exuberance that has overwhelmed other factors."[6] It is not that researchers aim to develop destabilizing weapons, but regardless of their motivations the result remains the same. This drive to produce and deploy technologically more efficient weapons is legitimized by the adversarial relationship that exists between the superpowers.

To date, the technological advancement of strategic weapons has to a great extent evaded the efforts of arms controllers, because arms agreements have focused upon aggregate numbers, definitions, and verifiable methods of counting strategic hardware. The more subtle elements of weaponry cannot be classified and counted easily. As Zbigniew Brezinski has said, "The realities of what goes on inside the payload of a missile do not fit a Procrustean rack of definitions."[7]

In fact, plausible evidence supports the claim that, as a result of the quantitative restrictions of arms control agreements, those agreements themselves have escalated the qualitative momentum of advancement. Political considerations cause the promoters of arms agreements to offer technological trade-offs for support. "Almost every governmental discussion of arms control measures tends to be accompanied by exhortations to increase research and development."[8]

Budgetary studies reveal this rechanneling effect of arms control accords. If an agreement precludes spending in one particular area of the defense budget, those funds tends to be redirected into other areas of defense spending. As a result of the prohibition placed on antiballistic missile systems in SALT I, monies earmarked for ABM were redirected to fund the Trident and B-1 programs. "In part this was done to appease SALT I opponents, but it was also driven by the technological imperative to use funds efficiently and fill in other gaps in the US–Soviet military competition."[9]

Qualitative Circumvention of Quantititive Limits

Modern arms agreements have failed to be comprehensive in scope. Critics therefore claim that partial arms limitations merely rechannel arms competition into areas that remain unrestricted by arms control accords. Conventional and strategic arms agreements demonstrate this phenomenon.

The Washington Naval Treaty of 1922 placed numerical limitations on battleships, battle cruisers, and aircraft carriers. It failed, however, to place any restrictions on auxiliary vessels (cruisers, destroyers, and submarines), thus

[6]Deborah Shapley, "Technology Creep and the Arms Race: Two Future Arms Control Problems," *Science*, **202**, October 20, 1978, p. 289.

[7]Harvey Brooks, "The Military Innovation System and the Qualitative Arms Race," from *Arms Defense Policy and Arms Control,* F. Long and G. Rathjens, eds., New York, Norton, 1976, p. 12.

[8]Roger George, "SALT and the Defense Budget," *Arms Control Today,* **8**, No. 10, November 1978, p. 4.

[9]Deborah Shapley, "Technology Creep and the Arms Race: A World of Absolute Accuracy," *Science*, **201**, September 29, 1978, p. 1192.

encouraging military competition to be concentrated in that area. Within two years after the conclusion of the treaty, the signatories were pursuing the development of cruisers at an alarming rate. "By 1929, the Anglo-American cruiser race became a source of serious friction."[10]

The fifteen-year moratorium on capital ship construction provided by the Washington and London Naval Treaties undoubtedly slowed qualitative improvements and reduced expenditures in the capital ship category.[11] But the concomitant emphasis placed on auxiliary vessels diminished the arms control impact of the agreements.

In 1972, the United States and the Soviet Union reached final agreement on the negotiations of strategic weapons that began during the Johnson administration. SALT represented the merging of the complexities of nuclear strategy, weapons technology, and the overall U.S. foreign policy toward the Soviet Union. The SALT I treaty was composed of two different accords, the ABM treaty and the Interim Agreement.

The ABM treaty has been hailed as the most significant arms control effort of the century. By ensuring that neither the United States nor the Soviet Union would be able to erode the threat to the civilian targets of the other nation through defensive means, the ABM treaty reinforced the doctrine of mutual assured destruction.

The Interim Agreement, which placed quantitative limitations upon ICBMs and SLBMs, became the focus of harsh domestic criticism from the conservative sectors of American opinion. The unequal static limits aroused cries for a vigorous research and development program in the area of strategic weapons. In an effort to assuage those demands, the Nixon administration emphasized negotiating from a "position of strength."

The quantitative restriction of SALT I did not abate the momentum in arms development. Again, the rechanneling of funds for unrestricted areas of the defense budget occurred. George Rathjens has called this the "displacement effect" of SALT. Weapons systems were developed and pursued both as bargaining chips for future negotiations and as a hedge against the possible breakdown of later strategic talks.

In a very real sense, the SALT I accords helped bring about the development of new weapons that have caused great consternation in SALT II and placed the survivability of the strategic triad in doubt. In 1972, the United States' interest in cruise missiles was revived—according to some as a bargaining chip for SALT, but according to others as a means to circumvent the quantitative limits placed on ICBMs and SLBMs in order to swing the strategic balance in favor of the United States.

The bargaining chip argument was also used for the decision to deploy MIRVs in the 1970s. MIRVs enable the accuracies of warheads to be reduced so low as to

[10]Hedley Bull, "Strategic Arms Limitation: The Precedent of the Washington and London Naval Treaties," in SALT: Problems and Prospects, Morton Kaplan, ed., New York, Praeger, 1972, p. 45.

[11]Ibid.

offer hard-target counterforce options. They also present a cost-efficient means of significantly increasing the potential "exchange ratio between the number of adversary missiles destroyed by each missile fired."[12] Yet MIRVs were considered a qualitative refinement of weaponry and therefore not subject to restriction. "In fact SALT I, by permitting the substitution of qualitatively improved weapons for older ones, provided powerful incentive for weapon system innovations on both sides."[13]

At the time of the SALT I agreement, the unfortunate ramifications of MIRV technology were not fully appreciated. Secretary McNamara even attempted to make the case that land-based MIRVs were a helpful contribution to arms control. He tried to persuade Premier Aleksei Kosygin that, because MIRVs offered a relatively inexpensive means of overcoming any conceivable antimissile system, the Soviets and the Americans should forgo the deployment of ABMs.[14]

The inability of SALT I to restrict MIRVs or to hobble the arms race in payload refinement has led to the development of technical capabilities that endanger the invulnerability of land-based ICBMs. These technical capabilities have also made it possible to pursue counterforce strategy and damage-limitation efforts. The recognition of the destabilizing ramifications of MIRVs has caused such former MIRV supporters as Henry Kissinger to regret their optimistic analysis of the implications of MIRV.

SALT I generated considerable debate concerning the best way to proceed with offensive force limitations. It was contended by some that the focus on quantitative limits was not the problem, but that the ceilings negotiated were so inflated that they could have no real impact on arms control. Others argued that the only way to arrest advancing technology was to make a frontal assault on the qualitative nature of arms race. The Stockholm International Peace Research Institute (SIPRI) is of the opinion "that the major test of SALT will be its effect on qualitative development."[15]

With the advent of the Carter administration, "the President's men were eager to do more than just finish the work of Henry Kissinger. Carter's inner circle had an idealistic commitment to real arms control and measures to halt the arms race."[16] This commitment was demonstrated in the Comprehensive Proposals presented by Secretary of State Cyrus Vance to Foreign Minister Andrei Gromyko in March 1977. The Soviets categorically rejected the proposal as "a cheap and shady maneuver aimed at achieving unilateral advantage."[17]

---

[12]Potter, *op. cit.,* p. 602.

[13]Brooks, *op. cit.,* p. 75.

[14]Ted Greenwood,*Making the MIRV: A Study of Defense Decision Making,* Cambridge, Ballinger, 1975, p. 131.

[15]"The Strategic Arms Limitation Talks," *Stockholm International Peace Research Institute [SIPRI] Yearbook 1975,* Stockholm, Almqvist and Wiksell, 1975, p. 78.

[16]Strobe Talbott, "Who Conceded What to Whom," *Time,* 113, No. 21, May 21, 1979, p. 25.

[17]*Ibid.,* p. 26.

The debacle of March forced the negotiators to work along the guidelines established in Vladivostok in 1974. The negotiations were laborious and lengthy, primarily as a result of the parties' desire to deal with sensitive issues that escaped the attention of SALT I. The major impasses were centered around MIRVs, cruise missiles, technological advancement, and verification.

SALT II is composed of three essential parts: a treaty, which will be in force through 1985, based on the Vladivostok guidelines; a protocol, which will cover certain contentious issues and qualitative limitations; and a Joint Statement of Principles, which will serve as a guideline for SALT III.

The SALT II treaty would set aggregate limits on the overall number of strategic nuclear delivery vehicles (2250). It would place sublimits on the total number of MIRVed ballistic missile launchers and heavy bombers armed with long-range cruise missiles (1320). The accord would establish a total number limit on MIRVs (1200) and call for a limit of 820 MIRVed land-based ICBMs, the most destabilizing of the two sides' forces. By doing so, SALT II offers the first quantitative limitations that will address all the central strategic systems.

After SALT I it was believed that meaningful limits on MIRVs would be impossible to attain, owing to the associated problems of verification. Yet by establishing counting rules for verification of MIRVed systems, SALT II is able to set limits that were unattainable in SALT I. "The new agreement does represent an evolution of SALT in the direction the United States has long argued it must if it is to continue to play a significant role in stabilizing the nuclear balance."[18] Modest qualitative improvements are provided for in the protocol.

The parties have agreed not to increase the launch weight or throw weight of their current strategic forces. SALT II continues the prohibition of construction of additional fixed ICBM launchers established in SALT I. When coupled with the provision limiting the fractionization of warheads, the magnitude of the land-based forces is predictable and therefore discourages worst-case analysis. A provision has also been negotiated that bans rapid reload systems, negating the cold-launch capability possessed by some fourth-generation Soviet ICBMs.

The protocol contains some unprecedented attempts designed to limit qualitative advancement. A provision banning the flight testing of mobile ICBMs has been included. Limitations on the number of long-range cruise missiles aboard heavy bombers have been set. And ground-launched and sea-launched cruise missiles have been banned for the duration of the protocol.

The most significant qualitative limit to be reached in the SALT II negotiations was a ban on the flight testing or deployment of "new types" of ICBMs, with the exception of one "new type" per side. This provision presented a definitional dilemma centered around what constituted a "new type." In early April of 1979, Ambassador Anatoly Dobrynin finally accepted the U.S. condition that a modification of 5% in increase or decrease in the "length, diameter, launch weight, and throw weight of an existing missile would constitute the development of a new

---

[18]Jan M. Lodal, "SALT and American Security," *Foreign Affairs,* Winter 1978/79, p. 264.

weapon."[19] This agreement was accompanied by provisions that require advance notification of ICBM tests and a mutual data bank for systems limited by the SALT II treaty. The Joint Statement of Principles establishes a framework for SALT III. It calls for further reductions in strategic offensive forces, further qualitative limitations, and a resolution to address the contentious issues of the protocol.

The accomplishments of SALT II are impressive when compared with earlier attempts at strategic restraint. Yet "the bane of SALT has been that the modernization of old weapons has had an insidious way of rendering arms control agreements obsolete if not equal."[20] The "new missile" provision will not necessarily result in the inability of either side to develop only one new generation of ICBM. "Since new missiles do not have to count as a new type if they are substantially similar to the older generation, the Soviet's fifth generation systems may in the end not be limited at all by this provision."[21]

The fact that the protocol limits systems that will not have initial operating capability until after the December 31, 1981, expiration date undermines the sincerity of the arms control proposals. The protocol is to serve as a powerful precedent for continuation of the restraints of SALT II in SALT III, but that is a dubious conclusion.

As in SALT I, SALT II is having to pay the political price of domestic acceptance. Exhortations for a presidential pledge to maintain an active modernization program have been heard from senators and military leaders alike. The MX is a "key component of the administration's stratagem to win support of moderate senators for the Treaty."[22] And the Pentagon has since proposed a large increase in spending for fiscal year 1980 to modernize the strategic triad.[23] The rechanneling or displacement effect of arms control agreements seems to be applicable to SALT II. "There is ample opportunity for rechanneling, with the most obvious areas being satellite and antisatellite research, greater air defense capabilities and more sophisticated and effective BMD techniques."[24]

Nevertheless, the Carter administration has attempted throughout the negotiations to slow the juggernaut of technology, and modest results will have been achieved if the SALT II treaty is confirmed. The treaty will provide predictable force deployment estimates for its duration, but it will not eliminate the inevitable pressure to modernize. SALT II is evidence that "the inability of political accords to keep pace with technological innovation is rendering strategic arms control agreements obsolescent almost before the ink dries."[25] It has been

[19]Talbott, *op. cit.*, p. 35.

[20]*Ibid.*, p. 30.

[21]Lodal, *op. cit.*, p. 78.

[22]"Upcoming Battle Over the M-X," *Congressional Quarterly*, **37**, No. 12, March 24, 1979, p. 539.

[23]"SALT II: Can It End the Arms Race?" *U.S. News and World Report*, **86**, No. 20, May 21, 1979, p. 21.

[24]George, *op. cit.*, p. 4.

[25]SIPRI, p. 80.

demonstrated that arms agreements have had little effect in slowing the qualitative arms race. The implications of the introduction of new weapons that will have initial operating capability during the 1980s demand the immediate attention of arms controllers and politicians for these weapons will threaten the basis of strategic deterrence and the survivability of retaliatory forces.

Destabilizing Technological Developments

The dilemma presented by the drastic improvement in the accuracy of ICBMs illustrates how arms control has been outwitted by technology in a contest reminiscent of the race between the tortoise and the hare. Subtle advances in guidance and control systems have resulted in a significant drop in the circular error probability of ICBMs.

Currently ICBMs use inertial navigations for guidance and control in order that the missile remain impervious to outside forces. Yet steady improvements in the gyroscopes used in ICBMs have led to the development of guidance systems that can direct a missile within meters of its target. Charles Draper, the foremost expert in guidance systems, has predicted that "fairly soon it will be possible to place a warhead on the silo door."[26]

By 1981, the U.S. Minuteman III force will be significantly improved as a hard-target kill weapon with the addition of the MK-12A warhead and the NS-20 guidance system. These improvements should reduce the CEP of the Minuteman III to less than 0.15 nautical mile.[27] The Soviets have also pursued the refinement of missile technology to increase accuracy. "From the data being analyzed, it appears the instrumentation on the SS-18 and SS-19 is as good or better than the Minuteman III with the NS-20 guidance system. In six tests conducted from Tyuratam, both ICBMs approached 0.1 nautical mile CEP."[28]

The United States is also funding further development of technologies that will decrease the inaccuracies of the ICBM below present capabilities. The advanced inertial reference sphere (AIRS) is being designed for application on the MX missile. "The AIRS is an all-altitude guidance system which can withstand the effect of movement of a missile along the ground before launch."[29] If the MX were equipped with AIRS and a new generation on-board computer, its CEP could be reduced to as low as 0.05 nautical mile, or 300 feet.[30]

---

[26]Graham T. Allison, "Questions about the Arms Race: Who's Racing Whom: A Bureaucratic Perspective," in *Contrasting Approaches to Strategic Arms Control*, R.L. Pfaltzgraff, Jr., ed., Lexington, Massachusetts, Lexington Books, 1975, p. 44.

[27]*Arms Control Impact Statements-FY 1979*, Washington, D.C., Government Printing Office, 1978, p. 4.

[28]Clarence Robinson, "Soviets Boost ICBM Accuracy," *Aviation Week and Space Technology*, April 3, 1978, p. 14.

[29]*Arms Control Impact Statements*, p. 12.

[30]Colin S. Gray, "The Future of Land-Based Missile Forces," *Adelphi Papers*, No. 140, London, The International Institute for Strategic Studies, 1977, p. 33.

Missile accuracies such as those contemplated in future development programs are more detrimental than beneficial to strategic security. Although improvements in accuracy will assist in reducing collateral damage, if associated with low-yield warheads, they certainly contribute to the anxiety of both the Soviet Union and the United States in regard to the vulnerability of their land-based ICBMs. "A strategic weapon's invulnerability, like a Victorian lady's good name, is its most precious asset."[31] And it appears evident that the land-based forces of the superpowers are accruing more liabilities than assets. By the mid-to-late 1980s, on current projections, neither nation will be able to retain sufficient confidence in the ability of its fixed-site ICBMs to ride out a surprise attack and reliably deliver a second strike.[32]

SALT II has proved unable to arrest the threat to land-based ICBMs. The possibility of block obsolescence of the most reliable leg of the strategic triad has encouraged technological solutions to eradicate the problem. The MX has received strong support among some officials as a means to eliminate the Soviet threat to our ICBMs. But although the MX may solve the military problem of land-based vulnerability, it will bring grave arms control problems in its wake.[33] Verification of a mobile ICBM system may prove to be an insurmountable obstacle, which would preclude future arms control efforts in that area. Recognition of this fact caused the United States to renounce mobile land-based systems during SALT I and the Soviets unilaterally to dismantle their SS-16 during the SALT II negotiations.

The quest for "absolute accuracy" and the concomitant vulnerability of land-based missiles may be the most immediate technological advancement with destabilizing effects, but the other components of the central strategic systems are undergoing modernization that could prove to be upsetting to the strategic balance as well. "Presidential guidance has established a requirement for an adequate survivable submarine force capable of successfully attacking all classes of Soviet targets."[34] This has led the Navy to seek development and deployment of an SLBM that will have counterforce capabilities. Trident II will embody the technological sophistication necessary for "absolute accuracy." MARVed missiles using three accuracy improvements have been considered. The first is an extrapolation of a stellar inertial guidance system, the second is a technique employing electromagnetic means of correcting trajectory error, and the third is a mechanism that will enable the re-entry vehicle to take a fix in the terminal stage, use its advanced maneuvering capability, and home in on the target.[35]

[31]John Newhouse, *Cold Dawn: The Story of SALT*, New York, Holt, Rinehart and Winston, 1973, p. 265.

[32]Gray, *Adelphi Papers*, No. 140, p. 47.

[33]Shapley, "Technology Creep...A World of Absolute Accuracy," p. 1193.

[34]U.S. Congress, *Hearings before the Senate Armed Services Committee on S1210*, 95th Congress, First Session, 1978, p. 3695.

[35]U.S. Congress, Senate Committee on Armed Services, *Prepared Statement of Rear Adm. Robert H. Wertheim, USN, Dir. of Strategic Programs, Hearings before Senate Armed Services Committee on S2571*, 95th Congress, Second Session, July 18, 1978, p. 6683.

The allocation for research and development of the Trident II in fiscal year 1980 will be ten times as great as that in fiscal year 1979—approximately 155 million dollars. The development of a terminal homing capability will decouple the relation between accuracy and launch platform characteristics.

Research is also being devoted to the establishment of a depressed trajectory capability. "Depressed trajectory is a term used to describe a reentry vehicle operating with a reentry angle of usually around 6 degrees over the horizon."[36] The advent of hard-target counterforce SLBMs in concert with efforts to achieve depressed trajectories will result in the serious erosion of warning time necessary for bomber forces to escape a preemptive attack. If these hard-target counterforce SLBMs, equipped with depressed trajectories, are coupled with ICBMs with nominal CEPs, the "synergistic relationship" between ICBMs and the strategic bomber force will be destroyed.[37] It has been argued by Thomas Downey that in such a situation the attacked party would have only a strategic monad to deter the aggressor.

The ramifications presented by the advanced technologies described above are not easily ignored. The vulnerability of any leg of the strategic triad arouses anxiety, prompts discussion of launch-on-warning contingencies, and generates a greater velocity of technical momentum aimed at reestablishing invulnerability. The block obsolescence of a major strategic system could redirect development efforts and possibly erode the invulnerability of another component of the triad. Reduced confidence in one's assured destruction capability fosters incentives for a preemptive strike during times of crisis.

Effects of Qualitative Improvements on Strategic Doctrine

The basis of strategic nuclear deterrence rests on the knowledge that one side's ballistic missile systems are capable of absorbing an offensive attack and in return inflicting unacceptable damage on the aggressor. It is claimed by some that the mutual assured destruction (MAD) orientation of strategic doctrine was an aberration—"the product of technological accident, strategic imbalance, and naive strategic theorising."[38] Gray suggests that MAD theorists have sought to freeze a moment in history, the technological plateau of the mid-to-late 1960s.[39] Yet the qualitative improvement of weapons has nudged the superpowers off their MAD precipice, into the gulch of counterforce doctrine.

The "Schlesinger shift" of 1974 was predicated on the belief that the best way to insure a credible deterrent was to maintain the capability of exercising nuclear options short of massive retaliation. "Technological innovation has made more

[36]Clarence Robinson, "U.S. Weighs SALT Offer to the Soviets," *Aviation Week and Space Technology,* September 4, 1978, p. 25.

[37]Thomas Downey, "How To Avoid MONAD and Disaster," *Foreign Policy,* June 24, 1976, p. 193.

[38]Gray, *The Soviet–American Arms Race,* p. 137.

[39]*Ibid.*

feasible limited or flexible responses to nuclear provocation."[40] Yet fear is expressed that the full exploitation of new technologies is better suited to nuclear warfighting.

Defense of the counterforce argument places heavy reliance on the apparent pursuit of counterforce by the Soviets. It is argued that the distinction between deterrence and war-fighting is representative of a bygone era. "In more recent years, as Soviet capabilities have become more formidable, this distinction has become blurred by the necessity of responses to attack at a variety of levels in order to maintain a credible deterrent."[41] Counterforce capability is useful only to the extent that it can erode the countervalue* response of an opponent nation. Thomas Downey convincingly contradicts counterforce logic by noting that "to respond to an opponent's counterforce capability by creating one's own is an exercise in irrelevance since it does nothing to reduce the impact of his counterforce on one's countervalue."[42] The United States should not respond to a Soviet counterforce with its own. It should ensure the invulnerability of its countervalue and reduce tensions by not adopting a counterforce strategy. "The trend in offensive technology that leads away from exploiting that technology to enhance deterrence, and towards improving the capacity to fight a nuclear war, inverts American security."[43] If a nuclear war becomes manageable, it also becomes more likely.

Counterforce strategy, made possible by technological advancement, encourages further innovation and refinement of strategic systems. "That the seeking of an improved counterforce capability might prove to be a stimulant to the arms race is difficult to dispute."[44] In fact, this is precisely what happened in the early 1960s after Secretary McNamara endorsed a damage-limiting counterforce strategy, and it was undoubtedly an important consideration in the abandonment of counterforce rhetoric.

Impact of Technology on Arms Control

The implications of technological development on strategic doctrine and the arms race are evident, yet possibly the most regretful ramification of qualitative

---

*"Countervalue" denotes a strategy of attacking targets other than strategic weapons launch systems and is therefore often used in contrast to "counterforce," which denotes attacks aimed only at an opponent's strategic retaliatory capacity.

[40] Paul F. Walker, "New Weapons and the Changing Nature of Warfare," *Arms Control Today*, **9**, No. 4, April 1979, p. 6.

[41] U.S. Congress, Senate Committee on Armed Services, *Prepared Statement of Dr. William Perry, Undersecretary of Defense, Research and Engineering, Hearings before the Senate Armed Services Committee on S2751*, 95th Congress, Second Session, 1978, p. 6675.

[42] Downey, *op. cit.*, 182.

[43] Walker, op. cit., p. 6.

[44] *U. S. News and World Report*, p. 22.

modernization is its impact upon the possibility of meaningful arms control.

In the debate that the SALT II treaty has prompted, verification of the agreement has become a central question. Exacerbated by the loss of monitoring facilities in Iran, the verification issue has occupied much attention in the Senate debate.

Technological innovation threatens the ability of nations to restrict weapon programs by "national technical" means of verification. As mentioned earlier, the subtle improvements in weaponry evade common definition and do not provide functionally related observable differences. Verification such as the kind demanded in the Senate debate may prove to be impossible to maintain if these elusive systems are not thwarted.

Cruise missiles offer a poignant example of this dilemma. Because of verification difficulties, the Russians maintained that the provisions of the SALT II treaty "should apply to armed cruise missiles: there should be no distinction between nuclear-armed and conventionally-armed cruise missiles."[45] Yet historically it has been the position of the United States that SALT would restrict only strategic and not conventional or tactical weapons.

During the 1960s, the demarcation between strategic and conventional weapons was convenient because the components of the strategic forces were intercontinental ballistic missiles, which it was believed played only a strategic role. "This distinction was institutionalized as the central organizing concept of SALT."[46]

The sophistication of cruise missile technology now promises unparalleled adaptability and versatility. Cruise missiles do not require identifiable launch facilities, such as silos, submarines, or airfields. Their versatility in mission performance makes them suitable for strategic or tactical roles armed with nuclear or conventional munitions.[47]

Weapons such as the cruise missile undermine the basic requirement of mutually agreed upon definition necessary for arms control agreements. Such weapons blur the distinction between strategic and tactical, between conventional and nuclear. The only way to limit such systems is to constrain their numbers regardless of their mission, an approach rejected by the United States at SALT II.

If the cruise missile used conventional munitions to attack strategic targets, the "territorial threshold attached to a nuclear threshold would become decoupled."[48] Although this would heighten the threshold for nuclear weapons use, it could effectively lower the inhibitions of leaders to use force. The ambiguous nature of such weapons will heighten uncertainty in times of crisis.

SALT has also relied on the fact that strategic weapons have been virtually the

[45]Talbott, *op. cit.*, p. 28.

[46]Richard Burt, "Cruise Missiles and Arms Control," *Survival*, January, February, 1976, p. 14.

[47]Kosta Tsipis, "Cruise Missiles," *Progress in Arms Control*, San Francisco, W. H. Freeman, 1978, p. 175.

[48]Burt, *op. cit.,* p. 15.

monopoly of the superpowers. Horizontal proliferation of nuclear cruise missiles threatens the bilateral nature of strategic talks. These weapons "would cut across the jurisdictional patterns established between SALT and the negotiations on mutual force reductions in Europe."[49] Attaining consensus between the superpowers concerning weapons limitation has proved difficult enough. Obtaining a consensus in Western Europe would prove to be exceedingly problematic, owing to differing perspectives and security needs.

So far, then, the efforts of the superpowers have not arrested the juggernaut of technology. And as the 1980s unfold, we shall be faced with the grim prospect of qualitative improvements that will be extremely destabilizing to security. Technological advancement is too incremental, diffuse, and difficult to categorize to be limited by traditional arms control methods. The destabilizing results of modernization can be minimized only if the qualitative arms race itself is slowed. Technical solutions to technically created problems will only create greater instability and complicate the nature of future arms control negotiations. Only mutual self-restraint will lessen the threat to stability, slow the arms race, and reduce international tensions.

Flight Test Quotas

The establishment of confidence flight test quotas offers a practical and verifiable solution to the problem of restraining qualitative weapons improvements. The road to deployment is composed of two stages of testing: research and development testing, and operational readiness or postdeployment testing. The United States and the Soviet Union both require testing to be done to determine if their deployed strategic systems are capable of carrying out their assigned strategic roles. The confidence flight test quota would curtail the total number of strategic missile tests conducted by each superpower, thereby diverting efforts centered on research and development testing to ensure that confidence could be retained in their presently deployed strategic systems. If the quota were severe enough, it could diminish confidence in highly complex systems and inhibit their deployment.

The flight test limitation would cripple the counterforce capability of the superpowers and reinforce countervalue retaliation as the basis of mutual deterrence. Strategic security would be bolstered, and international suspicions quieted.

A total ban of flight testing has been proposed in the past, but because of its uncertain outcome and drastic nature it has never overcome political resistance. Congressional testimony illustrates the need for confidence testing. "Minuteman is tested in its operationally deployed configuration over the lifespan of the system in order to avoid failure from things such as aging."[50] Funding for

---

[49] *Ibid.*, p. 16.

[50] Potter, *op. cit.*, p. 615.

Poseidon and Polaris missiles is also requested for testing, even though the weapons are no longer procured. It is deemed necessary in order to evaluate the readiness of deployed missiles in accordance with the test criteria of the Chiefs of Staff.[51]

One advantage to a missile flight test limitation is that it could be confidently verified with current reconnaissance capabilities. "Counting missile firings is what early-warning satellites, over-the-horizon radar, and fixed-land radar do best."[52] The task of reconnaissance could be substantially reduced if the flight test agreement provided that tests be preannounced and held at specific locations. This would be consistent with the precedent established in SALT II, calling for advance notification of certain ICBM tests.

Enumerated qualitative restrictions coupled with the test limitation would hobble the arms race even further. Destabilizing technologies would not be specifically banned by the flight test quotas. If one were willing to sacrifice confidence in the reliability of currently deployed systems, testing could be devoted to research and development in order to sustain a technological breakthrough. It is reasonable to assume, however, that this switch in emphasis could be detected and any agreement abrogated.

As demonstrated by previous arms agreements, verification of specific qualitative restraints is far more difficult than verification of missile test firings. It is important not to subdivide restrictions so as to attempt to distinguish between essentially indistinguishable missile testings. Examples are MRVs and MIRVs, AMARVs and terminally guided MARVs, and long-range and short-range cruise missiles. Special attention would have to be given to areas where technology transfers from tactical weapons to nuclear weapons could circumvent the restrictions. A case in point would be the Pershing II and MARV ban.

How much testing is necessary to ensure confidence in deployed systems, while retarding the development of newer ones, is a subjective question. It is further complicated by the disparities that exist between American and Soviet testing practices. The U.S. Minuteman force has never been fully tested from operational silos for safety reasons, whereas the Soviets conduct extensive testing of their ICBMs from operational silos.[53] And it has been suggested that the Soviets would require more tests, owing to the diversification of their strategic forces.

During the Department of Defense Authorization Hearings for fiscal year 1976, the Air Force indicated:

If we were required to reduce the level of testing for our ICBMs to only 5–10 missiles a year, it would have serious impact on our ability to flight test system changes, new

---

[51] U.S. Congress, Senate Committee on Armed Services, *Statement of Rear Adm. A. L. Kelln, USN, Dir. of Strategic Submarine Division and Trident Program Coordinator, Hearings before the Senate Armed Services Committee on S1210,* Part 5, 95th Congress, First Session, 1979, p. 3705.

[52] Ted Greenwood, "Reconnaissance, Surveillance and Arms Control," *Progress in Arms Control,* San Francisco, W.H. Freeman, 1978, p. 103.

[53] Potter, *op. cit.,* p. 613.

improvements... and most importantly our ability to detect deficiences resulting from aging.[54]

On the other hand, Harold Agnew, former Director of the Los Alamos Scientific Laboratory, believes that, "with adequate funding, it is possible to maintain stockpile reliability in the absence of nuclear testing."[55]

In testimony before the Senate Committee on Foreign Relations, Sidney Drell posited that an appropriate annual quota of roughly "10 to 20 missile tests for each country"[56] would prevent the attainment of destabilizing technology and be fully compatible with the assured destruction role of our ICBMs and SLBMs.

Evidence suggests that the testing quota number could be lower at present than at any earlier time. The technical sophistication of testing equipment and the weapons themselves have increased the success rate of system testing as much as ten times. After twelve flight tests in the Polaris A-3 program, there was only one completely successful flight, yet the Trident I has had ten completely successful tests out of twelve.[57] This suggests that in the future the quota level should be renegotiated to account for refinement of weaponry and testing techniques.

A missile flight test quota at a low-enough level would inhibit confidence in the attainment of strategic technology capable of obliterating fixed land-based ICBMs. Confidence would be retained in the ability of ICBMs to act as retaliatory instruments, but the quota would deny them the reliability necessary for confident "absolute accuracy." As John Walsh, Deputy Director of Defense Research and Engineering, stated before the Senate Armed Services Committee:

After you have done your initial testing... a large amount of testing can be done to refine your statistics, but if you go to a smaller amount of testing, you are no longer really adding to the statistics. Rather, what you are doing is looking to see if some aging phenomenon has occurred which results in major deterioration of your accuracy from that which you believe it is already, so that the confidence figure you are working with tends to be a reliability kind of confidence, rather than accuracy confidence.[58]

The reduction in the confidence in accuracy suitable to deliver a counterforce strike would be a deterrent for initiating an attack.

The political feasibility of a missile flight test limitation received considerable attention as a result of President Carter's March 1977 proposals. The flight test limitation would be consistent with the mutual advantages of both the United States and the Soviet Union. As William Perry has observed, "one of the

[54] *Statement of Rear Adm. Wertheim, Hearings before the Senate Committee on Armed Services on S2571*, p. 6867.

[55] D. G. Brennan, "A Comprehensive Test Ban: Everybody or Nobody," *International Security*, 1, 1976, p. 101.

[56] Sidney Drell, "Discussion Paper on Limitations on Missiles Test Firings," Stanford University, December 1976, p. 18.

[57] *Statement of Rear Adm. Wertheim, Hearings before the Senate Committee on Armed Services on S2571*, p. 6676.

[58] Statement of John Walsh, *Ibid.,* p. 6876.

considerations of our comprehensive SALT proposal in March was the extension of the survivability of Minuteman through limits on modifications.[59] Since the Soviets suffer from an even greater threat to their land-based ICBM force, it would appear that the logic would work in reverse. Both nations would then be spared the frightening specter of block obsolescence of their land-based missiles.

It is certain that President Carter obtained sufficient bureaucratic and military support to offer the March proposal. What is less clear is the extent of Soviet receptivity. Unfortunately, the quantitative restrictions accompanying the flight test proposal were clearly to the Soviet's disadvantage and therefore caused categorical rejection of the entire comprehensive plan. Denunciations of the proposal were leveled by Foreign Minister Gromyko on March 31, 1977. Conspicuous by its absence, however, was "any reference to the proposed limitation on missile test firings."[60] It is doubtful that the omission was the result of an oversight. More likely, it indicates a Soviet interest in that provision. The adoption of the flight test ban provisions of SALT II offers encouragement.

The major disadvantage of the missile flight test limitation would be the benefits inherently granted to the technologically superior party at the time of the agreement. If the proposal were coupled with other qualitative restraints, it would essentially freeze the technological levels of both nations. Bureaucratic drawbacks to the missile flight test approach have also been raised. The approach, as proposed here, calls only for limitations on the number of strategic tests made annually. Research and development would not be specifically limited by the proposal. In fact, they are encouraged, to avoid the possibility of technological breakthroughs by the Soviets. One should not overestimate the recalcitrance of conservative sectors of the bureaucracy and military to accept a missile flight test limitation, however. The reaction of the Soviet internal structure is difficult to predict, but it is reasonable to assume that United States resistance could be overcome, in light of the March 1977 proposal.

The advantages of a missile flight test limitation seem to overshadow the disadvantages. Both nations would be assured of their retaliatory capability, and neither would fear the rapid extinction of their land-based ICBMs. Advancement of the qualitative arms race, which threatens stability and the future of arms control, could be significantly decelerated. The flight test quota could be monitored confidently by both nations, and attempts to circumvent them could be detected. Coupling the flight test limitation with enumerated qualitative restrictions, such as a ban on MARVs and anti-submarine warfare limitations, would offer greater stability for the world.

It must be remembered that the quota to be established, the effectiveness of verification, and the merits of the proposal must be determined in the political arena. It has been demonstrated here to be technically and doctrinally prudent and feasible, but it must also be deemed politically prudent and feasible. It is

---

[59]Statement of Dr. William Perry, *Hearings before the Senate Armed Services Committee on S2751*, p. 985.

[60]Potter, *op. cit.,* p. 614.

hoped that the two will be congruent and that a missile flight test limitation will be implemented. Otherwise the 1980s will offer the depressing prospect of heightened strategic instability and mutual suspicion.

11. The Comprehensive Nuclear Test Ban

Herbert F. York, Jr., and G. Allen Greb

All American presidents since Harry Truman have called for comprehensive restrictions on nuclear testing. Indeed, this was originally thought to be the least difficult of all arms control measures to achieve. Yet the closest the superpowers have come to this modest goal is the Limited Test Ban Treaty (LTBT) of 1963, prohibiting explosions in the atmosphere, the oceans, and outer space. Leaders of the United States and the Soviet Union have signed three related pacts since: a nuclear nonproliferation treaty (NPT) in 1968, designed to prevent the spread of nuclear weapons to non-nuclear states; a threshold test ban agreement in 1974 that limits underground tests to 150 kilotons yield or less; and a treaty in 1976 restricting nuclear explosions for so-called peaceful purposes. Our purpose in the pages that follow is to analyze current arguments for and against a comprehensive nuclear test ban (CTB) and to trace their historical roots.

In Support of a Nuclear Test Ban

In addition to the nearly full-time job of answering their critics in the nuclear weapons establishment, proponents of a total test cessation have developed several positive arguments of their own over the past twenty years. The most important of these involves the political and military impact of a test ban. But the one that initially brought the matter to public attention and forced the first positive results is what we would now call its environmental impact. Extensive U.S. and Soviet nuclear tests, even in the face of a near-disastrous accident in 1954 and subsequent numerous fallout scares, led to a worldwide uproar against the superpowers during the late 1950s. At home, many American scientists and political leaders began to criticize the government's nuclear weapons test program openly for the first time. Even in Russia, scientists—including Andrei Sakharov, who helped develop the Soviet hydrogen bomb—began a much less vociferous yet significant campaign to stop their country's testing program.[1] Although the 1954 episode and the subsequent fallout "scares" played a particularly crucial role in stimulating widespread initial public interest in nuclear weapons and their potential harmful effects, no unilateral U.S. decisions or international treaty negotiations have been predicated on the fallout issue alone. Indeed, the prevailing pattern became one of "sudden interest in nuclear tests, intense debate and public discussion, and then the equally abrupt dropping of the issue."[2] Even most test ban proponents have regarded fallout as only a minor problem since the emotion-charged years of the late 1950s and early 1960s.

An issue of greater importance in the minds of many advocates has been the relationship of a test ban to the Soviet–American arms race. Even as early as the 1950s, many scientists who favored a test ban stressed this point above the fallout question. An example is the physicist Hans Bethe, who testified before a Senate disarmament subcommittee in 1958 that he believed the key point regarding a test suspension to be "a political one; namely to obtain a controlled disarmament agreement." Bethe further postulated that, "if we once get one controlled disarmament agreement, I believe that others may follow and that the principle will thereby be established. This, I think, is the overriding argument."[3] These early efforts resulted in a comprehensive test moratorium that lasted from 1958 to 1961. Attempts to turn that moratorium—which was initiated as a result of a roughly matched set of unilateral statements by Eisenhower and Khrushchev—into a formal treaty failed because the verification issue proved to be more intractable than was first anticipated. Eisenhower later remarked that his greatest regret about his presidency concerned his failure to convert the moratorium into a formal treaty prohibiting all forms of nuclear tests. Of course, his Atoms for Peace Program and Open Skies proposal also looked toward mutual Soviet–American arms reduction as one of their ultimate goals, but neither initiative was effective in that regard.

The Kennedy administration made arms control a core theme in its approach to the test ban question. In effect, Kennedy and his advisers presented the 1963 LTBT as a pivotal test case for all future disarmament agreements. And in retrospect the LTBT clearly did create a more favorable political climate, setting the stage for, among other measures, the 1968 NPT, SALT I, and SALT II.

[1]Lawrence S. Wittner, *Rebels Against War: The American Peace Movement, 1941–1960*, New York, Columbia University Press, 1969. pp. 240–247; Robert Gilpin, *American Scientists and Nuclear Weapons Policy*, Princeton, New Jersey, Princeton University Press, 1962, pp. 137–143; 155–161, 166–171; J. Rotblat, *Scientists in the Quest for Peace: A History of the Pugwash Conference*, Cambridge, Massachusetts, M.I.T. Press, 1972; Andrei D. Sakharov, *Sakharov Speaks*, Harrison E. Salisbury, ed., New York, A. A. Knopf, 1974, pp. 12, 32.

[2]Robert A. Divine, *Blowing on the Wind: The Nuclear Test Ban Debate, 1954–1960*, New York, Oxford University Press, 1978, p. 27. This new study provides the most thorough discussion of the fallout controversy and its relation to the test ban debate during the 1950s.

[3]Senate Committee on Foreign Relations, *Hearings on Disarmament and Foreign Policy,* 86th Congress, First Session, 1959, p. 178.

More recently, CTB advocates have turned their attention almost exclusively to a second political issue, the so-called nuclear proliferation problem. In this context, proliferation means the spread of nuclear weapons or nuclear weapons technology to states that do not now possess it. Kennedy administration spokesmen introduced the argument into the public test ban debate during the pivotal 1963 LTBT hearings. Although Secretary of State Dean Rusk considered fallout and the arms race to be issues of major importance at that time, he significantly placed proliferation ahead of both:

Among the dangers to the United States from continued testing by both sides I would consider the danger of the further spread of nuclear weapons to other countries of perhaps primary importance. Unlimited testing by both the United States and the Soviet Union would substantially increase the likelihood that more and more nations would seek the dubious, but what some might consider prestigious, distinction of membership in the nuclear club. The risks to the security of the free world from nuclear capabilities coming within the grasp of governments substantially less stable than either the United States or the Soviet Union are grave indeed.[4]

Arms Control and Disarmament Agency (ACDA) Director William C. Foster agreed that proliferation was "a problem of special concern to this committee, and...a grave threat to our future security." Secretary of Defense Robert McNamara also outlined the problem several times for anxious committee members, contending that one of the "great advantages" of an atmospheric test ban would be its positive effect on "retarding the spread of nuclear weapons." Similar pronouncements came from such divergent government sources as Glenn T. Seaborg, chairman of the Atomic Energy Commission (AEC); Harold Brown, then Director of Defense Research and Engineering; and General Maxwell Taylor, chairman of the Joint Chiefs of Staff (JCS).[5]

This argument gained particular urgency after 1968 with the signing of the NPT, which, like the LTBT, specifically committed the two major nuclear powers to achieve a comprehensive test ban (in the treaty's exact words, "the discontinuance of all test explosions of nuclear weapons for all time"). In 1971, Philip J. Farley, the acting director of ACDA, reported to Congress that a CTB "would have particular importance because such an agreement would involve most other countries of the world... so it would have a very wide application to the proliferation problem."[6] CTB proponents outside of government typically went beyond such moderate statements. They observed that half the nations of the world, including several major powers, still had not signed the NPT. As a

[4]Arms Control and Disarmament Agency, *Why a Nuclear Test Ban Treaty?* Washington, D.C., 1963, pp. 5–6.

[5]*Ibid.*, p. 11; Senate Committee on Foreign Relations, *Nuclear Test Ban Treaty,* 88th Congress, First Session, 1963, pp. 107–108, 144–145, 189, 259, 325, 562.

[6]Senate Subcommittee on Arms Control, International Law and Organization of the Committee on Foreign Relations, *Prospects for Comprehensive Nuclear Test Ban Treaty,* 92nd Congress, First Session, 1971, p. 33.

rationale for their refusal, these nations could and did point to the apparent unfairness and hypocrisy demonstrated by the superpowers, which asked others to abstain from all nuclear tests while continuing to improve their own nuclear expertise through extensive underground testing programs. A CTB therefore remained, as a Stockholm International Peace and Research Institute (SIPRI) report said, "a piece of *unfinished business* since the signing of the LTB in 1963 [emphasis in original]."[7]

Since the early 1970s, advocates both within and without government circles have escalated their agitation for a CTB, based on its relation to proliferation. Entire volumes of "special studies" have been devoted exclusively to the subject,[8] and it has been thoroughly discussed by Congress and in the public press. Wolfgang K.H. Panofsky, director of the Standard Linear Accelerator Center, made perhaps the most cogent recent statement of the proliferation argument before a Senate subcommittee in September 1977. The "principal motivation" for achieving a CTB, Panofsky said, related to this issue:

Here the linkage is both political and technical. By agreeing to the preamble of the NPT and as a signatory to the Limited Nuclear Test Ban, this country has undertaken a solemn obligation to strive in good faith toward the attainment of a CTBT. The lack of progress on the part of the Soviet Union and the United States in reaching a CTBT has added to the cynicism with which the NPT is viewed by some nations and individuals...

The technical reason why a comprehensive test ban agreement would serve the cause of nonproliferation is, of course, the fact that nonnuclear nations could not with confidence develop a nuclear explosive without nuclear testing.

The LTBT of 1963 currently has more than 100 signatories, and it would be.expected similarly that a comprehensive ban would be adhered to by a very large number of other nations.[9]

Significantly, President Jimmy Carter's administration had adopted this line of reasoning as official policy, despite mounting doubts about a CTB within the bureaucracy itself. As recently as March 1978, for example, Rear Admiral Thomas Davies of ACDA presented much the same arguments as Panofsky's in outlining the government's position on this question.[10]

[7] Robert Neild and J.P. Ruina, "A Comprehensive Ban on Nuclear Testing," *Science*, 175, January 14, 1972, p. 145; Herbert York, "The Great Test Ban Debate," *Scientific American*, 227, November 1972, p.23.

[8] See, for example, Stockholm International Peace and Research Institute, *Nuclear Proliferation Problems, Cambridge, Massachusetts, 1974;* Senate Committee on Government Operations, *Facts on Nuclear Proliferation: A Handbook*, 94th Congress, First Session, 1975; Jane M. O. Sharp, ed., *Opportunities for Disarmament: A Preview of the 1978 United Nations Special Session on Disarmament*, New York, Carnegie Endowment for International Peace, 1978, pp. 140–141.

[9] Senate Committee on Foreign Relations, *Threshold Test Ban and Peaceful Nuclear Explosion Treaties*, 94th Congress, Second Session, 1977, pp. 138–139, 143.

[10] House Intelligence and Military Application of Nuclear Energy Subcommittee of the Committee on Armed Services, *Current Negotiations on the Comprehensive Test Ban Treaty*, 95th Congress, Second Session, 1978, p.3.

Those who oppose a CTB generally contend that the connection between it and nonproliferation is somewhere between nonexistent and very weak. When the next nation decides to acquire a nuclear weapons capability, they argue, it will do so for reasons much more practical and immediate than whether or not the superpowers are still conducting tests. Robert W. Helm and Donald R. Westervelt, both of Los Alamos, recently stressed this point in an *International Security* article. "We are convinced," they maintain, "that individual states base their nuclear weapon decisions primarily, if not exclusively, on their perceptions of local—and largely conventional—threats, and that the arms race behavior of the superpowers is relevant only in that it can serve as a political excuse for pursuing already well-perceived national interests."[11] A few sometimes go further and contend that the connection is negative. They say that a cessation of testing by the United States would be an indication of failing will, and that nations that might otherwise look to U.S. nuclear strength for their own security would feel compelled to acquire weapons of their own.

In Opposition to a Nuclear Test Ban

During the early years of the nuclear test ban debate, opponents of a test cessation concentrated upon the issue of weapons development. At that time, nearly all informed scientists, technologists, military and government officials, and members of Congress believed that the United States had a real lead over the Soviets in nuclear weapons development. Many of them, however, also shared the conviction that a vigorous testing program was essential if the United States was to maintain its lead. The first explosion of the atomic bomb in 1945, the invention of the hydrogen bomb in 1951, and the development of physically much smaller but still extremely powerful variants of these warheads in the mid- and late 1950s came in such quick succession that it was only reasonable to think that further dramatic and politically significant technological breakthroughs would soon follow. In particular, further research was needed, experts said, both to perfect the design of existing nuclear devices—that is, to make lighter, smaller, high-yield warheads adaptable to special purposes such as an ABM system—and to develop new kinds of weapons, such as the pure fusion or so-called clean bomb, which would produce substantially reduced levels of radioactive fallout compared with either the fission or the standard thermonuclear weapons of that time. As the proponents of a test ban had concentrated largely on the dangers of radioactive fallout in their public arguments, it is perhaps no surprise that the proponents of continued testing placed special emphasis on a bomb that produced less radioactivity.

Edward Teller, along with several of his associates at the newly organized University of California Lawrence Livermore Radiation Laboratory, took the lead

---

[11] Robert W. Helm and Donald R. Westervelt, "The New Test Ban Treaties: What Do They Mean? Where Do They Lead?" *International Security*, 1, 1977, pp. 166–169, 177. See also Michael M. May, "Do We Need a Nuclear Test Ban?" *Wall Street Journal*, June 28, 1976.

in pressing this line of argument in the late 1950s. Supported by the chairman of the AEC, Lewis L. Strauss, and by certain members of the Department of Defense (DOD), these scientists worked hard and effectively in opposition to a test ban because of its possible impact on U.S. weapons development.[12]

During the 1963 public hearings on the LTBT, the notion of a quest for new knowledge—of the possibility of technological breakthroughs just around the corner—permeated the testimony of witnesses and the remarks of congressmen alike. Teller again became a major spokesman for the antiban scientists during these hearings. "Those of us who are worried about this treaty have a clear and strong obligation to put before you our worries," he told the Senate Committee on Foreign Relations. Missile defense, he stressed, required "proper and complete experimentation, experimentation that can be performed faithfully and in a relevant way only in the atmosphere." The extensive Russian test series of 1961 and 1962 in fact had included a number of explosions at the very high altitudes necessary for development of an ABM warhead. Several specialists believed that these tests in particular had given the Soviets a great advantage over the United States in the development of a workable ABM system. Teller went so far as to suggest that Khrushschev was willing and even anxious to have a test cessation precisely because of this hypothetical advantage. Teller's major worry, however, still related to the pursuit of new knowledge in general: "I think that this treaty really ties us, ties one hand behind our back in our endeavor to acquire new knowledge, and this is a field in which I want to work with both hands."[13] Several politicians and members of the military establishment echoed these ideas.

Under such constant pressure, members of the Kennedy administration who were pushing the treaty felt compelled to offer assurances about weapons development. Secretary of Defense Robert McNamara, after discussing the "ABM problem," declared that, "with or without a test ban, we could proceed with the development of an ABM system." Harold Brown had doubts about whether an antiballistic missile could be developed, but added that, "if it can, I do not believe that this treaty will greatly inhibit its development nor substantially reduce its effectiveness."[14]

Today, the argument over new weapons versus new knowledge persists, but those who oppose a comprehensive test ban no longer give it first place, largely because of a growing consensus, even among many within the nuclear weapons establishment, that nuclear weapons technology has "matured." In fact, the weapons laboratories have made no significant breakthroughs since Teller and other scientist–experts put forth their claims in the early 1960s that new advances

[12]Charles J.V. Murphy, "Nuclear Inspection: A Near Miss," *Fortune,* 59, March 1959, pp. 155–156; Edward Teller, *The Legacy of Hiroshima,* Garden City, New York, Doubleday, 1962, p. 68; Lewis L. Strauss, *Men and Decisions,* Garden City, New York, Doubleday, 1962, pp. 418–419; Harold Jacobsen and Eric Stein, *Diplomats, Scientists, and Politicians; The United States and the Nuclear Test Ban Negotiations,* Ann Arbor, Michigan, 1966, University of Michigan Press, pp. 27–28.

[13]Senate Committee on Foreign Relations, *Nuclear Test Ban Treaty,* pp. 373, 417, 423–424.

[14]*Ibid.,* pp. 103–104, 546.

were just around the corner. Even the "neutron bomb"—a highly controversial political issue in the 1970s—is a creature of the late 1950s and early 1960s. Moreover, it provides only marginal advantages over the weapons its advocates aspire to replace. Of course, substantial improvement in the physical characteristics of nuclear bombs have been made in the past twenty years, and more improvements are evidently in store, including the possible extension of the clean bomb idea to very small yields. Indeed, Livermore and Los Alamos now focus much of their attention on what were considered earlier to be peripheral issues—making bombs safe and secure against accidents, misuse, tampering and robbery—and modifying existing designs so that they make a better fit with new delivery systems. Many of these improvements involve considerable technological cleverness and expertise, but they have done relatively little to shift the nuclear balance, whether strategic or tactical.

Another issue raised repeatedly by opponents of a test ban over the past two decades is that of "weapons effects," broadly defined as the effects on people and equipment of the blast, heat, radiation, and fallout produced by a nuclear explosion. Since the beginning of the nuclear weapons program, this problem has been studied in great detail. In fact, the initial series of post-World War II experimental nuclear tests at Bikini Atoll, Operation Crossroads, were carried out specifically to determine the effects of nuclear explosions on naval vessels. From that time, the overall weapons development program has included a separately administered weapons effects program under the jurisdiction of a special Pentagon agency. Originally designated the Armed Forces Special Weapons Project in 1947, renamed the Defense Atomic Support Agency (DASA) in 1959, this office is known today as the Defense Nuclear Agency (DNA).[15]

As in the debate involving the safety of nuclear devices, members of the nuclear establishment have argued the pros and cons of testing weapons effects largely behind closed doors. It was not until 1963 that the public became significantly aware of this sensitive national security issue. During the limited test ban hearings of that year, several congressmen who supported the construction of an ABM system raised questions about the so-called blackout phenomenon. Their concern related to the fact that the ionization of the upper atmosphere produced by high-altitude nuclear explosions could "black out" ABM radar for a limited time. Both the heat of the nuclear explosion ("fireball blackout") and the decay of fission products ("beta blackout") could produce this effect. Representatives of the Kennedy administration spent much time reassuring committee members that the United States had not fallen behind Russia in this or any other sphere of weapons effects. "As for the blackout problem, "Secretary McNamara summarized,

Soviet and U.S. experience appears to be comparable although obtained in different ways....

[15]Richard G. Hewlett and Francis Duncan, *Atomic Shield, 1947-1952: A History of the United States Atomic Energy Commission*, University of Pennsylvania Press (University of Pennsylvania Press, University Park, Pennsylvania, 1969), Vol. II, pp. 131–132.

By theoretical analysis of presently available data, we believe we can adequately predict effects over the range of yields and altitudes in which we are most interested. We will be able to design around the remaining uncertainties.

Brown, then Director of Defense and Engineering, similarly disagreed with anxious senators that the Soviets stood ahead of the United States in weapons effects knowledge, adding that "we have done the [blackout] experiments both in 1958 and again [in] 1962. That is where we learned about this phenomenon, and the systems we are talking about are designed to survive in such a situation, to deploy additional radars, to be able to see, to allow for this blackout." On the same point, Los Alamos Director Norris Bradbury said he believed that "limiting our knowledge in this area... does not appear to be a great risk."[16]

As all the other difficulties involved in building and deploying an efficient ABM system became more and more apparent, the blackout argument gradually faded into the background. Discussion of other weapons effects did not, however. Since 1963, opponents of a comprehensive ban have focused on the need to know more about other exotic, difficult-to-calculate electromagnetic phenomena, including electromagnetic pulse (EMP), transient radiation effects on electronics (TREE), and various equipment-kill mechanisms—such as neutrons and x-rays— which are especially effective in outer space. In 1971, for example, Carl Walske, Assistant to the Secretary of Defense for Atomic Energy, argued that further testing was needed in part "to harden our weapons systems against nuclear destructive effects," such as those from neutrons and x-rays. "I am not saying that no progress at all could be made without nuclear tests," Walske hedged, "but rather that compromises of one sort or another would have to be made." In 1972, AEC Chairman James R. Schlesinger added that, if a CTB had been in effect in 1963, the United States would not have learned about the vulnerability of certain important effects.[17]

On the other side of the argument, there are those who believe that, although we do not know all there is to know about these phenomena, thirty years and hundreds of tests have provided enough data to enable weapons experts to design and plan adequately for the use of nuclear forces. In fact, they add, all the other fundamental uncertainties in a nuclear weapons battle—the detailed design of Soviet weapons, the number and types to be used, the general environment, the exact interweapons distances at the time of the explosion, and so on—will be much greater than the remaining uncertainties in the physics of weapons effects, so further elaboration of these uncertainties is of relatively little value.[18] The truth of the matter will depend on the finer details of some future battle pitting nuclear weapons against nuclear weapons. Unlike other analogous situations—tank

[16] Senate Committee on Foreign Relations, *Nuclear Test Ban Treaty*, pp. 103, 570, 581.

[17] Senate Subcommittee on Arms Control, *Prospects for Comprehensive Nuclear Test Ban Treaty*, pp. 100–101; Herbert York interview with James Schlesinger, December 11, 1972.

[18] See, for example, Jerome B. Wiesner and Herbert F. York, "National Security and the Nuclear Test Ban," *Scientific American*, 211, October 1964, pp. 30–31.

versus tank, aircraft versus aircraft—there are absolutely no experimental cases to which we can turn for guidance. It should therefore not be surprising that even within the nuclear weapons establishment there is argument over this issue.

Twenty years ago, in addition to their anxieties over unexplored areas of weapons technology, many nuclear scientists entertained doubts about a test ban because of the possibility of utilizing nuclear explosions for peaceful purposes. During the early 1950s, the major U.S. weapons laboratories made preliminary investigations into the technical feasibility of this idea. Members of the Livermore staff—Harold Brown, Gerald W. Johnson, Arthur T. Biehl, Edward Teller, and Herbert York—were sufficiently impressed by these initial studies to begin a formal research program, which they designated Project Plowshare (from the Bible passage in Isaiah that speaks of beating swords into plowshares.) By 1958, this group had discovered numerous potential nonmilitary applications of nuclear explosions, including construction of harbors and canals, mineral and geothermal extraction, creation of water storage reservoirs, and oil well stimulation. In the 1960s, the AEC sponsored extensive experimental and theoretical work in this field. At the same time, the initiation of test ban negotiations helped to assure that interest in peaceful nuclear explosions (PNEs) remained high. Influential scientists and officials repeatedly stressed that testing should continue in order to keep the Project Plowshare program alive. Teller, for example, conducted a public crusade in this direction, writing promotional tracts and books, appearing on television, and lobbying in Washington. A book that he co-authored with the Plowshare division leader at Livermore and other scientists, *The Constructive Uses of Nuclear Explosions*, contains a detailed discussion of the technology of the Plowshare program. Published in 1968, it envisions among other projects the construction of a new transisthmian Panama canal using a series of underground nuclear explosions, some as powerful as thirty-five megatons.[19]

Peaceful nuclear explosions persist as a test ban problem in large part because of ongoing Soviet interest in their utility. Originally indifferent or even hostile toward the idea of using nuclear explosions for peaceful purposes, the Soviets since 1965 have operated a very active PNE program. To date, they have made over forty PNE experiments, ostensibly to investigate such applications as extinguishing gas well fires, stimulating oil fields, and creating underground storage cavities for water and gas. They also envision a major excavation project involving the digging of a canal between the Kama and Pechora Rivers in northeastern Russia.[20]

Because of this vigorous program, the Soviet government has steadfastly

[19]Edward Teller *et al., The Constructive Uses of Nuclear Explosives*, New York, McGraw Hill, 1968. See also Hans A. Bethe and Edward Teller, "The Future of Nuclear Tests," Foreign Policy Association, *Headline Series*, 1961, p. 145.

[20]On the Soviet PNE program, see Glenn C. Werth, "The Soviet Program on Nuclear Explosives for the National Economy," *Nuclear Technology*, 11, 1971, pp. 280–302; Milo Nordyke, "A Review of Soviet Data on the Peaceful Uses of Nuclear Explosions," *Annals of Nuclear Energy*, Oxford, 1975, Vol. II, especially pp. 657–673.

insisted on excluding PNEs from any CTB agreement. By contrast, the United States, despite signing a draft protocol with Russia in 1976 regarding PNEs, now desires to close firmly what it regards as the "PNE loophole" or "escape hatch." As Assistant Secretary of Defense David E. McGiffert explains: "We maintained, and continue to maintain, that PNE technology is indistinguishable from that required for weapons tests and that a state undertaking PNEs inevitably derives weapons-related information." Rear Admiral Thomas Davies of ACDA similarly asserts that "the continuation of a PNE program of any sort would provide such a substantial weapons testing infrastructure that it simply cannot be considered as acceptable in any way if weapon testing is banned."[21] Prospects of finally settling this issue have improved somewhat because of recent Soviet indications that they are willing to suspend their PNE program temporarily. "Over the past several months," McGiffert states, "negotiations have made considerable progress... and the Soviets have agreed in principle to a moratorium."[22] Soviet negotiators have emphasized, however, that such a moratorium would be acceptable only if it ran concurrently with a fixed-duration CTB. It is still uncertain whether the USSR would agree to extend the PNE moratorium if the CTB were extended beyond its initial period.

When Congress reopened its investigation into the test ban question in 1971, proponents of continued testing again raised the familiar issues of weapons development and weapons effects. Significantly, however, they centered their attention on the relatively novel question of stockpile reliability, or stockpile confidence. "The basic problem," as Walske put it in his 1971 remarks before Senator Muskie's subcommittee on arms control, "is the stockpiles of nuclear weapons much more than it is the development of nuclear weapons." Observing that since the mid-1950s "we have had five principal cases in which a nuclear test was an integral part of a corrective program for a nuclear weapon," Walske went on to explain how a comprehensive test agreement might result in the critical deterioration of U.S. stockpiles:

This would come about... because of seemingly acceptable changes that would in time be made in nuclear warheads without the check that comes from a performance test of the modified warheads.... Even if an effort is made to be prudent by avoiding changes in an existing stockpile, changes would sometimes prove necessary anyway because an aging weapon required a modification to correct a developing defect.... After years of no testing, the degree of uncertainty in the reliability of stockpiled wepons might be quite large. If this uncertainty were equal for all nuclear powers, it could be beneficial by serving as an added factor in deterring the first use of nuclear weapons. However, it would at the same time create anxieties for each power about the reliability of its second-strike, retaliatory capability. Furthermore, if the uncertainty in stockpiled weapons were not

---

[21] House Military Appplication of Nuclear Energy Subcommittee, *Current Negotiations on the CTBT,* pp. 4,8; "Opening Statement of David E. McGiffert, Assistant Secretary of Defense for International Security Affairs before the House Armed Services Committee on Comprehensive Test Ban Hearings," August 14, 1978.

[22] *Ibid.*

about equal for all—that is, if one power could with confidence exploit weaknesses in another power's stockpile—then the uncertainty would be quite destabilizing.[23]

Current nuclear weapons technologists regard stockpile reliability as an extremely serious question, emphasizing it above all others in their efforts to forestall a comprehensive test ban. Representatives of the nuclear weapons laboratories generally present a twofold argument. First, like Walske, they maintain that a vigorous research, development, and test program is required both to protect against the aging problem and to facilitate possible planned design changes in weapons. Second, and more important, they view a CTB as a threat to the very existence of the laboratories. Without at least a modicum of testing, they reason, technical expertise will eventually atrophy ("within a few years," according to the laboratory directors), leaving no one to deal with emergencies or special problems that might arise. "In summary," John C. Hopkins, director of field testing at Los Alamos says, "the basic argument for not stopping testing is that the present stockpile requires experts for its maintenance and experts cannot stay proficient without an R&D program that includes full-scale nuclear testing."

Laboratory representatives also raise the question of "asymmetry" in relation to stockpile reliability. They challenge the assertion of some CTB advocates that any stockpile degradation that occurs under such an agreement will affect the United States and the USSR equally, leaving the strategic balance undisturbed. Because of the autocratic nature of its society, they argue, the Soviet government could maintain a pool of technical experts and solve stockpile problems through clandestine testing, whereas the "virtually absolute self-verification imposed by our open society" would preclude this option for the United States. Sometimes they add that the problem of asymmetry is compounded by the greater complexity of American nuclear systems compared with those of the Soviets and other nuclear powers.[24]

Nearly all knowledgeable officials in the Department of Energy (DOE) and the Department of Defense, and many in the international community, support the position of the weapons laboratories. "It is our belief," Donald Kerr, DOE's acting Assistant Secretary for Defense Programs, asserted in March 1978, "that in the long run without testing, we could not maintain the same confidence in our nuclear weapons stockpile that we have today.... We don't believe that in the long run you could rebuild the stockpile without having access to designers and engineers who had experimental verification of their thoughts and design judgment."[25] Europeans understandably worry primarily about NATO de-

[23] Senate Subcommittee on Arms Control, *Prospects for Comprehensive Nuclear Test Ban Treaty,* pp. 101–106.

[24] Senate Committee on Foreign Relations, *TTB and PNE Treaties,* pp. 68–73; John C. Hopkins, "Despite All the Bombs, We Still Need Testing," *Washington Star,* May 8, 1977; Michael M. May, "Do We Need a Nuclear Test Ban?" *Wall Street Journal,* June 28, 1976; Donald R. Westervelt, "Candor, Compromise, and the Comprehensive Test Ban," *Strategic Review,* Fall 1977, pp. 33–44.

[25] House Military Application of Nuclear Energy Subcommittee, *Current Negotiations on the CTBT,* pp. 22–23.

fenses, which rely heavily on tactical nuclear weapons (TNWs) supplied chiefly by the United States. They believe that these weapons require even more "proof-testing" than do strategic nuclear weapons. "The problem of reliability and confidence would be much more serious...," one writer in the *London Daily Telegraph* explains, "for under present strategy NATO is very likely to want to use these first. The plausibility of doing so... depends very much on the weapons being used in a controllable fashion to achieve precise effects on the battle-field."[26]

Others take the opposite view. Three outstanding experts—Norris Bradbury and J. Carson Mark, who were director and head of the theoretical division, respectively, at Los Alamos for a quarter century, and Richard Garwin, a highly respected general consultant on military technology—recently wrote a letter to the President in which they said: "We believe that the Department of Energy, through its contractors and laboratories, can through the measures described provide continuing assurance for as long as may be desired of the operability of the nuclear weapons stockpile." The "measures described," which are simple and straightforward, include forgoing minor improvements "in order not to sacrifice stockpile reliability."[27] Most nuclear experts in ACDA and many of the expert consultants to ACDA and to the Office of Science and Technology Policy hold similar views. All this creates a dilemma for the national security leadership, which must, of course, give substantial weight to the opinion of those currently responsible for nuclear weapons design while at the same time recognizing the intellectual commitment and the vested interests of these scientist–technicians.

Other Technical Issues

The discussion to this point has been concerned with the potential consequences of a CTB for the national security of the United States. But besides the political and military implications of a total test ban, and just as important to advocates as well as to opponents, is an entirely separate category of issues related to technical aspects of the problem. Since the mid-1950s, debate has focused on two such issues. Verification, or how to monitor compliance with agreements, has been a critical problem from the outset of test ban talks. And, more recently, the nuclear community has devoted time and effort to the problem of how precisely to define what constitutes a nuclear weapons test. Of course, throughout the great test ban debate, authorities on both sides have not always found it easy or expedient to separate clearly these technical questions from purely political–military ones.

"Verification of compliance with a comprehensive test ban is one of the most difficult issues to understand and one of the most important," Senator Dewey F. Bartlett, the first congressional observer at the current Geneva CTB talks,

---

[26] *London Daily Telegraph,* October 3, 1977.

[27] Norris E. Bradbury, Richard L. Garwin, and J. Carson Mark, letter to President Jimmy Carter, August 15, 1978; released through Senator Edward Kennedy's office, August 17, 1978.

declared in 1978.[28] This statement could just as easily be attributed to any of the Eisenhower administration officials who first seriously considered a test ban twenty years ago. In fact, the sharp intragovernmental debate over a possible ban during the Eisenhower years can be said to have hinged on this issue alone. Because of their fundamental distrust of Soviet intentions, those in the administration who opposed a ban assumed that the Russians would take every opportunity to conduct clandestine tests (in particular, underground tests), which would be difficult if not impossible for the United States to detect. As Edward Teller put it in 1958: "In the contest between the bootlegger and the police, the bootlegger has a great advantage."[29]

In early 1958, President Eisenhower charged his new science adviser, James Killian, to investigate this thorny problem. Killian appointed Hans Bethe, a member of the President's Science Advisory Committee (PSAC), to chair a special interagency panel for this purpose. From a study of the first U.S. underground nuclear explosion, the Rainier test of 1957, the Bethe Panel estimated that a reliable control system based largely on a network of seismic observation stations could be devised. That summer, the Geneva Conference of Experts, a series of meetings arranged by Eisenhower and Khrushchev which brought together representatives from eight nations, arrived at much the same conclusion.

But shortly thereafter data gathered from new U.S. underground tests, part of the Hardtack II series of October 1958, threw the conclusions of the Bethe Panel and the Conference of Experts into doubt. These tests led to the formation of two new PSAC study groups in 1959—the Berkner Panel on Seismic Improvement, and a panel headed by Robert F. Bacher of the California Institute of Technology—which were much less optimistic about the reliability of seismic monitoring, even if combined with the right to carry out some on-site inspections. Interestingly, George Kistiakowsky, who took over for Killian as Eisenhower's science advisor in the middle of this turmoil, says in his diary that the Bacher Panel reached its negative conclusions about on-site inspections primarily because of the "aggressive arguments" of Harold Brown. Then deputy director of the Livermore Laboratory, Brown, again according to Kistiakowsky, was "violently opposed" to a test cessation that did not provide for a high-yield threshold.[30]

To compound these difficulties, antiban forces advanced two new theories at this time about how nuclear tests might be concealed, thus raising once more the specter of Soviet evasion. First, Albert Latter and other Rand scientists, following a suggestion of Teller, studied the possibility of muffling ("decoupling") nuclear explosions in large underground cavities. Their report on the "big-hole" decoupling theory, published in March 1959, maintained that "a yield of more than 300 KT could be made to look seismically like a yield of 1 KT." Second, test ban

[28] Senate Committee on Armed Services, *The Consequences of a Comprehensive Test Ban Treaty: Report of Senator Dewey F. Bartlett,* 95th Congress, Second Session, August 11, 1978, p.7.

[29] Edward Teller, "Alternatives for Security," *Foreign Affairs,* 36, 1958, p. 204.

[30] George B. Kistiakowsky, *Scientist at the White House: Private Diary of President Eisenhower's Special Assistant for Science and Technology.* Cambridge, Massachusetts, Harvard University Press, 1976, pp. 6, 38, 146–147, 169.

opponents suggested that tests might be concealed by exploding devices at extremely high altitudes or in outer space, even behind the sun. What opponents failed to emphasize was the both of these concealment methods involved great expense and tremendous engineering and practical difficulties. In retrospect, Killian calls them "bizarre" technologies "contrived as part of a campaign to oppose any test ban." He regrets that he and others "who spoke for science never succeeded in making clear the difference between *probability* and *possibility* . . . in such fields as test detection and monitoring negotiations."[31]

The verification issue continued to create problems for the Kennedy administration. In 1961, arguments over on-site inspection of suspected underground tests became the major stumbling block to concluding a comprehensive agreement. The United States initially insisted on a large number of such inspections— between twelve and twenty per year. At first intractable, Khrushchev eventually offered to permit as many as three inspections per year. The United States in turn reduced its requirement to seven, but here the negotiations stalemated. Despite extensive discussion of the possible modalities of on-site inspection during the Geneva Conference on the Discontinuance of Nuclear Weapon Tests between October 1958 and January 1962, neither the United States nore the USSR ever made concrete proposals about how *any* such system (whatever the number of inspections allowed) might work. Some observers interpreted this as meaning that each side harbored   less than serious intentions of reaching a viable compromise, and that each merely used this issue as a convenient pretext or rationalization in order to continue nuclear weapons research through underground testing. Jack Ruina, in 1963 director of the Defense Department's Advanced Research Projects Agency (ARPA), later wrote that "it is difficult to see how U.S. security could have been dependent on precisely seven on-site inspections or how Soviet security could have been more compromised by seven (or five) inspections per year than by three. But this numbers game became a politically inseparable obstacle to an agreement."[32]

Several expert studies conducted both during and since the Kennedy years have greatly advanced our knowledge about verification problems. The VELA research program, a direct outgrowth of the Berkner Panel recommendations, and other programs sponsored by ACDA's Science and Technology Bureau have demonstrated significant improvements in teleseismic detection methods.[33] To summarize this research, the yield separating those tests that can be both detected and remotely identified from those that cannot is centered at a few

[31] *Ibid.*, pp. 10–12, 22–23; Gilpin, *op. cit.*, p. 236; James R. Killian, Jr., *Sputnik, Scientists, and Eisenhower: A Memoir of the First Special Assistant to the President for Science and Technology*, Cambridge, Massachusetts, M.I.T. Press, 1977, pp. 150–168, 172.

[32] Neild and Ruina, *op. cit.*, p. 141.

[33] See Cecil H. Uyehara, "Scientific Advice and the Nuclear Test Ban Treaty," in *Knowledge and Power: Essays on Science and Government*, Sanford A. Lakoff, ed., New York, Free Press, 1966, pp. 124– 133, 136; Herbert Scoville, Jr., "Verification of Nuclear Arms Limitations: An Analysis," *Bulletin of the Atomic Scientist*, **26**, October, 1970, pp. 6–11; Henry R. Myers, "Extending the Nuclear Test Ban," *Scientific American*, **226**, January 1972, pp. 13–23.

kilotons, but the amount varies up or down from that figure by at least an order of magnitude. The variance in detection reliability depends on such details as the availability of continuous seismic data from stations within the Soviet Union, the distance from these stations to the site of the event, the seismological nature of the test region itself, the effort put into evasion, and whether one asks, "Can our side detect a single test with high confidence?" or "Can the other side get away with an entire series of test with high confidence?"

The necessarily "noisy" and imprecise nature of the basic data and variations in the emphasis placed upon the "details" above permit totally opposit perspectives on the verification problem today. In 1971, test ban proponent Jerome Wiesner of MIT, former science adviser to Kennedy, argued:

The feasibility of an underground test ban is even greater [now than it was in 1963]. It was recently announced that a scientists' panel at a test detection conference of the Advanced Research Projects Agency... concluded that progress in seismology now makes it possible to distinguish all but the smallest tests from earthquakes. A test ban agreement without on-site inspection, therefore acceptable to the Soviet Union and practical to implement, would now appear possible.[34]

Drawing from the same pool of technical data, test ban foe Congressman Samuel S. Stratton recently maintained that "we are getting a pig in a poke, if we think that we can really tell whether there is going to be any violation [of a comprehensive ban]." Former DNA Director Johnson agreed: "To me, it is quite clear that the technical problems associated with reliably detecting small-yield detonations are so great that thinking that we can verify adequately is silly."[35]

In sum, in the past two decades we have learned a great deal about the seismology of test detection, and we have simultaneously increased the sensitivity of our detection systems. In addition, we have achieved fairly broad agreement as to the basic scientific facts. But difficulties in determining which environmental and political conditions apply make the interpretation of these facts as blurred and uncertain today as it has ever been.

At the very beginning of the nuclear era it seemed easy to distinguish between a nuclear weapons test and any other event. Within just a few years, however, the distinction became blurred in at least one dimension, and as the years went by many other such situations arose. Senator George D. Aiken neatly summarized the growing scope of this problem when he asked at the 1971 hearings on a CTB whether "anyone had any idea how... [to] tell the difference between one bang and another bang."[36] We are not much closer to answering this question today than we were in 1971.

One important gray area of definition involves the purposes for which nuclear explosions are intended. In 1949, when the United States observed and an-

[34]Senate Subcommittee on Arms Control, *Prospects for Comprehensive Nuclear Test Ban Treaty,* p.5.

[35]House Military Application of Nuclear Energy Subcommittee, *Current Negotiations on the CTBT,* pp. 47, 76.

[36]Senate Subcommittee on Arms Control, *Prospects for Comprehensive Nuclear Test Ban Treaty,* p. 19.

nounced the first Russian tests, the Soviet government at first attempted to confuse matters by claiming that U.S. sensors had detected not weapons tests but rather some "experiments" in the use of nuclear energy for "moving mountains."* The Soviet news agency Tass made the following classic statement of obfuscation about the tests on September 25:

In the Soviet Union, as is known, building work on a large scale is in progress—the building of hydroelectric stations, mines, canals, roads, which evoke the necessity of large-scale blasting work with the use of the latest technical means.
. . . As for the alarm that is being spread on this account by certain foregn circles, . . . [there] are not the slightest grounds for alarm. . . . The Soviet Government, despite the existence in its country of an atomic weapon, adopts and intends adopting in the future its former position in favor of the absolute prohibition of the use of the atomic weapons. . .[37]

Since then, confusion has grown about how to differentiate between nuclear weapons tests and tests and applications of nuclear explosive devices for engineering or commericial uses. In 1974, India labeled its first underground nuclear test a PNE. The Indians have since consistently denied that they tested a nuclear weapon then or at any other time. The Soviets currently insist that any long-term CTB must allow for economically oriented tests and applications. They have offered only a short-term moratorium on such explosions in order to create the proper atmosphere for working out a permanent arrangement. Most American scientists and arms control experts have grave doubts about the practicality and viability of such a distinction. The United States takes the position, therefore, that a provision in a CTB permitting PNEs would constitute a huge loophole and would therefore be unacceptable. Even so, U.S. and Soviet negotiators worked out a Threshold Test Ban Treaty in 1974, which recognized the problem created by the USSR's desire to conduct PNEs. Two years later, they concluded talks on a specific PNE Treaty. As described in an ACDA publication on arms control agreements, the statement agreed upon, which precedes the PNE Treaty, specifies that "no weapon-related benefits precluded by the Threshold Test Ban Treaty are [to be] derived by carrying out a peaceful nuclear explosion" and that "a 'peaceful application' of an underground nuclear explosion would not include the developmental testing of any nuclear explosive."[38] Accordingly, the treaty established identical yield limits for PNEs and nuclear explosions at weapons test sites. Both the Threshold and the PNE treaties have been signed, but neither has

*The first study of what later became Project Plowshare was stimulated by this event. See Frederick Reines, "Are There Peaceful Engineering Uses of Atomic Explosives?", *Bulletin of the Atomic Scientists*, 6 June 1950, pp. 171-172.

[37] Herbert F. York, *The Advisors: Oppenheimer, Teller, and the Superbomb*, San Francisco, Freeman, 1976, pp. 34–35.

[38] Arms Control and Disarmament Agency, *Arms Control and Disarmament Agreements: Texts and History of Negotiations*, Washington, D.C., 1977, p. 167.

been ratified. Extension of the limitation embodied in them to any CTB will obviously be a very difficult matter to iron out.

Two areas where nuclear weapons tests may overlap events of a different type are (1) very-low-yield "nuclear experiments" and (2) the so-called inertially confined fusion (ICF) device. The first case can be illustrated best through hypothetical examples. Suppose, for instance, that one builds and explodes a device that looks very much like an atomic bomb (both inside and out) but contains an amount or kind of fissile material such that its nuclear yield is only one one-millionth of its chemical yield. Would that be a nuclear weapons test explosion? Or suppose that the nuclear yield is one-half the chemical yield, or equal to it? Or perhaps it is a matter of style: What if a low-yield test is conducted in an above-ground containment facility at a laboratory instead of at an underground nuclear test site? There seems to be no clear and simple answer to these questions, but they are important nonetheless.

The case of the ICF device is quite different. Here the goal of the current development program is to focus a large amount of power and energy (in the form of a laser beam or an electron beam) onto a capsule containing a very small fraction of a gram of deuterium and tritium. When this is accomplished in just the right way, the capsule should implode and its contents would reach the high-density and high-temperature conditions necessary for a thermonuclear explosion of relatively tiny proportions. If the capsule contains some tens of milligrams, the resulting explosion could be equivalent to that of a ton of TNT. Is this a nuclear explosion? Certainly. It it or could it be considered a nuclear weapon? Not in its present form, which typically involves very complex and exceedingly heavy ancillary equipment for generating the driving beams. For this reason, a consensus exists that the possible future overlap between these devices and nuclear weapons should be ignored for purposes of current CTB. Most expert also agree, however, that the situation should be monitored carefully in case something more ominous develops in the future.

This presentation of arguments and counterarguments explains why, on the one hand, advocates of a comprehensive test ban continue their efforts, and why, on the other, such a ban has proved so difficult and elusive to achieve. The President, the top national security leadership, and, ultimately, the Congress must seek to weigh the various incommensurate risks and benefits inherent in each of the major issues raised by test ban proponents and opponents. Only then can we arrive at a final conclusion about a suitable future course.

III. Biomedical Research and Applications

These papers treat a variety of issues. Salk takes the most global perspective, arguing that human survival now depends upon the development of a new respect for wisdom—wisdom drawn from an understanding of the laws of nature, and of the needs of a new epoch in which human knowledge faces its severest test. Like Salk, Kieffer contends that the future must be planned, especially to take account of new dilemmas posed by advances in biomedical knowledge. Shapiro and Lefebvre discuss the social and professional problem of allocating scarce resources, in research as well as in health care. Reiss and Murtagh present differing views of the effort to regulate recombinant DNA research; Grobstein looks ahead to the likely future of genetic and epigenetic intervention and considers its moral and policy implications.

12.   Toward a New Epoch*

Jonas Salk

An unprecedented explosion of interest and movements concerned with the survival of the species is now taking place. The idea of the extermination, by Man, of various forms of life on the planet, and the danger to human life, induces a fear that preoccupies increasing numbers of individuals, especially of the generations now maturing. Those who are ecologically oriented and those who are profoundly concerned about the quality of life for the species as well as for the individual appear to stand in opposition to others less aware of such problems, who are more concerned with themselves in their own life spans. The fundamental difference between these two attitudes is that the first expresses concern for *the individual and the species;* the second reveals principally, and perhaps exclusively, an interest in the *individual* and the *particular group* of which he is a part. The more broadly concerned *(i.e. with the species and the individual)* fall into two categories. One consists of those born after such threats came into full evidence; the other, of those born earlier but who, having witnessed the change, are now reacting to previously prophesied dangers which have become realities. Those preoccupied only with their own problems are either unaware or unperturbed in the face of a process in human evolution to which others are sensitive and, if aware, feel frustrated, helpless, or apathetic.

A major threat to the species is attributed to the increasing size of the human population, which, in turn, is ascribed to successes in science and technology. This "explanation" has evoked an attack upon science and the exploitation of its technology, to the development of which are attributed many adverse effects upon the human species and upon other forms of life. "Polluters" who befoul the planet affect the "quality of life" and are regarded as a threat to the present and future equilibrium of the species and of the planet. Those who consider themselves *on the side of* Nature, and therefore of the human species, see others in

*From *The Survival of the Wisest*, Copyright 1973 by Jonas Salk, Harper & Row, New York.

opposition to both Nature and Man. Hence we are to be concerned not only with Man's relationship to Nature but with Man's relationship to himself.

Viewed in this way we realize how much blindness to Man's true nature actually exists. This may be understandable in the young, who have not lived very long, but it is equally true of those who have lived longer. How we grapple with our blindness is of the greatest importance for the present and the future; it is the central problem of our time.

If human life is to express as much harmony, constructiveness, and creativity as are possible for fulfilling the purpose *of* life, as "required" by Nature, and the purposes *in life* as "chosen" by Man, an attitude will be needed, not of Man "against" Nature, but of Man "inclusive with" Nature. A more reasonable attitude would be for Man to "serve Nature" in order to serve himself, rather than to "serve himself" without regard for, or at the expense of, Nature and others. By recognizing and respecting the natural "hierarchies of purpose," Man would be better able to gauge his latitude to select and pursue his own "chosen purposes" without coming into conflict with the "purpose of Nature," which appears to be the continuation of life as long as conditions on the planet permit.

As a process, evolution seems to be Nature's way of finding means for extending the persistance of life on earth. This involves the elaboration of increasingly complex mechanisms for problem-solving and adaptation. The ability of the human mind to solve the problem of survival is part of this process. In this respect Man has evolved so successfully that he is now to be tested for his capacity to "invent" appropriate means to limit the harmful or lethal excesses of which he is capable. The conflict in the human realm is now between self-expression and self-restraint *within* the individual, as the effect of cultural evolutionary processes has reduced external restraint upon individual expression and increased opportunities for choice.

The fork-in-the-road at which Man *now* stands offers either a path toward the development of ways and means for maximizing self-expression *and* self-restraint, by means of external restraints that are not suppressive or oppressive, or an alternative path of limitless license which would unleash destructive and pathological greed at the expense of constructive and creative individuals. In the latter case, a strong reaction can be expected to develop in response to the sense of order upon which their survival is based. The challenge is to establish an equilibrium between *self-expression with self-restraint* on the one hand, and *self-protection with self-restraint* on the other. If Man is to take advantage of opportunities to remedy diffulties that have arisen as a result of his evolution, then he needs to understand his relationship to the evolutionary process which plays with and upon him.

As a logical overture, let us look at the growth of the human population on the face of the earth and the present reasonable projection over the next few decades to the year 2000. It raises vast and complex implications for the character and quality of human life that concern relationships as well as resources for the present and the future. It raises questions as to the means that Man or Nature will invoke to deal with the excesses that have developed and the insufficiencies that persist. Will Man create his own procedures to deal with them or will Nature's simple ways come into play, some of which may prove quite undesirable from Man's point of view? This, in fact, may already be occuring. Before turning our attention to the questions and consequences of the rapidly mounting curve of population increase as drawn in Figure 1, or to the implications of its curtailment or of its continuation, let us look at patterns of growth in other living systems. For example, Figure 2 shows the growth curve of a fruit-fly population in a closed system as observed by Raymond Pearl in 1925.

The S-shaped, or sigmoid, curve that describes the growth of fruit flies is also seen in curves of growth of microorganisms and of cells or molecules. Since the planet earth can be considered a closed system and *since the sigmoid curve reflects the operation of control and regulatory mechanisms that appear to be associated with survival of the individual or of the species*, it would seem reasonable to expect that the pattern of future population growth in Man will tend to stabilize at an optimal level described by an S-shaped curve. It is possible, of course, that an alternative pattern might resemble that of the lemmings (Fig. 3), in which periodic catastrophe occurs with enormous loss of life. However, Man's attitude toward human life would have to alter significantly for such patterns to be endured; he is more likely to choose *other ways than catastrophe for maintaining optimal numbers on the face of the earth while remaining within the limit of available resources.*

As Man has still to complete a cycle of growth on this planet, he has not yet fully revealed the pattern biologically programmed in him, or the way it will be influenced by factors he is responsible for, or by natural forces beyond his control. Therefore we are unable to know the pattern of his trajectory in the short-term or longer-term future. The "catastrophists" and harbingers of doom *are in themselves evidence that Man possesses a signaling mechanism for sounding warnings*

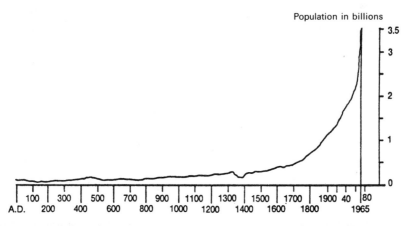

Figure 1. World population estimates, A.D. 0-1965. Adapted from *World Facts and Trends,* by John McHale.

of danger sensed more acutely and more clearly by some who alarmingly represent the problem of population increase as shown in Fig. 1.

If we assume, however, that Man has the power of choice and can influence the course of his growth curve on this planet, then it is of special interest to look carefully at the sigmoid curve in terms meaningful for him. Since our deeper purpose is to try to discern the nature of order in the human realm in relation to the nature of order in the realm of life in general, it is interesting to explore the possible meaning of the similarities observed in the human population growth curve as manifested thus far, and the first portion of the growth curve of the fruit-fly population and similar curves in the subsystems of other living systems.

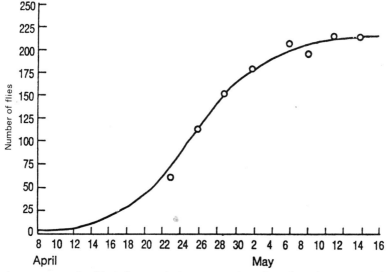

Figure 2. Growth of fruit-fly population. From *The Biology of Population Growth*, by Raymond Pearl. Copyright 1925 by A.A. Knopf, Inc., and renewed 1953 by Maude de Witt Pearl. Reprinted by permission of the publisher.

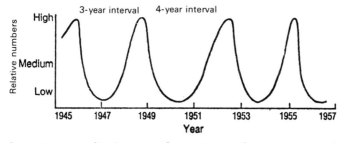

Figure 3. Generalized curve of the three-to-four-year cycle of the brown lemming population. From CRM Books, *Biology: An Appreciation of Life*, 1972 by Communications Research Machines, Inc.

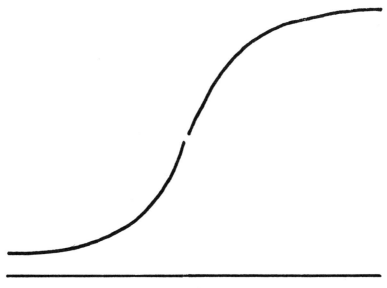

Figure 4

Since, through the process of natural selection, living organisms that have survived have revealed their fitness for persistence thus far in the evolutionary scheme, we would like to have some prevision of Man's program. Is he programmed for relatively short-term survival in which his end may come of his own doing? Or is he programmed for a life in which only those who have lost the power to discriminate, or who are otherwise degenerate, will continue to inhabit the planet as long as reproductive activity continues to supply "victims" of life, struggling to preserve itself in the "human" form? And what other alternatives exist?

It is likely that Man's brain has developed as it has, in the course of natural selection, partly in response to exogenous forces active against survival. Does that same brain also possess the capacity to tame and discipline those inner forces that act against long-term survival, in opposition to a life of high quality? The struggle for survival once manifest principally *between Man and Nature now seems to be taking place within the human species itself, between Man and men* and within the individual himself.

My purpose is to elucidate the factors and forces affecting the quality of human life through ideas that emerge while "playing with" the growth curve and reflecting upon the developmental and evolutionary processes of Man in the critical stage in which we seem to be at this point in time.

As we study the curve in Fig. 4, consideration of the lower portion only gives the impression of continuous, even explosive expansion, whereas consideration of the upper portion gives the impression of modulation and control of this expansion, so that finally a limit is established. At the junction of the lower and

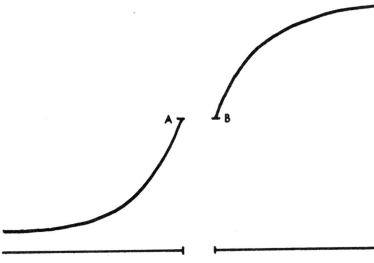

Figure 5.

upper portions of the curve is a region of inflection at which there is a change *from progressive acceleration to progressive deceleration* and at which the influence of the controlling processes is clearly visible. The break apparent in this region suggests that a "signaling" mechanism of some kind must operate to bring about this change, producing an effect that, judging from the shape of the curve, indicates the existence of a uniform process, reflecting the operation of some kind of ordering principle in response to "signals" both from the environment and from within the organisms themselves. At different points in time along the curve, latent qualities and reactions are evoked appropriate to survival, the program for which is coded in the germ plasm, which also contains an accumulation of control and regulatory factors essential thereto.

At the plateau stage of numbers, the individuals in the fly population would be expected to "behave" differently as compared with those alive earlier in the growth curve, before the zone of inflection when different "problems" prevailed. The extent to which circumstances differ, at different points in time along the curve, is graphically suggested in Figure 5 by breaking the continuity at the point of inflection so as to create two curves, A and B.

These curves are intended to emphasize the difference in attitude and outlook in the two periods and help create a visual image of what can be sensed "intuitively." They also convey concretely what might be appreciated "cognitively" by means of an objective analysis of the increasingly complex problems generated by the growing numbers of individuals. In the discussion to follow, curves A and B will be used as symbols of the "shape" of the past and of the future, as we attempt to characterize each. When we speak of the fruit fly in anthropomorphic terms, it is to suggest, using this caricature, the nature of the forces operating in the human realm. For example, if we speak of flies as possessing, individually or collectively, a "sense of responsibility" and "insight and foresight," it is to suggest the existence of the equivalent of conflicting forces by which they would, were they human, be impelled to "judge" and "choose." Such judgment would be exercised according to the contesting "value systems" that would be in operation during periods as different as those suggested by curves A and B.

The fact that the fruit flies are a product of a long evolutionary history, whose survivors react according to their genetic programming, leads us to think that Man, who is of more recent origin and, moreover, at or near the point of inflection in his present curve of population growth on the planet, is about to find out whether he is programmed to behave in ways leading to a population growth curve similar to the fruit fly's, or to a curve of another shape. He has still to find out about the nature of the quality of life under circumstances that remain to be experienced. In being tested for survival, he still has a way to go, not only quantitatively but qualitatively. The curves, however, provide some insight, their shapes suggesting the character of the problems that prevailed in the past, those now existing, and those likely to be encountered as Man continues to move through evolutionary time.

Man differs from other living organisms in possessing another "control and regulatory" system, for response to environmental and other changes, in addition to that genetically coded and automatically operative as in the fruit fly, which has been tested and selected in the course of its evolutionary history. Man is able to exercise learned behavior. He also possesses individual will, which can be either in accord or in conflict with genetically coded patterns of response. In this sense Man is more complex and more unpredictable than the fruit fly. He can learn to behave in ways that are anti-life as well as pro-life, anti-evolution as well as pro-evolution. He remains to be tested for this pattern of response to all that is implied in the need for changing values to make the transition from Epoch A to Epoch B. In view of the greed and ideologies of Man as causes of his conflicts, attitudes as well as values will be put to test in the transition from Epoch A to Epoch B.

Genetic programming does not change as rapidly as the attitudes and values that also guide human behavior. Since genetically as well as culturally determined responses are "environmentally" linked, the circumstantial differences implied by the dissimilar "shapes" of the curves symbolizing Epoch A and Epoch B will be expected to evoke different sets of genetic as well as cultural potentialities. In Epoch B those attitudes and attributes that are of the greatest value will determine the "real" and not merely the "presumed" shape of the population growth curve and the quality of life. Value systems such as these that prevailed in Epoch A will, of necessity, have to be replaced by those appropriate for Epoch B, and new concepts will emerge about the nature of Man and his relationship to all parts of the cosmos. Since the conditions into which future generations will be born are not yet determined, it will be of interest to see how men in different cultures, with

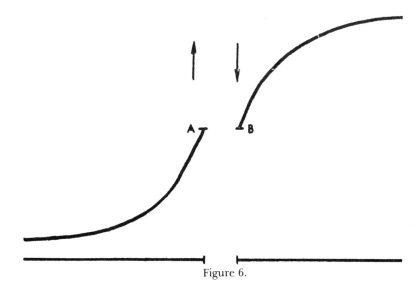

Figure 6.

different genetic backgrounds and capacities, will respond to the human and planetary changes now well under way. It is not yet possible to see how Man will deal with attributes that dominated in Epoch A nor to foresee very clearly precisely what attributes will emerge in Epoch B.

Thus Man is being subjected to a new and possibly more severe challenge than ever before, for which he needs perspective and insight. He must become aware of the opportunities and the dangers that he will have to face when confronted by the conflicts resulting from a necessary inversion of such magnitude as is implied by the diagram in Fig. 6. The profundity of the change in values required for survival and for quality of life in the periods described by curves A and B makes it seem not only that what was of positive value in A may, in fact, become of negative value in B; but also, if "B values" had prevailed earlier they would have been of opposite value in the A epoch. From this point of view it is not difficult to understand the depth and meaning of the change that Man is now experiencing in the various forms that have already become manifest under the specific historical circumstances in different cultures in all parts of the world.

This change is of such magnitude and significance that it may well be judged to be of major import in the course of human evolution. At this time Man seems to be seeking tolerable levels quantitatively and is being called upon to develop qualitatively satisfying ways and means for living with himself and with others that fit what might be thought of as the scheme of Nature. Man's choices will be "judged" by Nature, thus revealing the wisdom of his selections from among many alternatives.

We need to discern Nature's "game," as well as Man's. The choices that Man makes from the alternatives available to him will profoundly influence his own evolutionary destiny. The outcome will reveal the extent to which he will have succeeded in understanding the workings of Nature, at a time in his own evolution when he is being tested for his capacity to accomodate himself to change, and for his ability to create the possibilities for existence under circumstances as different from those of the past as suggested by the shapes of curves A and B. Until this point in evolutionary time Man has been selected for characteristics that were of value for survival during the A epoch. Now, quite abruptly, a new "selection pressure" has appeared, for which he is ill-prepared by experience but for which there may exist within him a reservoir of potential appropriate to the new circumstances such as are now developing.

In the course of evolution many more species have become extinct than have survived, each perhaps for particular causes very different from those which might cause the extinction of Man. For in Man's case, at this point in his evolution, his extinction might well arise for internal reasons. The way he deals with unresolved conflicts within himself individually and collectively might lead to his own destruction. The process of natural selection has developed survivors resistant to various infectious diseases and to some of the vicissitudes of the environment. It has also led to the selection for survival of those successful in escaping the ravages of war and those ingenious enough to escape human tyranny. Thus until now the qualities that have been selected for survival reflect the conditions and circumstances that prevailed as much as the potentialities that exist in Man. As Nature continues its game of biological mutation and selection, and as Man plays his own games of selection of ideas and of cultural innovations, Nature will have the last word. Therefore it is up to Man to look closely and deeply into Nature's workings, not only at the molecular and cellular levels but also at the consequences of advancing knowledge and cultural practices as these bear on the question of survival and the quality of life. It is in this respect that wisdom will be required for which a balanced creative center for judgment is needed.

We must look to those among us who are in closest touch with the unfathomable source of creativity in the human species for an understanding of the workings of Nature and for insight into Nature's "game," as we enter upon an epoch in which new values are required for choices of immediate need as well as for those with longer-range implications. This is especially important when, as now, the number born in each new generation exceeds the number born in each of the earlier generations. For this reason, the character and quality of the individual that will survive and predominate in our period will have a very profound effect upon the character and quality of human life for a long time to come.

To what extent will we be able to affect the course of Nature, in the short or in the long run? That remains to be seen. Nevertheless, we are fully conscious of this problem. How will we deal with this opportunity and this responsibility knowing as much as we do? What more do we need to know, being as aware as we are now of our limitations and our capabilities?

To help our understanding of man's relationship to Man, and Man's relationship to Nature, in terms of their inherent complementary character, analogies have been drawn between the "games" of Nature and of Man. The point that has already been made is that the laws that govern Nature's game require, under certain circumstances, "double-win" rather than "win-lose" resolutions which Man must also develop.

Although each of us would like to be "winners," at least in terms of individual satisfaction and fulfillment, it is also clear that more luck, knowledge, and wisdom are required than are possessed by very many. Some are more fortunate than others, but none are born either fully knowledgeable or infinitely wise; hence Man's search for perspective and guidance in dealing with the unknowns and uncertainties of life in all its complexity.

Since many of the problems for which Man seeks solution are an inherent part of the process of human development itself, and since he is both a contributing cause as well as a sufferer, his position as both patient and physician is a difficult one. And yet he must be both. Fortunately, means do exist for self-correction, for self-cure, and for prevention even of those potentially harmful or lethal effects that are self-induced. The approach employed here in thinking about this dilemma has been to seek useful analogies in the self-correcting processes that are an essential and integral part of living systems. The assumption is that if individual man were aware of the existence of such processes and the way they operate, knowing that they are an integral part of his own self as well, he might develop the desire to learn how to use them consciously and deliberately not only for survival but for fulfillment in his lifetime. Such clues can be found in the way in which control and regulation operate in living Nature, where success is evidenced in the persistence of life in spite of vicissitudes, difficulties, and seeming impossibilities. On the assumption that metabiologic evolutionary problems (that is, those surpassing the ordinary limits of living matter) are similar to problems encountered and solved in biological evolution, analogies are sought to serve as models helping us to deal more realistically, and therefore more appropriately, with some of our unresolved problems, and even, possibly, to accept the existence of insoluble enigmas.

Referring to Man as a metabiological entity implies that he possesses self-correcting, self-controlling, and self-disciplining mechanisms, as well as biologically governed balance-mechanisms for each of his two distinct yet related evolutionary purposes—that is, for improving the quality of life as well as for survival. It implies, also, that change in human behavior that will serve both biological and metabiological aims requires many steps and stages involving both error-making and error-correcting. In spite of our prior limitations, due to ignorance of the character or details of the processes involved in evolution, can we, now, with our increased knowledge of the nature of living systems, and of Man, apply ourselves to conceive of ways and means of influencing the course of future events toward fulfilling Man's evolutionary potential? In Epoch A it appears that greater success has been achieved in reducing premature death than in improving the quality of life in terms of individual satisfaction. Hence, to the gains made in Epoch A, facilitating survival by better hygienic conditions and other measures for the prevention of disease, new challenges will have to be accepted in Epoch B, testing Man's ingenuity in developing the means to enhance the degree of fulfillment in the life of the individual and in the quality of life generally.

The difficulties and complexities involved in such a challenge are considerable; the mere existence of innate mechanisms for meeting them does not mean that the odds are in favor of success. Human history is replete with evidence that *de*volutionary processes also operate, with deterioration of the human condition, unless foresight, imagination, ingenuity, determination, and wisdom are brought to bear, to increase self-awareness and self-discipline in the choice of ends as well as means. To be able to prevent such deterioration, principles will be required by which to live, and by which to intervene judiciously in the process of biologic and metabiologic evolution with knowledge of the *de*volutionary as well as the evolutionary consequences of each action or nonaction when we face issues that affect our well-being individually and collectively.

The hypothesis has been proposed that if the mind of Man is exposed to the economy of Nature, as revealed in the workings of living systems, he will become sensitized to recognize the necessity of balancing values. Thus measure is established as the source of wisdom. By improving the quality of life, wisdom, thus, can influence the processes of metabiological evolution, just as the enhancement of physical fitness functioned in the struggle for survival in biological evolution. If Man can come to recognize that the use of wisdom in the game of life leads to the reward of a greater measure of fulfillment and satisfaction, then he will value the development of such special skills; nowadays more individuals have the opportunity to do so for more years than was generally true heretofore. In this, everyone has much to gain.

If, in course of this quest, a struggle arises between the wise and the nonwise, the conquest or elimination of either one would result in loss to both, just as if Life and Death were to "conquer or eliminate" each other. The wise must avoid a "win-lose" conflict with the unwise, just as it was necessary in biological evolution for Life and Death to arrive at a "double-win" resolution in order for either one, and hence both, to persist. Even though Death eventually wins over Life so far as the individual is concerned, Life wins over Death in the perpetuation of the species. This is to say that Life "wins immortality" for the species and Death, mortality for the individual; the individual may be unwise, but not the species. For the quality of life to be improved, and for survival, Mankind will have to respect those who are wise and expect the individual to behave as if he were. If wisdom is, in fact, a new kind of fitness for survival, the operation of the equivalent of natural selection in the metabiological evolutionary processes will have been guided by the choice of human values.

In Epoch A Man acted effectively on the side of Life, both of the individual and of the species, by reducing the incidence of disease and the frequency of dying prematurely. Correspondingly, in Epoch B Man may be able to devise ways of improving the quality of life of the individual and of the species by reducing unwisdom or its adverse effects and by respecting and applying wisdom for increasing the possibility of personal fulfillment. Among individuals who now have less to struggle for personally in order to survive, as a result of the changes brought about in Epoch A, a new syndrome has developed, manifest in seeming purposelessness, for the treatment of which new experiences will be needed, possibly leading to new motivations.

Judgment is required in larger measure than ever before if Man is to succeed in balancing the adverse effects, both upon the species and upon individuals, resulting from the increased knowledge and improved technology that reduce the need for struggle and also the opportunity to learn how to experience a sense of satisfaction. This is seen among the increasing numbers of individuals whose lives have been prolonged and made more secure by the metabiological evolutionary developments that have occurred in recent times, without effort on their part.

There is a further undesirable side effect of the benefits brought about in Epoch A. Among the individuals who feel purposeless, some become wantonly and pathologically destructive, threatening and interfering with the development and achievement of fulfillment of others. For resolving such problems as these, far more insight is needed than has as yet been activated.

By suggesting the idea of survival of the wisest, I mean not only that the more discerning will survive but also that the survival of Man, with a life of high quality, depends upon the prevalence of respect for wisdom and for those possessing a sense of the *being** of Man and of the laws of Nature. These are necessary for choosing from among alternatives, for fulfillment as well as for survival. Man's metabiological questions and problems still need answers.

The idea of wisdom as a pro-health, pro-life, pro-evolutionary influence still leaves open and unresolved the question as to how this might be developed and applied. Man's capacity to bring this about is also not known. The role of religious and political organizations has been to enlighten and to guide. Now new means are needed for inner self-regulation based upon naturalistic rather than on arbitrary moralistic formulations. In spite of the difficulties involved in devising and developing such formulations, this could provide an important purpose *in* life and serve the purpose *of* life individually and collectively.

The extent reached by Man in his capacity to create, to destroy, and to move in space as well as over the surface of the earth has indeed been remarkable. To what extent does he have the ability to invent new ways to act wisely as a species even if his aptitude to so behave individually is relatively limited?

Exposing Man's mind to the laws of Nature may help him discover and apply whatever insight and foresight he possesses for dealing with the problems of relationships to himself and to others, and to the universe. This way of thinking about Man and Nature and relationship and wisdom is new to most, and to be of value will require modern patterns of perceiving one's self and others. New attitudes and behavioral patterns will follow.

---

*"Being" is the center in which exist the possibilities that, when unfolded, reveal the essence of the person, both as a member of the species and as an individual.

It is simpler to conceive such notions than to apply them in everyday life. Nevertheless, it is far easier to reach objectives based upon sound concepts and hypotheses than upon those without basis. Hence the challenge with which Man is generally confronted at this point is to see himself as a biological and meta-biological entity, possessing attributes capable of reversing some of the *de-volutionary* trends. These attributes can also be directed and disciplined to facilitate and increase the probability of achieving a greater measure of fulfillment in life than has been possible until now.

Paradoxically, this challenge and hope exist in the face of enigmas more difficult to overcome than ever before, because greater opportunities for fulfillment are matched by correspondingly greater obstacles. For this reason, wisdom, understood as a new kind of strength, is a paramount necessity for Man. Now, even more than ever before, it is required as a basis for fitness, to maintain life itself on the face of this planet, and as an alternative to paths toward alienation or despair.

13.   Whither Bioethics?

George H. Kieffer

Science is now in a vastly more powerful position to influence world events than it ever has been before. The most striking example of this is the fact that the theories of atomic physics developed by Einstein, Fermi, Schrödinger, and others resulted in the development of atomic weapons, so that now we are caught up in an arms race that threatens to bring about the demise of modern civilization. Of course this is not the only instance in which scientific discovery applied to technology has revolutionized our ways of thinking. Science, in a profound way, has brought about an interdependence in our society. People today depend upon science, and science, if it is to flourish, must rely upon the people for its support.

The scientific and technological capabilities with which we are endowed are part of our continuing history. Throughout time, the human mind has attempted to understand, control, and harness nature for its own use. The condition that sets us apart from earlier times, though, is that at this point in our history the broad effects of science on people's lives must be accorded a high priority, for it is an axiom of this science that risks must be taken if we are to move on in the world. Scientific and technological advances already have had major impacts on our individual and collective lives; therefore, they merit serious attention. These same developments have and will continue to have implications for ethics, too. Some of our most basic conceptions of life and death, for instance, or of what it means to be human, are difficult to justify with the new conditions resulting from advances in biology and medicine. We have already entered a new era medically, morally, and ideologically. These aspects all go together.

There are several reasons for believing that we can no longer keep our system of morality and our system of scientific expertise in separate compartments. Among these is the fact that science has called attention to the significance of an entirely new range of good and bad behaviors. These new possibilities have forced theologians and ethicists back to the drawing board to reconsider the older conceptions of life, death, and humanness, of what things are really beneficial and what things are harmful. We have learned from history that moral rules that are out of touch with reality encounter at least two problems: (a) They may be harmful to the very values that ethical norms are established to safeguard; and (b) they run the risk of diminishing returns in terms of human compliance. Let me give some examples.

Regarding the first, the Catholic Church today recognizes the actuality of overpopulation in some parts of the world, but it is forbidden by a moral rule to act effectively against this problem. The Church is in a moral bind; the problem is clear, but the solution requires action contrary to previously determined moral positions—in this case, the widespread implementation of effective birth control programs. The same kind of moral bind arises in regard to euthanasia, genetic engineering, truth-telling in medicine, environmental matters, and the like. Over and over again we see the same circumstances—namely, a conflict between what we recognize as being necessary, and moral rules that block appropriate action. The very values we want to protect—the right to life, individual freedom of action, the quality of life—are harmed by our inability to enact effective responses to perceived problems.

The situation in Colombia, as described recently by a release of the World Watch Institute in Washington, D.C., exemplifies the second problem—a diminishing of human compliance with stated policies when those policies are out of touch with the real world. Colombia is a staunchly Catholic country, and the state actively forbids abortion, even to save the life of a pregnant woman. Yet the largest maternity hospital in Bogota devotes half of its beds to cases resulting from complications caused by illegally induced abortions.

The same World Watch report describes a similar situation in Brazil, where it is estimated that half of all conceptions end in abortion, which is illegal there, too. The practice of contraception in the American segment of the Catholic Church, reports of suspected cheating in recombinant DNA research, physicians practicing illegal euthanasia, and so forth, again illustrate that legitimate authority may be overridden or simply ignored when values clash. A transition in ethical standards seems mandatory, for, whether we choose to accept it or not, society is changing, and biology is contributing much to this change.

To gain a better perception of the new conditions, let me construct a rough history that describes an evolution in our view of the world and our relationship to it. At one time in our past, humans learned to relate to the gods. The gods controlled everything. They were the cause, and humans were merely part of the vast scheme of things, possessing little if any value. The whims of the gods ruled earthly as well as heavenly matters. All values derived from the gods. This was probably quite satisfying inwardly and may have provided a measure of security in a period when knowledge about the world was very limited. However, as time went on, humans became more conceited. They could still accept the view that they lived in a planned universe, but they came to believe that they themselves had value. As a result, people started talking about such things as duties and responsibilities toward fellow persons. Then another shift occurred, reflecting the judgment that humans have not only value but a capacity for control of their lives. The belief spread that we could direct events through our own efforts, and we entered the age of science and technology. We left behind the old world view, which was based on a unifying vision of wholeness. The new culture could no longer look to the oracles for help with difficult problems. There was an empty place left by the diminution of religious authority. But still, we were not satisfied. In those earlier days of the scientific age, there were limitations on what we could do, because nature still set the boundaries on our activities, and science was not powerful enough to overcome those limits. But now we have placed ourselves in the center of things. Science and technology grew up, and the result has been that humans have become a cause. Paul Ehrlich calls our species a geologic force. The powers at our disposal are comparable to those found in nature, and we are at the controls. Cause coming from without has completely disappeared; we are the cause, and we decide! As a result, our sense of human significance has also been enhanced; we count for a great deal more than we formerly did.

And what are the consequences of this new world view? Plenty. First, we have a new concept of freedom; we can freely choose those policies and applications that contribute to what *we* judge to be in our own best interest, both for the short term and for the long term. We choose what we are or what we should like to become. This has placed upon us tremendous responsibilities, for our decisions can result in either good or evil. We have a responsibility to make the right choices according to our own standards.

Second, we have a new concept of relationships. Interdependency is the word for this new concept, and it reaches into all segments of our world. We can no longer make decisions in isolation. Both in time and in space, we are starting to talk about intergenerational responsibility: the separations of "I and thou, here and now" of classical ethics are being replaced by the view that we are bound by moral ties to generations still to be born, and we can do irreparable harm to them by what we choose to do now. Along with the concept of interdependence, we have a new idea of unity. We are aware of threats to the global community, such as ecological threats to our physical survival, and threats that can degrade our humanity.

Lastly, we have a new concept of resources. we now know that physical resources are depleted by use, and we are concerned about limits—limits to what we can do, and limits to the physical resources available to work with. We also know that human resources are depleted by nonuse. We talk about the aged, those members of our society who have been formally isolated. How do we involve them in action?

The lesson of this short list is that we live in a world where futures are limited. No longer is the utopian concept of a limitless future, promoted until several years ago, a viable option. The future must be planned. The choices we make today are conditioned by many factors, including past events and future needs. If there is any advantage in our present position, it is that we have historical experience on our side. By using history wisely, we may learn from the past. We know what powerful technologies can do, and we hope that we need not repeat all the mistakes of our predecessors.

What does this have to do with the problem at hand—biomedical innovation and its implications in our technological society? I think that the relationship is quite clear. The perplexing circumstances into which we have been thrust is that we find ourselves to be technological giants but ethical infants. We have an almost infinite capacity to uncover new knowledge, but only limited wisdom to apply this knowledge. We have much maturing to do ethically, with many growing pains and a lot of stumbles and falls ahead of us.

We are almost continuously exposed to bioethical problems. A few of these have been momentarily alleviated, but fewer still have been corrected. In fact, we have yet to resolve a major biomedical ethical issue at the societal level. To be sure, we are not quite so indiscriminate as we once were about the pollution of lakes and watersheds or wide-scale release of toxic and persistent chemicals into the environment; and the use of fetal tissue for research is not so freely engaged in, nor is the manipulation of genetic material as easily practiced. Study commissions on a national level have investigated a number of biomedical ethical problems; they have made their recommendations, and now we have guidelines governing a number of our activities. But in none of these cases have the ethical problems been solved. These I call "first-generation" responses—measures that were initiated to respond to immediate crises. The objective of these responses has been to avoid biological or ethical calamity. Now we are entering a new phase, one that is aimed at searching out effective and long-term resolutions. It is in these areas that a transition in ethical thinking is mandatory. We need something in our world view to fill the empty space left by the weakening of religious authority. This I call "second-generation" responses. Some examples may be instructive.

Should welfare mothers be provided elective clinical abortions at government expense? With this question, we are caught between two morally valid principles. One is distributive justice, which demands that what is available to some should be made available to all in the same class or category. The other is that it is wrong to insist on making those who believe that abortion is immoral finance it with their taxes. This defense, you may recall, was the rationale used during the Viet Nam War by those who refused to pay their taxes as a symbol of their nonsupport for what they judged to be an unjust war. Or take the example of the Supreme Court decision on abortion in 1973, which changed our country overnight from one of the most restrictive in Western society to one of the most liberal. The decision did not settle the moral question. Are fetuses really human beings, entitled to equal protection under the Constitution, or are they of inferior legal status? Does the freedom of women apply to those instances in which they may make personal decisions on reproduction, including the elimination of an unwanted fetus?

Health care raises more questions. Since 1950, as all of us know, the cost of health care has risen over a thousand percent, so that today the average wage earner is devoting well over one month's salary each year just to pay these costs. How much of our income can we as individuals and as a nation devote to this single service? No one knows. Is the ability to pay an adequate criterion for obtaining necessary medical services? On one hand we hear that health care is a right, and on the other that it is a privilege. Others insist that it is a service that the physician is free to give or withhold. No resolution is in sight for this problem. And still more difficult is the question of euthanasia, or elective death—an issue that arouses heated controversy wherever it is raised. California broke new ground when it enacted the Natural Death Act in January of 1977, and at least seven other states have followed suit since then. To some, it seems an inconsistency in our logic, as well as in our ethics, that we allow fetal euthanasia of the grossly deformed unborn but prohibit the taking of a life judged to be helpless and hopeless when it occurs at the end of a sojourn on earth. Can and should the two be separated? Are there social and moral dangers associated with legal euthanasia?

One last example: Do we as an affluent people have the right to dictate population policy to less wealthy nations, even to the point of refusing aid if they fail to comply with our demands? "Lifeboat ethics" holds that it is much more heartless to feed starving people who have no long-term hope of surviving than it is to let nature take its course. The utilitarian principle of "give if it helps but not if it hurts" has been applied to this situation. But still, the essential tenet of a responsible ethic is the unqualifiable worth of every human being. Who can judge which humans are to be saved and which allowed to die? By what standards?

The new biology is replete with unfamiliar problems for which traditional ethical principles and historical moral standards simply are ineffective. Somehow we have to narrow the gap between the power of science to discover new knowledge and our prescriptions for applying this new knowledge wisely. The moral bind threatens to overwhelm us, cutting off options for rational actions. How can the really hard questions be resolved? What ought we to do?

To begin with, settling ethical issues by flipping coins is less than acceptable. It is not true that any choice is as good as any other, nor that it is simply a matter of preference whether this or that value is selected. To be sure, all ethical views are matters of individual decision, but to allow a moral free-for-all simply is incompatible with a well-ordered society. The cases of people whose brain function has ceased but whose remaining vital functions can be mechanically sustained are concrete and tragic instances of conflict between best interests of the patient, the family, the medical profession, and society. When pressed to its limits, complete ethical freedom carries with it the potential for its own destruction. If there is any lesson to be learned from past history, it is that total concern for oneself leads to socially disruptive behavior.

Ethical problems are notoriously open-ended. It is a central feature of contemporary bioethical debates that they are impossible to settle and are interminable. Bertrand Russell bluntly claimed that in ethics there are no facts. David Hume's critique of the naturalistic fallacy insists that one can never derive an "ought" from an "is," for somewhere along the way a value judgment has to be made that is independent of the facts. Those trained in the sciences find it difficult to resolve the hard questions of what ought or ought not to be done. They are disturbed by nonsolutions. They have a different way of approaching a problem. The scientist sees the problem and wants to get on with it, whereas the philosopher says, "No, you cannot solve ethical problems by attacking them head on, you first must have a philosophy; from a philosophy you derive an ethic, and then maybe you will be ready to attack the problem." So philosophers and scientists have a devil of a time getting together.

The word, "trans-science" was coined by Alvin Weinberg several years ago to identify those issues that fall somewhere between the usual domains of science and society. The questions are raised by science, yet science cannot answer them because they do not depend solely on facts. So what then are we left with? What criteria can we apply to the solution of problems?

Daniel Callahan, the director of the Hastings Center, notes that there are three moral premises that dominate the debates on "ought" questions in bioethical matters. The first of these is the principle of individual liberty, which simply means the maximization of individual freedom. This is interpreted to mean that one has the right to be left alone and the right to seek out what one desires, if, in doing so, there is no demonstrable harm to others. It is the moral premise of the radical psychiatrist Thomas Szazz, who advocates the right of a person to self-medication, for whatever reason. It was the argument of the Supreme Court in 1973 in its abortion decision, when it ruled that the right of women to legal freedom in matters of personal concern overshadows the right of society to make laws restricting that freedom. It is in part the defense for continued recombinant DNA research, which involves the right to seek out knowledge as guaranteed by the First Amendment. It is the moral axiom on which the case for *in vitro* fertilization is being argued.

The second principle is the risk–benefit approach. When there is uncertainty about a particular policy, costs are weighed against benefits to be gained, and the morally correct action is determined by the outcome of this arithmetic. Cost–benefit is, of course, the ethical principle derived from utilitarianism; strictly speaking, it is a technique and not a moral principle. Lately, however, it appears that it has been given moral status. It is this principle that has dominated the recombinant DNA debate leading to a continuing relaxation of the National Institute of Health guidelines.

The third principle is that it is always better to try do good than to avoid doing harm. It is better to do something if that something could result in a good end, than to do nothing for fear that some harm might come. This is a principle much used in experimental medicine. This moral premise is taken a step further by those who insist that failure to pursue good is, in effect, a form of doing harm. For example, it is argued that, if recombinant DNA research is not carried out, untold benefits will be lost to future generations, and that loss in itself is morally reprehensible.

Callahan criticizes these principles, since they always lead to the same conclusion: Proceed with whatever it is that you are doing. The deck seems to be stacked every time in favor of what it is that you intend to do. If outcomes are preordained, he argues, then these cannot be the only right moral principles, if, indeed, they are the right ones at all. To be valid, moral principles must have a degree of ambivalence about them so that when they are applied to real problems different judgments result in different instances.

And so the question must be asked, Whither bioethics? What has bioethics provided us with as a way to resolve the perplexing "ought" problems of biology and medicine? At this point in our history the judgment must be that it has contributed very little. Nothing presently exists that clearly shows how we might make "ought" choices–nothing, at least, that meets with any sort of consensus. To emphasize the point, I should like to read to you from a recent editorial in *Chemical and Engineering News:* "No decision-making process exists. We make do with disparate efforts of individuals, special interest groups, self-appointed interest groups and legislative, judicial and regulatory systems. Thus we have not yet developed the philosophical network, the societal traditions or the institutional mechanisms to deal holistically with the risks that we now increasingly identify or create." The message of this editorial is clear, but can the issues be settled more smoothly? We are talking about second-generation responses, shifts in ethical standards that can humanize biotechnology. The questions that must be answered are not easy ones. They may turn out to be unanswerable. But this doesn't absolve us of the obligation to try. In the long run, bioethics may be the most important body of knowledge we have ever attempted to define, for in order to do so we must also define ourselves. If moral philosophy does indeed reflect the society and culture of which it is part, and at present moral philosophy in our society is ill-defined, we can reach a better understanding only by thinking through the concrete questions that confront us. It is a hard but vital challenge.

14. Social Priorities in Biomedical Research

Craig Shapiro

Since World War II, scientists have had a unique position of authority in the public mind, and until recently they have had largely unquestioned financial support from the government. This came about mainly because of the results achieved by research groups during World War II, which gave scientists credibility among government policy makers in claiming that science could meet societal needs. It was due also to the prestige accorded scientists because of their education and expertise, to the political advantages of the support of scientific research, and to the fact that at that time research funding did not represent a large proportion of the gross national product. As a result, the government could be relatively flexible in meeting demands for funding. During the 1950s and the early 1960s, the number of scientists and the funds for research grew at such a rapid rate relative to the rest of the economy that if this rate continued unchecked, by 2010 everyone in the United States would have been a scientist and the entire GNP would have been devoted to research.[1]

During the late 1960s, the demands, activities, and goals of scientists, as well as of other professionals and institutions, began to be questioned by the public and by Congress. A widespread feeling developed that the advancement of knowledge for its own sake was not necessarily desirable, that it drew resources away from other national needs, and that it sometimes led to technologies that were potentially dangerous to the individual and society.

Congress became more knowledgeable about the workings of science as more scientists and engineers were employed on its staffs, and it began to view scientists as another costly special interest group.[2]

[1] J. Katz, "Scientific Mystifications," *Society*, 15, 1978, p. 84.

[2] B. Bozeman, "Reflections on the End of Carte Blanche: The Inevitability of Conflict between Congress and the Scientific Community," *Policy Studies Journal*, 5, 1977, p. 175.

The conflict between science and the government and public interests can be considered more systematically within the theoretical framework of accountability to government presented by Robinson.[3] Three different kinds of accountability in government contracting for research are identified: (a) Program accountability, or whether the researcher fulfills through performance the goals of the awarding agency. In this type of accountability, good performance is often rewarded by grant renewal; (b) Process accountability, or whether the research is carried out according to rules specified by the awarding agency. Since program accountability is often hard to assess, in practice process accountability often substitutes in the evaluation of research; (c) Fiscal accountability, which focuses on the actual expenditure of funds.

Several aspects of accountability that are most prevalent in the public concern about science can be added to this framework. One is safety accountability: to do no harm, either to the individual or, on a larger scale, to society. As a result of the disclosure of certain harmful incidents and of ethically questionable practices of scientists who were regulating themselves, there has been a clamor by the public and government for more active roles in the regulation of research. This has resulted in the establishment of guidelines for experimental research on human subjects, for fetal research, and for recombinant DNA research, and in the establishment of the HEW Ethics Advisory Board for the consideration of research involving *in vitro* fertilization. Another form of accountability is professional accountability: whether the research is conducted by using the scientific method, whether evidence is used without experimenter bias, and whether results are disseminated to the scientific community. A third form of accountability not covered by Robinson's framework, although it is included indirectly in the agency goals involved in program accountability, is social accountability: Should research be required to be directed toward a desired social end? In the face of limited resources, priorities on funding and setting research agendas must be determined. Social benefit is the most commonly used priority, and it is often the basis on which the researcher justifies his project to the awarding agency.

In all these areas of accountability, scientists until recently have not had to account directly to the government or to the public. Questions have been raised, however, about scientists' ability to regulate themselves. There is much debate on the extent of internal and external constraints, and on institutional mechanisms. All the areas of accountability involve ethical issues. With respect to the ethical principles of nonmalevolence and beneficence, one could say that safety accountability is concerned mainly with nonmalevolence—for example, the concern that recombinant DNA research not lead to hazardous outcomes. Social accountability deals with beneficence: how to ensure that scientific research, most of which does not involve apparent safety issues, is ultimately socially beneficial.

[3]D. Robinson, "Government Contracting for Academic Research: Accountability in the American Experience," in B. Smith, and D. Hague, eds., *The Dilemma of Accountability in Modern Government,* London; Macmillan, 1971.

Some people argue that all research should have direct social relevance. Others believe that the very nature of science demands that it function independently to be productive. In the middle are debates on such topics as the balance between applied research and basic research, and the evaluation of research objectives and results. This paper will examine some issues raised in consideration of the social accountability of science, and in particular the accountability of biomedical science and cancer research.

Accountability of Science

In discussing the issues associated with the social accountability of science, the first question one must address is whether science *should* be socially accountable. The public provides funds for research with some expectation of a future return. It is really the only reason that the public funds research—for the ultimate social utility of its products. This is often the reason given by scientists to granting agencies to justify the funding of their research. (There is a joke among researchers that a grant application has little chance of success if it does not contain in its introduction the words cancer or heart disease.) In accepting public funding, the scientific community tacitly agrees to public proscription and prescription of research. Because the public supports scientific activity, the public has the right to demand something in return for its investment. The question then shifts to how and at what level scientific activity should be directed and managed to ensure that it is socially accountable.

One must consider whether science can survive if its directions are determined, to any significant extent, by society. It has been argued[4] that the informal agreement between society and scientists, which has existed for the past several decades and which allows scientists to operate relatively freely, is necessary for the functioning of science, and that any government intervention to make science accountable will be counterproductive; science, in its quest for truth, cannot be suject to ideology or time schedules. It has also been suggested[5] that a principle anologous to the "invisible hand" of the market holds for scientific research. When researchers are left free to do the projects they as qualifed scientists consider desirable, the results are socially optimal—that is, they provide the most benefit at the least cost.

Jonas,[6] however, challenges the view that science has ever operated free of external constraints. Part of his argument involves the concept of technology pull, or the notion that science relies on feedback from and receives its assignments from sectors that utilize its technological applications. Furthermore, science uses

---

[4]D. Baltimore, "Limiting Science: A Biologist's Perspective," *Daedalus,* **107**, 1978, p. 37.

[5]H. Brooks, "The Problem of Research Priorities," *Daedalus,* **107**, 1978, p. 171.

[6]H. Jonas, "Freedom of Scientific Inquiry and the Public Interest," *Hastings Center Report,* August 1976, p. 15.

technological advances as tools in its day-to-day operations, and the funds to acquire these tools come from the public.

It probably can be safely said that science, while requiring a certain degree of self-government to be productive, has always functioned under external constraints imposed directly or indirectly. Social accountability involves constraints imposed through the funding mechanism. The fact that there are more research proposals than there are funds available for awards and that there is concern for the efficient functioning of the research establishment in the pursuit of socially beneficial results requires that priorities in funding be set, which necessitates bringing in considerations external to science itself.

Setting Priorities in Basic Research

The institutional compromise evolving out of the need for science to function independently and yet be accountable in the peer review system. This system is fundamental to determining which projects will be funded in a given field and to overseeing the quality of research. It relies totally on judgment by scientists and on criteria internal to the scientific establishment. Perhaps suprisingly, the latest studies[7-9] indicate that the peer review system is extremely equitable, efficient, and accurate in providing judgment on the scientific merit of applications to agencies such as the National Institutes of Health, the National Science Foundation, and the National Institute of Mental Health. As Bondurant points out, however, "the peer review step in research decision making at the NIH does not address national goals in health research because the study sections do not articulate for this purpose with the legislative and executive agencies where goals and budgets are developed." The system of setting of goals and appropriation of funds for biomedical research in the specific programs of the NIH involves negotiation between the executive branch and Congress with advice from the NIH staff and consultants. The peer review system plays a large but indirect role in this process, however, and there are indications that it may not be the best for this purpose.

As Brooks explains,[5] the peer review system is "frequently used as a signaling mechanism to indicate to program administrators where the most significant opportunities lie within a much broader domain of science," through proposal pressure, "which represents what the scientific community has pluralistically decided what is worth doing." But this representation is far from ideal. Proposal pressure is dependent on the availability of money; researchers apply for funds where the money is and justify their research in terms of a specific program. For

[7]S. Cole, L. Rubin and J. Cole, "Peer Review and the Support of Science," *Scientific American*, **237**, 1977, p. 34.

[8]C. Wilkinson, and V. Reus, "An Examination of the Federal Psychiatric Training Grant Peer Review Process," *American Journal of Psychiatry*, **194**, 1977, p. 6.

[9]S. Bondurant, "Peer Review of Research Project Grants by NIH Study Sections," *Clinical Research*, **25**, 1977, p. 297.

example, a proposal in immunology can be phrased in terms of immunology itself, or in terms of developmental biology, biochemistry, genetics, allergy or a variety of other topics. Secondly, the level of activity in a certain area of science, and thus the number of proposals in that area, reflects grants awarded previously. A step-up in funding of virology projects three or six years ago means that there will be a host of researchers reapplying to continue their work at the present time, somewhat independently of the present popularity of the field or its prospects of beneficial outcomes. Prior funding will have also educated graduate students in virology and thus produced more applicants in the field. Trends are easily created in biological science, and they do not necessarily represent the best direction. It is my opinion that real discoveries, real breakthroughs that change our idea about biology, are very rare, but when they do happen, researchers, each of whom has his own experimental system, look for the new discovery in their system. Thus, biology research is a matter of follow the leader until another breakthrough occurs. The general direction of research does not necessarily represent what science collectively feels is the best direction for itself; it is where the latest big discovery has occurred. For example, for the past two years everyone has been sequencing the RNA and DNA of their systems, whether they be adenoviruses or Chinese hamster ovary cells, looking for intervening sequences. Next year it may be something different. Finally, as a critique of the peer review system, it has been said that the system promotes the generation of safe proposals, in fields where issues are already well established or in which quick results can be obtained. In part this is due to the nature of the award. Most NIH grants are for three years—a short time for a basic research plan, since science progresses slowly, and the development of ideas and experiments often requires a long-term program.

Thus, peer review, because of its distortion of scientific consensus due to the dependence of proposed pressure on the availability of funds, due to changing trends in science, and due to the creation of inertia in certain directions of research, does not appear to be a satisfactory way to establish priorities among different fields of basic science.

Another approach to the setting of priorities is the economically based extrinsic value theory.[10] This theory claims that research can be justified only if it has a recognizable and specific value to the society that supports it. Some research, particularly applied research, delivers an intrinsically valuable social product, such as increased agricultural productivity or analysis of social conditions—products that can be judged on their merits, assigned direct values, and viewed as investment goods that will ultimately be inputs to production. Basic research is viewed as having no readily identifiable investment function or intrinsic value as a service, and thus it should be judged by the services and products that accompany its activity: teaching, for example. Research could then be considered as having extrinsic value as an anti-obsolescence device by

[10]D. de Solla Price, "An Extrinsic Value Theory for Basic and Applied Research," *Policy Studies Journal,* 5, 1977, p. 185.

allowing the researcher to maintain or increase his level of knowledge in order to perform his teaching duties better, and levels of funding could thus be determined by this value. His teaching duties would be viewed as the ticket a researcher must buy in order for society to let him pursue research freely. This view has also been proposed by scientists. Nossal[11] describes how a main function of research is the updating of teaching and states that, without research, universities could not maintain a level of excellence.

In a sense, the coupling of teaching with research has been a major tool of the scientific community in procuring funds from the government. Price[12] describes how the research community used this tactic to justify funding of projects and to maintain research free from political pressures and economic self-interest. However, although teaching and research may be intertwined and inseparable activities, to utilize teaching or some other criterion extrinsic to research seems a replacement of one arbitrary process with another process just as arbitrary. For, although teaching and other research-related activities are socially beneficial, they are not the primary benefits that society seeks in funding research, and to maximize a variable other than the variable of major interest seems an inefficient way of achieving societal goals.

Other allocation schemes have been proposed, such as the manpower model,[5] which bases its funding on the availability of talent in various subsections of research; but this model also suffers from the fact that the distribution of talent within the scientific community does not necessarily reflect which research areas will ultimately provide socially beneficial results.

The essence of the problem is that, although social utility may be the primary reason for federal support of research and for making a choice among areas of support, there appears to be no good procedure for evaluating research projects with reference to this quality. Biologist David Baltimore[4] believes that it is a fallacy to assume that there are enough clues about the future to provide a good basis of prediction about what effects on society, negative or positive, outcomes of scientific investigation will have. This may be so, for predicting the future can be problematic. A biologist of a different opinion, Robert Sinsheimer,[13] dismisses the issue. "In a fundamental science such as biology," Sinsheimer suggests, "most of the overt phenomena of life have long been known.... It would seem likely that only within the central nervous system may there be potential for wholly novel—and correspondingly wholly unpredictable—process." Further research will elucidate underlying mechanisms, but will not yield "novel processes." Thus, "the unpredictability of a research outcome is not an absolute."

It is hard to reconcile such an opinion with fact. Even if most of the novel processes in biology are known—a point debatable in itself—the revelation of underlying, previously unknown mechanisms will certainly yield techniques with

[11] G. Nossal, *Medical Science and Human Goals,* London, Edward Arnold, 1975, p. 97.

[12] D. Price, "Endless Frontier or Bureaucratic Morass?" *Daedalus,* **107**,1978, p. 75.

[13] R. Sinsheimer, "The Presumptions of Science," *Daedalus,* **107**, 1978, p. 32.

tremendous and previously unpredictable applications. One has only to point to the discovery of restriction enzymes and the subsequent techniques of recombinant DNA, or the elucidation of the hormones responsible for ovulation and the subsequent techniques of *in vitro* fertilization, to question Sinsheimer's belief.

But because research outcomes are unpredictable, because basic research is termed basic for the very reason that one has little idea of how its outcomes will be applied, and because advances in one field are often based on results generated in a totally unrelated field of research, the possibility is raised that no assessment of the ultimate beneficence of basic research can be made. Brooks[5] claims that because the outcomes of research are so uncertain, and because any "one practical consequence is so improbable or speculative that it would be dishonest to use it as justification for supporting the research, it is therefore best to assume that increases in knowledge will on the average be beneficial." He goes on to argue that the setting of priorities at the fundamental research level will not have a great effect on social and political priorities, because research and development have "little leverage on social priorities."

Without an adequate procedure to evaluate ultimate social utility, scientific merit would appear to be the least arbitrary criterion for the awarding of grants. Thus, however inadequate it may be, the peer review system, in combination with scientific advisory councils for each research category, appears to be the best indicator of research opportunities and thus of choices for funding of basic research. Clearly, improvements must be made in our ability to assess the consequences of research so that the ultimate goal, social utility, can be a basis for setting our priorities.

Basic versus Applied Research

The predominant controversy in the management of research to yield socially beneficial results is the still unresolved issue of the relative merits of basic and applied research.

Prior to the 1960s, the prevailing opinion within the government and the scientific community, largely established by Vannevar Bush's 1945 report, *Science, The Endless Frontier,* was that "basic research was a necessary foundation for practical progress and that applied results would flow automatically from it."[12] It was believed that basic research supplied the information necessary for applied research to progress, and thus defined the limitations and potentials for applied research. This belief was used by scientists in their bargaining with Congress for funding of basic research. But starting in the 1960s, the importance of basic research began to be questioned, partly because, as resources became scarce and other social needs began to compete, basic research became a low-priority issue and a political liability. Some believed that only a few leading researchers are responsible for most of the scientific advances and that a cutback in basic research would not affect the rate of advancement of scientific knowledge.[1] It was also argued that scientific advancement is more a result of "technology pull" than of its own intrinsic momentum: "Thermodynamics owes more to the steam engine than ever the steam engine owed to thermodynamics."[10] This viewpoint started to be translated into policy. Said Lyndon Johnson in 1966, "Presidents need to show more interest in what the specific results of research are. A great deal of basic research has been done... the time has come to zero in on the targets by trying to get our knowledge applied." Published at the same time was a Department of Defense study, Project Hindsight, which concluded that, for the advancement of military technology at least, i) the contributions of university research were minimal, ii) scientists contributed most effectively when their efforts were mission oriented, and iii) the lag between initial discovery and final application was shortest when the scientist worked in areas targeted by his sponsor."[14] The level of federal funding for basic research fell in constant dollar terms 20% between 1966 and 1976.[15,16]

One cannot confidently say that basic research has yielded important contributions to the improvement of health and health care. According to David Baltimore, a basic research biologist, "Basic research in biology has yet to have a major impact on the prevention and treatment of human disease."[4] Donald Frederickson, Director of the NIH, has observed that "The ready translation of health research into benefits for patients is limited; a great deal of research cannot be directly coupled to the solution of health needs and only a fraction of the yield at any one time is convertible to useful technology."[17] The reduction in the infant mortality rate and the extension of the average life span, as well as the reduction in the incidence of many diseases, can be attributed more to widespread public health measures than to breakthroughs in basic biology. The belief is growing that traditional medicine and research are reaching their limits in the treatment of chronic diseases, and that improvements in the state of health will be obtained through public education, changes of life style, and elimination of hazardous elements from the environment, rather than in advances in the ability to cure specific diseases. It has been questioned whether the tremendous progress in the understanding of diseases is relevant to health. Before, when resources were not limited, the frontiers of biological knowledge could be pushed ahead indiscriminately. Now a call must be made for accountability. Thus, Frederickson foresees in the next few years a continued decrease in the money allocated to science and an increase in funds for the study of social and environmental factors important in health.

This questioning of basic biomedical science has led to a funding trend in the past decade stressing applied, targeted, "mission-oriented" research. The categorization of federal support through NIH into organ systems has increased over

[14] J. Comroe and R. Dripps, "Scientific Basis for the Support of Biomedical Science," *Science*, **192**, 1976, p. 105.

[15] B. Ancker-Johnson, "National Science and Technology Policy—Current Policies and Options for the Future," *Research Management*, **20**, 1977, p. 7.

[16] P. Handler, "Basic Research in the United States," *Science*, **204**, 1979, p. 474.

[17] D. Frederickson, "Health and the Search for New Knowledge," *Daedalus*, Winter 1977, p. 159.

the years because of specific legislative mandates from Congress. Legislation arising from political pressure, specifying cancer, cardiovascular, lung and blood diseases, was passed in the early 1970s. There has been a shift from project grants to individuals to program and center grants, and an increase in contracts that "throw greater responsibility for details and strategy of the research concept on the staff of the supporting agency"—tools that facilitate the targeting of research. In priorities for the decades ahead, Frederickson emphasizes that, although investment in research must be continued, scientists "must assume a greater degree of responsibility in... developing ways for applying research findings to the health care system."

Recently, however, owing to certain perceived failures of mission-oriented applied research, the efficiency of this type of research in satisfying societal goals is being questioned. Rettig and Wirt[18] argue that science policy is dominated by "moon-shot" mentality—the philosophy that mobilized efforts on a national problem will yield the most tangible benefits—and the public politician can view things only in this context. This methodology exists mainly because there are few metaphors, or theories in science policy, to replace the moon-shot metaphor. They claim that such large-scale science lacks economic efficiency restraints, and reflects an unrealistic, politically motivated policy that does not take into account the limits of science and engineering. Additionally, this type of science does not conduct its problem-solving with reference to the social constraints of health, safety, and the environment, and concerns for the consequences of technology.

Besides criticism, other social demands such as increasing health care costs are forcing reexamination of this type of research. It has been argued[19] that the creation of "big science" within biomedical research has been due to the failure to pass a national health insurance program, that the establishment of the NIH (and the NSF) was a political compromise between Congress and the powerful American Medical Association, which blocked national health insurance, and that the step-up in biomedical research was the government's justification to the public for its efforts to improve health care when it could not pass a bill for national health insurance. Thus, expectation that medical research would "ultimately produce better health for the American people" led to additional targeting of research. Now, as the government intervenes in health care and assumes more health care costs, research, which was used as a substitute solution for health care problems, must compete for funds within the health sector with delivery, prevention, and health care financing.

It is instructive to examine the National Cancer Program as a case study of targeted research and an example of the moon-shot metaphor. It is important because of the magnitude of the funding going to this area, the seriousness of the disease, and public concern. The call for a national program to conquer cancer was initiated not by the scientific community, but by Congress or the White

---

[18]R. Rettig and J. Wirt, "On Escaping the "Moon Shot" Metaphor," *Policy Studies Journal,* 5, 1977, p. 168.

[19]S. Strickland, "Integration of Medical Research and Health Policies," *Science,* 173, 1971, p. 1093.

House, but by an organized group of private citizens led by Mrs. Mary Lasker and referred to as the "cancer crusaders" by Rettig.[20] Many groups and factors were subsequently involved in the establishment of the National Cancer Program, but the beliefs of the cancer crusaders that cancer was a categorical disease problem and that applied and clinical research was the best approach toward conquering cancer prevailed in the final legislation, the National Cancer Act of 1971. The passage of this law provided for a fourfold increase for federal funding of cancer through the National Cancer Institute (NCI). By 1976, the NCI budget was 762 million dollars, or 35% of the total NIH budget, and 14% of all federally funded health research and development.

The NCI relies heavily on research contracts rather than on research grants, the former being much more structured and objective-oriented, with continuous evaluation of results. The emphasis on this type of award mechanism, when originally debated during legislative consideration of the National Cancer Act, was largely opposed by medical scientists but reflected more the interests of the cancer crusaders and the public, who felt that targeted research was the way to go. It was essentially a debate on applied research versus basic research. The scientists did not disagree with the objectives of the cancer crusaders, but rather with the means of achieving those objectives. But their political attack was decentralized, and, as Comroe and Dripps point out, the argument for basic research is difficult to justify analytically. Rettig claims that the legislation was ultimately passed without a thorough consideration of the relative merits of the NIH peer review process versus the contract system, of whether cancer research did in fact need a large funding initiative, and of whether there were major scientific and clinical advances to warrant such an initiative.

It was soon after the enactment of the National Cancer Act that the contract system began to draw much criticism, which prompted the NCI to undertake an internal study of the largest contract program, the Virus Cancer Program, so as to determine whether this was the proper way to support viral oncology research. The results of the study were that "much more would have been accomplished if equal support had been provided on a competitive basis to many more laboratories with greater capability and experience in particular areas."

The National Cancer Program has been criticized on other grounds. The emphasis on curative treatment is said to be misdirected because of the belief that most cancer is environmentally caused, and it is claimed that other areas of research are being deprived of funds. The prevailing view is that too little was known about the disease for cancer research to have been appropriate for a mission-oriented approach. NCI Director Arthur Upton stated recently: "There are still vast areas of research where we are unable to lay out a blueprint and a timetable."[21] This realization by NCI officials has recently changed the emphasis from contract research to independently proposed projects.

---

[20]R. Rettig, *Cancer Crusade: The Story of the National Cancer Act of 1971,* Princeton, New Jersey, Princeton University Press, 1977.

[21]T. Maugh II, "Canning Concentrated Cancer Research," *Technology Review,* March/April, 1979, p. 77.

In retrospect, the optimism with which the National Cancer Act was conceived seems foolish. As stated in the House Concurrent Resolution 675, passed by both the House and Senate in 1970 (as quoted by Rettig,[20] p. 82),"It is the sense of the Congress that the conquest of cancer is a national crusade to be accomplished by 1976 as an appropriate commemoration of the two hundredth anniversary of the independence of our country." Rettig points out that the administration's vacillation, its rapid changes of policy during 1971—first favoring just an increase in appropriation without specific legislation,then shifting to an internal reorganization of the NIH and NCI, then proposing its own legislation to create a new "conquest of cancer" agency within the NIH—indicated a general lack of health policy within the administration. Instead, the guiding principle was the moon-shot metaphor, as expressed by Richard Nixon in his 1971 State of the Union Address: "I will ask for an appropriation of an extra $100 million [which ultimately was included in the National Cancer Act] to launch an intensive campaign to find a cure for cancer... the time has come in America when the same kind of concentrated effort that split the atom and took man to the moon should be continued to conquer this dread disease." Although the moon-shot approach to cancer increased our understanding of the disease and provided improved forms of treatment for some victims, it is now rightly being questioned whether it is the best approach toward the desired goal.

Conclusion

The strong political factors behind the National Cancer Program do not make it a unique case of large-scale targeted research. Jackson[22] asserts that "big science" in general tends to have politically determined goals masked by secondary goals justifying it on grounds of practical utility. He claims that by "virtue of the political determination, the goals set for much of [big science] are so extravagant that even the most unprecedented advances would fall short, or alternatively the questions set [big science] by the public policy are often very broad."

This statement appears to be an accurate assessment of the National Cancer Program, and it points to the problems inherent in the logically desirable but practically difficult ambition to direct research to be socially beneficial. Out of the debates and discussion presented here, no clear theories or frameworks have emerged for judging between basic and applied research or for setting priorities within these categories of research to achieve social goals. One is faced with the situation that, even if goals are agreed upon, the means are not clear, and no good form of analysis exists to choose among alternatives.

The nature of science—its unpredictability, its sequential, cumulative strategy—is possibly such that no institutional mechanism or form of analysis can be designed that would permit correct judgments to be made on how to achieve

---

[22]M. Jackson, "Achieving Social Benefits through Science Policy: Is Big Science the Way?" *Australia-New Zealand Journal of Sociology,* **12**, 1976 p. 213.

social goals. Lindblom[23] speaks of the formulation of policy in terms of "muddling through." Most problems in policy formulation are too complex to be approached in a rational, comprehensive manner. They are beyond the intellectual capacities and sources of information that man possesses. Instead, most policy is developed incrementally; some possible consequences of the alternative policies and some relevant social values must inevitably be ignored through the "successive limited comparisons" method.

It is debatable whether science policy and implementation should be conducted by the successive limited comparison method, but, on the basis of past experience, attempts to conduct science policy by the rational comprehensive approach have proved unsuccessful, mostly because science itself, especially biology, reaches its solutions by muddling through.

It would thus seem that, in view of the uncertainties concerning the outcomes of science, the best strategy to achieve a socially beneficial goal would be a diversified, fragmented approach, involving a mix of both basic and applied science. Deciding the balance of this mix is difficult, since both are potentially beneficial. Comroe and Dripps,[14] disappointed with the anecdotal, "let-me-give-you-an-example" justification of basic biomedical research, made a detailed study and classification of the contributions responsible for major advances in the treatment of certain cardiovascular and pulmonary diseases. The majority of the contributions later judged essential for the clinical advances came from basic research, and 41% of the work was not clinically oriented at the time it was undertaken. Yet, many examples can be given of beneficial results from applied research. In setting priorities for basic and applied research, in areas where judgments can be based on scientific criteria they should be used, and in areas where social merit can be judged, these judgments should be employed to determine the allocation of resources.

There are institutional mechanisms that can change the nature and direction of research, principally through funding. Basic research in areas can be promoted or inhibited through decisions on funding. Resources can be concentrated in specific areas, such as cancer. But when one is confronted with ignorance or uncertainty about the limitations and potentials of science, it should be recognized that, although these institutional mechanisms can change the direction of research, they may not be able to provide the answers as to how research should be directed.

[23] C. Lindblom, "The Science of Muddling Through," *Public Administration Review*, 19, 1959, p. 2.

15. Allocations of Scarce Medical Resources

Diane Lefebvre

Medical scarcities of some kind have always been a fact of life, whether it be in the form of medical personnel, supplies, or equipment. The popular view of scarce medical resources may be illustrated by the example of the hemodialysis unit, which, until recently, was so limited in quantity that the number of patients needing hemodialysis treatment greatly outnumbered the amount of machines available. Yet this view of medical scarcity is somewhat confining; a more common example would be that of the well-known specialist refusing to accept any more patients, owing to his already enormous workload. The main problem to be addressed in this paper is the microallocation of scarce medical resources, which considers the idea of distributive justice, or how these resources should be allocated. After three ethical views that offer methods of allocation are examined, the problem of resolving ethical and economic perspectives will be considered. Finally, some possible trends in the allocation of scarce medical resources (SMR) will be reviewed.

Three main methods of determining the distribution of SMR have been proposed. The first and simplest is the view that since a scarcity exists, and not all can be given the needed resource, then none should have it. The second view, considered to be utilitarian, is that the determination of who gets the scarce resource is to be made by considering various characteristics of the patient, such as family role, future contributions, and past services rendered. The third view can be called the "randomness theory," according to which each potential patient has an equal chance to gain access to the treatment, since selections are made totally at random.

The position that when not all can receive the same medical treatment then none should receive it is derived from a belief that all people have the same right to medical care. In this sense, allowing one person to have the SMR would be an infringement on the rights of those who did not receive it. In cases where the

SMR is a lifesaving treatment (as in hemodialysis), this idea could be extended to the point where one could argue that allocating an individual this treatment would be, in effect, denying the right to life of another.

Although this view would be equally fair to all, and very simple to put into effect, it is not a view that most would hold, for two main reasons. First, it seems intuitively unreasonable to have a certain quantity of beneficial medical treatment available for use and not utilize it simply because that quantity is limited and not accessible to all. Second, if this position were to be taken to its extreme, most of the medical services available in the United States today would have to be abandoned, since our present level of health care would be considered a scarce medical resource in most underdeveloped countries. Many people would be more willing to take their chances with the possibility of receiving the SMR, rather than being equally certain with everyone else of not obtaining it.

The utilitarian theory, as a method of determining allocation of SMR, would take into account personal attributes of the patients to find which would most "merit" receiving the medical scarcity. There would first be an initial screening process in which, by medical evaluation, some prospective patients might be rejected on basis of factors that would put them at a severe disadvantage medically in comparison with other patients (such as those having other complicating diseases or injuries in addition to the one being considered for treatment. After this screening, there would more than likely still be an excess of patient-candidates in relation to the amount of SMR available. It is here that the attempt to evaluate social worth in order to justify allocating scarce medical services to individuals comes into play.

The most common method of making this sort of determination has been the use of an anonymous panel. The idea of using a panel or committee to decide who is to receive the SMR creates its own specific problem—that of deciding who is to serve on this panel. Some argue that it is strictly a medical problem and should be handled solely by medical experts. Others believe that, although the physician's knowledge is necessary to determine which patients would be most likely to benefit medically from the treatment, the fact that the panel takes into account the principles of societal worth (an undefined and subjective consideration) indicates that more than just the medical profession should be involved in the decision-making. An argument in favor of the lay panel is that physicians are already too close to the problem of allocating SMR, and they might be inclined toward a more confining view of who is to receive the benefits. "The use of lay members—public-interest lawyers, community and minority group representatives—in decision-making positions in medical areas helps ensure that decisions will not be made by scientific experts with vested interests."[1]

A technique of screening and selecting patients on the basis of social merit was actually employed in 1961 at the Swedish Hospital in Seattle, Washington, to determine who should receive hemodialysis treatment. A panel of seven anony-

[1]J. Jan Stein, *Making Medical Choices,* Boston, Houghton Mifflin Co., 1978, p. 236.

mous members (two physicians and five lay members) agreed to base all determinations on a number of social and economic factors such as age, sex, marital status, number of dependents, income, net worth, emotional stability, educational background, nature of occupation, past performances, future potential, and references. The results of such efforts caused one observer to comment, "On the basis of the past year's record, the candidate who plans to come before this committee would seem well-advised to father a great many children, then to throw away all his money, and finally to fall ill in a season when there will be a minimum of competition from other men dying of the same disease."[2] It appears that this panel was assessing something other than social worth in choosing patients for hemodialysis tratment; in this case, societal dependency seemed the real criterion necessary to obtain treatment. According to Robert Young, "the supposed dilemma: the greatest future contribution versus greatest future burdens imposed on others, should in my view always be resolved in favor of the latter horn."[3] Although those families who would be most dependent on society (upon the possible loss of the family member needing the specific treatment) certainly have the most urgent needs for that treatment, it seems terribly unjust to penalize those who have managed to attain some degree of independence and propority for precisely those achievements. If the panel considers the social burden created by withholding life-giving treatment from a patient the most important factor in their decision-making, it follows that a prospective patient would then be encouraged to rely as heavily as possible on society, as opposed to assisting positively in its advancement.

One of the major problems encountered in developing a system for evaluating social merit as the determining factor in distributing medical scarcities lies in the practicality of this method. Clearly, with so many variables to take into account, it becomes necessary to assign specific values to each in a sort of point-system fashion. The difficulty occurs when panel members do not agree on which criteria are to be held as most important. The likelihood of such disagreement is high, since each individual has his or her own views as to what is considered a valuable characteristic to possess in society; a consensus is improbable. The problem is further aggravated by the fact that several of these factors seem at odds with one another. For example, the future-contributions factor would favor a younger person, the past-performances factor, the older individual. The number-of-dependents factor would favor the parents of large families, but the income criterion might be to their disadvantage. In these cases, it appears that, instead of singling out certain individuals as conspicuously valuable members of society, this system would be more likely to equalize most candidates to the point that all would fall within a fairly homogeneous range of scores.

[2]Shana Alexander, "They decide Who Lives, Who Dies," *Life,* **53,** November 9, 1962, p. 125.

[3]Robert Young, "Some Criteria for Making Decisions Concerning the Distribution of Scarce Medical Resources," *Theory and Decision,* **6,** No. 4, November 1975, p. 439.

The true issue underlying the problem of practical applications of the social-worth method is an ethical one. Panel members cannot agree on a way of rating patients, mainly because no set definition of social worth exists. Since they attempt to judge social value, their decisions are necessarily highly subjective. Which patient is most deserving of SMR—the mother with six small children, or the Catholic priest? the famous artist, or the unemployed bus driver with dependents? What panel is capable of making such decisions without being prejudiced by personal preferences?

One might argue that, if the social-value theory is too biased a method to be employed when human lives are at stake, then any process that entails decision-making would suffer from the same faults. Even the initial screening phase, using medical criteria only, would be inappropriate, since the doctor might attempt to make some sort of value judgment about the prospective patients. Physicians may counter that their decisions are not and cannot be without some subjectivity, for, along with all the physical elements involved, any good practitioner knows the importance of the patient's personal attitude, especially when concerned with treatments requiring self-discipline in diet, exercise, hygiene, etc. Still, it is advisable to reject for this reason only those patients who have a clear record of refusal or inability to follow procedures outlined by their physician regarding their treatment, since estimates of "cooperation" can easily become value-laden.

Another problem that would possibly arise from granting of SMR on the basis of societal worth or dependence involves further decisions regarding the patient once he receives the treatment (provided this treatment is continuously needed). An example is the hemodialysis patient who needs to be on the machine every few days. What happens if, during the course of the treatment, this patient contracts another disease, which lowers his overall chances for success? Or, what if his morals take a turn for the worse and he is no longer the socially valued patient he was originally thought to be? Does the physician (or anyone else, for that matter) have the right to remove him from the machine, or is the patient exempt from any changes upon being chosen for treatment? Far from simplifying the situation, this "social-value" theory only serves to make a complex dilemma even more complex.

The theory of social worth and others like it imply that society invests this SMR in the patient with the hope of receiving some sort of return on its investment. "This approach stands out as a glaring inconsistency with the now-popular notion of medical care as a right, not a privilege. Which is it? If medical care is a right, and dialysis is medical care, committees have no more right to inquire into the social worth of the recipient than a public health nurse would have asking a person if they are Democrat or Republican before giving him a polio booster."[4] The idea that the patient is only a commodity whose main function is to serve society is not a particularly flattering one to many people; it also conflicts with the traditional view in medicine that the rights of the individual supersede the needs of society. "The physician–patient relationship takes place in a societal frame and

[4] Howard Brody, *Ethical Decisions in Medicine,* Boston, Little, Brown and Co., 1976, p. 190.

neither party can totally evade social responsibility. But when that responsibility is in conflict with what both conceive to be in the interest of the patient, that interest can be overridden only in the most exceptional circumstances."[5] With the social-value theory, it is society's "need" for the individual, which ultimately takes the place of the patient himself, that is considered to be of fundamental value.

The third method of allocation proposes the use of a lottery to choose those patients who will receive the medical scarcity. There is some screening involved in that patients with the best medical chance for success are the ones who will participate in the lottery. Physicians make this selection, using their medical knowledge of the groups that might have the highest probability of success with the specific treatment. This screening process should be as lenient as possible, excluding only those patients who, for some strictly medical reason, would be clearly at a disadvantage as compared with other patients on a relative scale of success. As few value choices as possible should be made by the doctor concerning the patient's supposed abilities to cope with the treatment. Finally, all the patients who pass the screening process would have an equal chance to obtain the SMR, as it would be distributed randomly among possible candidates, perhaps in the form of a lottery.

The advantages of using a lottery system to allocate SMR correspond to some of the disadvantages of the social-worth system of allocation. The first is its relative simplicity. Compared with the idea of using a panel to devise a complex and subjective system of scoring a potential patient's worth, a lottery would make allocations of SMR as rapid and as unbiased as possible. This factor becomes especially important when a lifesaving treatment is to be distributed to critically ill patients; these patients may not have the time to wait for a decision to be reached among a panel of anonymous members, who may be having trouble deciding among themselves which candidates to choose. This system also relieves the strain placed on this group of individuals faced with the awesome task of determining who will have access to the SMR, for these members realize that their decisions will affect all those patients in need of treatment, and some might feel burdened by the responsibility of participating in this decison-making process.

Random allocation of medical scarcities would also preserve the trust that is so important in determining the success of a good working relationship between physician and patient. The idea of a doctor's having to make fine medical distinctions between patients needing SMR would tend to undermine this relationship to some degree. The patient might not have as much confidence in his physician under these circumstances, thus decreasing his own chances for success.

The most important result of using the random method of distribution is that the dignity of the patient is maintained. "The individual's personal and trans-

[5]Edmund D. Pellegrino, "Medical Morality and Medical Economics," *Hasting Center Report,* August 1978, p. 9.

cendent dignity, which in the utilitarian approach would be submerged in his social role and function, can be protected and witnessed to by a recognition of his equal right to be saved."[6] Whereas the social-worth theory contends that one individual "deserves" the SMR more than another (implying that the former individual is the more valuable of the two), the randomness theory begins with the idea all men have an equal right to health care. In the event of the scarcity of a needed commodity, all men should have the same opportunity of obtaining that commodity.

Through this preservation of dignity, the randomness theory allows for a less traumatic perspective of possible rejection. To be refused medical treatment on the grounds that one is not as "worthy" to receive it as another would be much more degrading and detrimental to a patient's self-image than if one were to be refused treatment simply because one did not have the luck to be selected through a lottery. Although it would be unfortunate that the patient was not allocated the SMR, he would not be as apt to take this bad fortune as a personal affront to his integrity.

Random allocation of SMR is not without its problems. Some feel that it is too irrational a method of handling such an important issue. One author comments: "The crisis involves stakes too high for gambling and responsibilities too deep for destiny."[7] Certainly, random distribution, by its own definition, is not arrived at by rationalizing. But, as has been seen, selection via the societal-worth theory is just as subjective and much more biased; and no distribution at all of available resources seems a desperate attempt to solve a problem by ignoring its existence. None of the theories discussed is perfect; still, one must determine which is most fair and least subjective.

The main problem with the practice of random selection is that, by its refusal to make any moral judgments about needy patients, there may be times when it seems unjust that one patient receives treatment rather than another. If an Adolf Hitler were to acquire lifesaving treatment by a random selection, at the same time that an Abraham Lincoln was rejected under the same method, the outcome would be disagreeable but nevertheless in accord with the underlying principle of equality of opportunity. Critics of the randomness theory believe that it permits men to be irresponsible by rewarding "socially disvalued qualities by giving their bearers the same medical care opportunities as those received by the bearers of socially valued qualities."[8] In such a case, it is not clear that social value is the true issue; perhaps the term "moral value" should be used instead of social value.

These considerations lead to the following problem: Are there instances when

[6]James F. Childress, "Who Shall Live When Not All Can Live?" *Soundings*, 53, No. 4, Winter 1970, p. 349.

[7]Edmund Cahn, *The Moral Decision: Right and Wrong in the Light of American Law*, Bloomington, Indiana University Press, 1955, p. 71.

[8]Leo Shatin, "Medical Care and the Social Worth of a Man," *American Journal of Orthopsychiatry*, 36, 1967, p. 119.

individuals "forfeit" their right to equal access to a medical scarcity? Intuition and a glance at the hypothetical case above might prompt one to reply yes to this question; yet it becomes extremely dangerous to make exceptions to the rule of randomness. The difficulty lies in the attempt to determine what is to be considered "morally disvalued." It may seem easy to exclude murderers from the lottery pool, but where does one stop? Should rapists and burglars also be excluded? What about embezzlers or juvenile delinquents? Making these types of exceptions to inclusion in the lottery would reduce randomness to just one more form of social control; convicted criminals pay for their crimes through their prison sentences and should not be further penalized by exclusion from this lottery. Therefore, until concepts of social or moral merit can be clearly and universally defined, the most just method of allocation is that of randomness, without exceptions on the basis of moral standards. "In allocating sparse medical resources among equally needy persons, an extension of God's indiscriminate care into human affairs requires random selection and forbids god-like judgments that one man is worth more than another."[9]

The only foreseeable exception to the rule of randomness is the case where supplying an individual with the needed treatment will work to *everyone's* benefit. Suppose there has been a flood in a small, isolated, rural district, and the only available physician has been seriously injured and is in need of a scarce medical resource. It is to everyone's benefit, sick and well alike, to have the doctor able to lend his assistance and knowledge to the situation. In this case, everyone will profit from the nonrandom allocation of scarce treatment to the physician, and it can be justified on these grounds.

Using randomness as a method of distributing available medical scarcities gives an effective and justifiable way of coping with a problem for which there is no obvious solution. Yet this issue is too critical not to make some attempt to close the gap between what would be an egalitarian health care system accessible to individuals everywhere and the realistic perspective of what is both feasible and reasonable to expect to occur in the near future. This gap may never be completely closed, but it is essential to recognize that it can be considerably narrowed. Attempts to limit or decrease the number of medical scarcities become problematic in two often-conflicting respects. The first problem is an economic one, touching the queston of the way in which scarce medical services may be most efficiently distributed in the face of economic realities; the second concerns the ethics of the problem, or how medical services can be arranged to provide all men with as equal a health care system as is possible.

The dilemma presented by these two views is that of trying to fashion a quality health care system for all, when economics inherently limits the range of options and causes choices to be made among them. "The current slogan 'quality health care at a cost we can afford' epitomizes the difficulty of implementing this economic goal... any debate on the issues of resource allocation amply illustrates

[9]Paul Ramsey, *The Patient as a Person,* New Haven, Connecticut, Yale University Press, 1970, p. 450.

the sharply divergent meanings of the key words of the slogan—quality, care and affordability."[10]

Quality health care becomes increasingly more expensive, and thus more limited, because along with rises in standards of living and medical care comes a heightened perspective of what constitutes healthiness. People in countries with advanced medical care systems no longer want to live with the illnesses and disabilities their forefathers accepted as a normal part of their existence (high infant mortality rates, loss of eyesight and hearing, etc.). In the meantime, those unfortunates who live in underdeveloped areas with little or no medical care struggle to survive.

One result of the progress of medicine and its increasing technological triumphs has been the successively increasing complexity and expense encountered with each new development. Thus the problem of medical scarcity is amplified with each new highly technological complen treatment discovered or developed. Critics blame most modern-day scarcities on the "principle on which medical practice has been based, that one should do everything possible for the individual patient…few constraints are placed upon the introduction of new medical practices. I believe this is a luxury we can no longer afford."[11] As the expectations of the public become increasingly elevated by the impressive innovations of medical science, while economic resources are continually depleted by these same innovations, a cost–benefit analysis of this type of high-technology medicine is necessary to determine if the public is receiving the greatest return on its enormous investment in health care.

An alternative to this difficulty may be a shift from a "medical services" approach twoard the promotion of health through a variety of methods, of which medical care is only one possibility. Improvements in sanitation, diet, education, and safety features, for example, have often proved to be at least as effective as the curative medical techniques so heavily emphasized today. Although undoubtedly less dramatic than crisis-directed medicine, these measures are often much cheaper and of more benefit to the public at large. The most pressing need today in dealing effectively with the health care crisis is that of informing the public of the choices it now faces. People must be made aware that the development of new life-saving or life-enhancing innovations does not necessarily make such innovations universally accessible. Only by careful evaluation of all options available in the promotion of quality health care can responsible decisions be made.

---

[10]Edmund D. Pellegrino, "Medical Morality and Medical Economics," *Hasting Center Report,* August 1978, p. 8.

[11]Howard H. Hiatt, "Protecting the Medical Commons: Who is Responsible?" *New England Journal of Medicine,* **193**, July 31, 1975, p. 235.

16. The Regulation of Recombinant DNA Research: A Critique

Stephen L. Reiss

This paper deals with the reactions of scientists and governing bodies to the development of particular experimental techniques in microbiology that have gone under the general title of recombinant DNA technology.[1] More popularly known as "gene-splicing," this relatively new tool of basic research has prompted intense debate among the members of the scientific community and in political forums in almost every industrialized nation.

To a great extent, technologies have been independent variables in our society. Although they are the product of our labor, we, in turn, are shaped by them. Our technologies determine our ethics, our esthetics, and our physical well-being. The classic example is the automobile, which has brought us landscapes of interstate highways, the economics of extensive credit, and the social artifact of the drive-in.

In recent years, however, there have been inklings of what has been derogatorily termed an anti-intellectual backlash against technology. It is evident in the fights against nuclear power, the supersonic transport, and recombinant DNA. Unlike the unguardedly optimistic outlook of the turn of the century, there is now a tendency to look at technology and science with a jaundiced eye, to ask where it will lead and whether new technologies are really needed. We have lost faith in miracle cures and technological fixes.

The flowering of the biological sciences in the last twenty years has unavoidly brought them up against this spreading dissatisfaction. This is illustrated by the public alarm over fetal research, over controversial research on the genetic basis of intelligence, over experimental animal studies, and over the costs and benefits of the "Green Revolution." The motivations, the methods,and the aims of the

---

[1]This paper is adapted from my undergraduate senior thesis for the Woodrow Wilson School of Public and International Affairs, Princeton University.

developers of the new technology are under increasingly hostile examination, both by the public at large and by dissident members of the scientific and technical communities.

This paper will use the controversy over the creation of *in vitro* recombinant DNA molecules as a case study to examine this conflict critically and as a template by which to observe how our society grapples with sophisticated technological developments.

The debate over recombinant DNA technology is enmeshed in a whole spectrum of issues that cannot be adequately treated in this format, ranging from the desirability of public participation to the validity of risk–benefit analysis for newly developing fields. These issues can be loosely classified into structural concerns growing out of the nature of the technology leading to such questions as whether recombination of genes across natural barriers could wreak evolutionary havoc and procedural ones, having to do with how the structural issues are addressed (leading to such questions as what the role of the expert should be in forming public policy).

I shall be concerned primarily with the procedural issues. These may not be more important than the first kind on an absolute scale, but they are capable of being applied to other technologies and are therefore more interesting questions in the context of this analysis. Owing to space constraints, I shall not be able to give a history of the process.[2] Instead I shall present a menu of the major procedural problems associated with the decision-making on recombinant DNA. This will be followed by a more abstract treatment of the problems of public control over innovative technologies, and finally, a suggested framework for an appropriate social decision-making process for questions associated with innovative technologies.

Regulatory Problems

At this Council of Asilomar there congregated the molecular bishops and church fathers from all over the world, in order to condemn the heresies of which they themselves had been the first and the principal perpetrators. This was probably the first time in history that the incendiaries formed their own fire brigade. The edict published in due course, which lists the various forbidden items, reads like a combined curriculum vitae of the conveners of the conference.[3]

ERWIN CHARGAFF

The regulatory saga of recombinant DNA can be roughly divided into two periods—one of self-regulation by scientists, and another characterized by governmental involvement. In surveying the experience with self-regulation, a

[2]Nicholas Wade, *The Ultimate Experiment: Man-Made Evolution,* New York, Walker, 1977; Michael Rogers, *Biohazard,* New York, Knopf, 1977; John Lear, *Recombinant DNA: The Untold Story,* New York, Crown, 1978.

[3]Erwin Chargaff, "Profitable Wonders," *The Sciences,* August/September 1975, p. 26.

number of points stand out as significant. First, the decision-makers all had roughly the same backgrounds. At the 1975 Asilomar conference, the uniformity of the participants was embarassingly obvious, for aside from Andrew Lewis, who issued a subdued demurral, there was no one attending who was opposed to the research. Jonathan Beckwith, one of the early critics, was invited at the last minute, but he declined because he felt that his invitation was an instance of tokenism. Beckwith thought that the conference should have included specialists in public health, epidemiologists, labor unionists, OSHA officials, and social historians, and that one critical scientist could not compensate for their absence. The organizers hastily transferred his invitation to another well-known dissenter, Jonathan King. He had prior commitments and suggested a California member of the Science for the People group. Paul Berg and David Baltimore, two of the organizers, vetoed the suggestion because the Californian was only a post-doctoral student.

The resulting uniformity of background among the participants, which is alleged to have encouraged a uniformity of outlook, was equally prevalent at the local level until the revised National Institutes of Health guidelines governing the research (issued in December 1978) required public participation. A study of the composition of the Institutional Biohazard Committees found them to be totally nonrepresentative of the communities whose safety they were protecting. The people on these committees, which ruled on the safety of experimental protocols, were almost exclusively white males with graduate degrees. Within that range, 60% of the members were scientists in disciplines utilizing recombinant techniques, and an additional 13% were scientists in peripherally related fields.[4]

Although the composition of the committees does not necessarily affect the quality of their work, it does affect whether or not the work of the group is perceived as legitimate. It is this doubt about the legitimacy of the decisions made, regardless of whether they are right or wrong, that underlies the attacks on professional self-regulation as a strategy for making public policy without public participation. Scientists may be best qualified to calculate risks and to predict benefits, the argument runs, but they have no such special qualifications when it comes to deciding whether the risks should be assumed. As the biologist Ruth Hubbard has put it,

Obviously one wants to balance risks against benefits, but one has to also balance to whom the risks accrue and to whom the benefits accrue, and as long as these decisions are made by the people who are most likely to reap the immediate benefits and the people who are equally likely to reap the risks are not involved in the decision-making process, then the process itself is flawed.[5]

[4] Nancy Pfund, "Testimony for HEW/NIH Public Hearing on Revised Guidelines for Recombinant DNA Research," September 15, 1978.

[5] United States Congress, House Committee on Interstate and Foreign Commerce, *Recombinant DNA Research Act of 1977. Hearings before the Subcommittee on Health and the Environment,* 95th Congress, First Session, 1977, p. 93.

Senator Edward Kennedy convened his Subcommittee on Health for hearings on the Asilomar Conference only a month after it took place. The hearings took on a definite adversarial tone as Kennedy questioned the propriety of such strictly scientific decision-making processes. In a speech that May at the Harvard School of Public Health, he said:

It was commendable that scientists attempted to think through the social consequences of their work. It was commendable, but it was inadequate. It was inadequate because scientists alone decided to impose the moratorium and scientists alone decided to lift it. Yet the factors under consideration extend far beyond their technical competence. In fact, they were making public policy. And they were making it in private.[6]

Richard Roblin, one of the organizers of Asilomar, argued that the criticisms were valid only if one viewed the conference as settling the questions once and for all, rather than as the beginning of an ongoing process:

Given the burgeoning use of these techniques in research, I believe that the conference had no alternative but to devise an interim policy that would guide the increasing number of researchers entering this area until more comprehensive guidelines are established by appropriate national bodies.[7]

The lack of public input also opened the participants to charges of being self-serving. A widely noted example of this was the decision to allow experiments with DNA from cold-blooded animals under more permissive conditions than those with DNA from warm-blooded animals, although the rationale for the difference was entirely speculative. Critics noted that David Hogness, a member of the NAS committee that called the conference, had already begun working with fruit-fly DNA, and a different classification would have meant extensive delays in his work. Although no explicit accusations were made, it did not escape critics of self-regulation that, in the words of a professor of medicine at Yale, "while the long-term benefits of recombinant DNA research to mankind may be arguable, the short-term benefits to scientific careers are not."[8]

The progression of events and decisions was admittedly ad hoc, and the scientists cannot be blamed for the lack of a proper forum. Still, as Stephen Toulmin has noted,

[6]Barbara Culliton, "Kennedy: Pushing for More Public Input in Research," *Science,* June 20, 1975, p. 1188.

[7]Richard Roblin, "Reflections on Issues Posed by Recombinant DNA Molecule Technology," in *Ethical and Scientific Issues Posed by Human Uses of Molecular Genetics,* Marc Lappe and Robert Morison eds., New York, The New York Academy of Sciences, 1976, p. 62.

[8]P.T. Magee, letter to Adlai Stevenson, November 16, 1977, in United States Congress, Senate Committee on Commerce, Science and Transportation, *Regulation of Recombinant DNA Research. Hearing before the Subcommittee of Science, Technology and Space,* 95th Congress, First Session, 1977, p. 382.

the fact remains that what they did was a piece of institutional improvisation, and in consequence, whether or not the safeguards they arrived at really did proper justice to the legitimate interests of the wider public (and they very possibly did), that justice was not and could not have been seen to be done.[9]

The conflict between promotion of a technology and its control is inherent in the process of self-regulation. Regulators and promoters are expected to have different perspectives arising from different motivations. Some scientists have argued that the recognition of this conflict of interest has led to overregulation. In this case, they contend, precisely because there was no participation by laymen, the scientists bent over backward to avoid any appearance of being self-serving and therefore produced an unwieldy, overly detailed, overly restrictive system.

The specter of public regulation probably did influence the behavior of the scientists. This bogeyman was used within the scientific community in an attempt to quiet dissident scientists and led to resentment (and even possible retaliation) against those who tried to bring the issue to the public. The "outside regulation as big stick" attitude was articulated more delicately by an NIH official:

I am fearful that if dissident scientists, in public forum, engage in irresponsible attacks on the rules we have so painstakingly tried to write, they will stimulate political intervention that will block free inquiry...[10]

An editorial in *Science* warned DNA researchers to be especially careful while the debate was on, for any mistake would increase the likelihood of regulation. "A scientist who furnished the pretext for restrictive legislation could count on the ill will of many of those he or she most wants to impress."[11] The threat of unpopularity had already been manifested against opponents, according to historian Rae Goodell of MIT:

The scientific community seems to have a tremendous sense of living in a hostile environment, of being a little enclave of rationality in a hostile world.
There are rigid rules about what a scientist should and should not do. It's fine to be critical in private but not in public. If you want to express social responsibility, it is fine to do so in Washington but not on the street.[12]

Erwin Chargaff, one of the early pioneers in DNA research, said he has not been even more outspoken in his opposition to recombinant work because of the

[9] Stephen Toulmin, in *Research with Recombinant DNA, An Academy Forum,* Washington, D.C., National Academy of Sciences, 1977, p. 106.

[10] Leon Jacobs, "The Role of the National Institutes of Health in Rulemaking," in *Recombinant Molecules: Impact on Science and Society,* R.F. Beers, Jr., and E.G. Bassett, eds., New York, Raven Press, 1977, p. 453.

[11] Philip H. Abelson, "Recombinant DNA Legislation," *Science,* January 13, 1978, p. 135.

[12] Nicholas Wade, "Gene-Splicing: Critics of Research Get More Brickbats than Bouquets," *Science,* February 4, 1977, p. 469.

pressure of scientific majority opinion. When a scientist of his stature admits to feeling cowed, it is not surprising that there should be persistent rumors that scientists who opposed the establishment view of recombinant work would have difficulty getting grants approved or being promoted. Although it is hard to prove that such things actually occur, the almost total absence of young, untenured scientists from the debate on almost any level shows that at least they believe it is a possibility (or that they all favor the stance taken by their senior colleagues). For example, even though there were members of Science for the People at the University of Michigan, none of them would participate publicly in the debate there, and opponents had to be flown in from Cambridge to take part in the forums.

On the positive side, it is important to realize that the scientists were willing to stop and question their work without any outside prompting and without any proved hazard. They did so even though there were large costs involved. These costs ranged from being criticized to neglecting their work so that they could attend conferences debating topics like safety, which were only incidental to their main research interests. They did not, at least in the beginning, anticipate to what extent the safety of their research would become a public issue and the large costs that would follow (for example, lobbying, and testifying here, there and every-where).

In view of all this self-initiated activity, one might argue that self-regulation is a workable system, and if the risks prove to be greater than is now believed, scientists will again act responsibly to protect themselves and the public. The thundering of James Watson in the *New Republic* might give rise to doubts, though. He argues that the Berg committee, which called for a moratorium on some recombinant experiments in 1975, reached its decision in large part because of the national mood of Watergate and the desire to "come clean" and that, in retrospect, it was not a wise decision.

Unfortunately, there is support for this more pessimistic view of scientific responsibility. As David L. Bazelon, a distinguished federal judge, has observed, some scientists

say that they would consider not disclosing risks which, in the scientist's view, are insignificant, but which might alarm the public if taken out of context. This problem is not mere speculation. Consider the recently released tapes of the Nuclear Regulatory Commission's deliberation over the accident at Three Mile Island. They illustrate dramatically how concern for minimizing public reaction can overwhelm scientific candor.[13]

Two other points deserve to be mentioned. The first, although possibly obvious, is that where compliance is difficult to determine, regulation depends for its success on the support of those being regulated. A system would not be

[13]David L. Bazelon, "How Much Is Too Much, and Who Decides?" *Legal Times of Washington,* April 30, 1979, p. 32.

workable if researchers did not believe it was necessary or realistic. The scientists were limited in what they could do by what their colleagues would accept.

Second, the whole process moved very quickly and as a consequence gave short shrift to ethical questions in order to concentrate on issues of safety. The amount of consideration given at any point in the self-regulatory process to the possibilities of biological warfare, human genetic engineering, or the value of this field of research as compared with others was small and superficial, as a critical biologist, Robert Sinsheimer, points out:

Why was it, for instance, that the first fully public discussion of recombinant DNA was already focussed on draft guidelines? A decision even to prepare and discuss guidelines seemed to assume that the basic ethical, political, and social questions about recombinant DNA research had all been raised and fully discussed. But they had not been.... It should hardly be astounding that some felt vested interests were pulling the strings and that, once again, the scientific establishment was slipping one over on the public.[14]

The period of government regulation shared many of the problems character-istics of self-regulation: the lack of a proper forum, the speed of consideration, and the relative inattention paid to nontechnical questions. It did not suffer from a problem of illegitimacy in the eyes of the public (although a number of scientists saw it in that fashion).

Characteristics of Technological Change

We are becoming creators, makers of new forms of life, creations that we cannot undo, that will live on long after us, that will evolve according to their own destiny. What are the responsibilities of creators for our creations and for all the living world into which we bring our inventions?[15]

ROBERT SINSHEIMER

The problems associated with the development of recombinant DNA are not unique to this technology. Indeed, they are inherent in technological innovation. Issues like balancing potential risks and benefits and free inquiry versus for-bidden knowledge will undoubtedly crop up again and again. Although they will occur in different contexts and with different degrees of emphasis, those issues derive from a common technological–social archetype of change. This section will examine some of the characteristics of this archetype and how they shape the technical–social issues, and then make some suggestions as to what the bench-marks are of an appropriate legal–regulatory response on the part of the public.

One of the most striking aspects of technological innovation is that we really do not know where it will lead, how long it will take to get there, or what mitigating

[14]Daniel Callahan, in *Research with Recombinant DNA,* p. 33.

[15]Robert Sinsheimer, in *Research with Recombinant DNA,* p. 79.

factors will come to bear along the way. As the historian Daniel J. Boorstin has written:

No inventor can know an invention's precise period of gestation or the time required to bring an invention to maturity. Nor can an inventor even begin to imagine the outcome of his success. Eli Whitney surely was not trying to make a civil war. Cyrus McCormick was not intent on depopulating our farms.[16]

Unlike political change, which, although varied, occurs within recognizable frameworks, technological change seems to create a new world with new rules every time. Although recombinant techniques were developed by microbiologists to solve their own particular problems, there is no way to restrict them to this sphere of application. "They are available to entrepreneurs, to flower fanciers, to the military, to subversives."[17] The possible applications multiply geometrically, with numerous first- and second-order effects, and the shape of the resulting society is extremely difficult to forecast.

This inability to predict outcomes severely hampers any attempts at technology assessment, particularly when the technology has only recently come into being. Such difficulties have been obvious throughout the debate over recombinant DNA in discussions over balancing risks and benefits. Freeman Dyson, a member of the Princeton Community Biohazards Committee, dismisses out of hand the utility of risk–benefit analysis. "Any attempt to measure the risks or the benefits analytically is an attempt to predict the history of the next hundred years, including the scientific discoveries that we have not yet made."[18] Discussing the 1976 NIH draft environmental impact statement on the guidelines before the House Subcommittee on Science, Research and Technology, NIH Director Donald Fredrickson admitted it had only limited value (and not because it was a shoddy piece of work). "I think that we really cannot make a useful technology assessment beyond imagining various scenarios."[19]

Despite an uncertain future, one cannot responsibly advocate taking no action at all, since to do so is to abdicate any possibility of beneficial control. "We are forced to make some guess as to the comparative consequences of banning or permitting experimentation, but in these circumstances that guess is necessarily in some degree at least an act of faith."[20]

A new technology is irreversible. It cannot be uninvented, although it may

[16]Daniel J. Boorstin, *The Republic of Technology,* New York, Harper & Row, 1978, p. 93.

[17]Robert Sinsheimer, "An Evolutionary Perspective for Genetic Engineering," *New Scientist,* January 20, 1977, p. 151.

[18]United States Congress, House Committee on Science and Technology, *Science Policy Implications of DNA Recombinant Molecule Research. Hearings before the Subcommittee on Science, Research and Technology,* 95th Congress, First Session, 1977, p. 839.

[19]*Ibid.,* p. 287.

[20]*Report of the University Committee To Recommend Policy for the Molecular Genetics and Oncology Program (Committee B),* Ann Arbor, Michigan, University of Michigan, March 1976, p. 27.

become obsolete. "There is no technological counterpart for the political restoration or the counter-revolution."[21] Recombinant DNA is here to stay. A decision by any body to completely ban the work would be impossible to enforce. It might have limited success for a time, but there has yet to be a type of technology that has been successfully repressed. Even in the area of nuclear weapons, where the potential evil effects could not be more evident, research is actively encouraged.

A new technology acquires an internal momentum as it develops from the first exploratory research to eventual industrial applications. The further along it is in the process, the harder it becomes to control. The momentum in recombinant DNA research is striking. Before 1974—only five years ago—it was an esoteric biochemical research tool used by only a few academic laboratories in California. It has now spread across the world and is one of the fastest growing areas of biological research. Practical applications, such as the production of interferon and insulin, are already under way. Within twenty years, the results of recombinant techniques will begin reshaping not only our medical care but to some extent our economic, political and social systems.

If control over a technology is to be significant and effective, it must be exerted early in the developmental process. It can perhaps be most effective, although necessarily tentative, at the stage of basic scientific research. Thus, while microbiologists were arguing for freedom of inquiry and attempting to distinguish between research and its applications, potential regulators recognized that their effectiveness depended on interfering with that freedom by explicitly and pessimistically considering the potential hazards. Control also becomes progressively more costly, the later it is applied, as Arthur Schwartz has noted:

The cost of slowing technological development and building considerably more prudence into the basic designs, as well as the cost of an intense and widespread research into the hazards of the technology will be more than balanced by the benefits of less damage to people, our environment, by the reduced need to combat these effects and by the reduction of the costs of redesigning, which is always more expensive than designing correctly in the first place, various components of the technology.[22]

Technological momentum coupled with our general respect for innovation has placed the burden of proof on the opponents of any new technology. Over and over, various groups have accepted the premise that, since there was no positive proof of danger, the work should proceed.

New technologies create needs and problems. While social and economic needs contribute to the acceptance and dispersion of new technologies, they are not primarily responsible for calling them into being. As Boorstin points out,

We will be misled if we think that technology will be directed primarily to satisfying "demands" or "needs" or to solving recognized "problems." There was no "demand" for

[21]Boorstin, *op. cit.,* p. 30.

[22]Arthur Schwartz, in *Regulation of Recombinant DNA Research,* p. 307.

the telephone, the automobile, radio or television. It is no accident that our nation—the most advanced in technology—is also the most advanced in advertising.[23]

Already we have found needs to be fulfilled by recombinant DNA techniques: a need for nitrogen-fixing plants, a need for mass production of hormones like insulin, a need for genetic cures.

Although these accomplishments are still outside the realm of the possible, problems attendant upon them can easily be foreseen. For example, creating nitrogen-fixing plants could severely damage the chemical/fertilizer industry while giving a boost to pharmaceutical firms. What will it do to food prices and to farmers' incomes? How will our diet change? (Will nitrogen-fixing corn and "normal" corn taste differently?)" Will a stigma come to be attached to products that have been genetically treated? Who will know about it? What institutions will have records of it? The point is clear—these "benefits" do not come unadulterated or without social implications. Whether any of these possible social effects are desirable is not a question any public body has confronted in a rigorous manner.

It is important to realize that controversies over many technical issues involve questions of value and are therefore political issues. Although clothed in esoteric jargon and apparently concerned only with the interpretation of "facts," science and technology are not neutral processes. The products of science are used by society, and their creation is a matter of political, economic, and social concern, and not simply the end result of a disinterested search for truth. As Primack and von Hippel point out,

The citizen must realize that important political issues are almost always present when a debate of an apparently technical nature bursts into the public arena. Although it may often appear that the partisans involved are asking the public to make determinations in areas where even the experts disagree, the experts are often talking past one another; in reality, the debate revolves around unspoken political questions.[24]

The unspoken political questions in the recombinant DNA debate concerned issues like whether scarce grant money should be spent on medical research or preventive medicine, or whether preventing an unknown but probably small amount of risk was worth the slowing of scientific research.

As was mentioned earlier, recombinant techniques were developed to solve a certain set of scientific problems, but the importance of gene expression in the monkey virus SV40 was a product of the Nixon administration's "war on cancer" and the immense amount of money flowing into biological research. The use of these techniques as opposed to controls on environmental carcinogens was determined as much by perceived public resistance to such efforts (for example,

[23] Boorstin, *op. cit.*, p.8-9.

[24] Joel Primack and Frank von Hippel, *Advice and Dissent, Scientist in the Political Arena*, New York, Basic Books, 1974, p. 5.

the FDA ban on saccharin) as by their inherent intellectual appeal, as Marc Lappe points out:

What scientists choose to study is conditioned as much by values as by heuristic appeal or scientific merit. *When* scientists select an area of research, their interests may be piqued as much by political considerations as by the timeliness of discovery. And, *How* scientists go about doing science involves political judgements as well as abstract, rule-following procedures.[25]

Once the political nature of the scientific process is made obvious, the legitimacy of handling technological problems through the political process gains ground. What stands in the way is the supposedly esoteric nature of scientific/technological concerns and the need to master a "foreign language" to discuss them. Yet problems in other equally esoteric fields such as economics and defense policy are treated via the political process. The problem of jargon is dealt with by translating technical verbiage into everyday language.

An Appropriate Public Response

It is the political system through which man aggregates and allocates the resources needed to develop and implement large-scale technology. Further, it is through the political system that the cost and benefits associated with technologies are distributed throughout society. Thus, if we are to alter significantly the present patterns of technology–society interaction, it is likely that the changes involved will begin with the political system.[26]

ALBERT TEICH

A frequently repeated complaint during the controversy over gene-splicing was the lack of an efficacious forum for discussion and debate of technical issues. The most complete discussions were at the congressional level, but not all issues can command congressional attention for a sustained period of time or deserve to do so. Mechanisms for local involvement in technical issues are conspicuous by their absence. The citizen review boards in Cambridge and Princeton, for example, were ad hoc and experimental. The city with an established procedure, San Diego, utilized an expert process with little public input.

If the technology had not presented what were perceived as obvious and immediate threats to public health, almost every governing body would have been at a loss as to how to proceed. Without that "peg" they would have had to struggle for a pretext to regulate, because the present system does not allow for prior restraint where there is no responsible basis to suspect immediate harmful effects. As a legal study points out,

[25]Marc Lappe, in *Science Policy Implications,* p. 1017.

[26]Albert H. Teich, ed., *Technology and Man's Future,* New York, St. Martin's Press, 1972, p. 195.

The common law reflects traditional laissez-faire attitudes which deter public input into enterprise decisionmaking; members of the general public may challenge a given technology, if at all, only if it directly affects them in some adverse way.[27]

If the pace and direction of new technologies are to be controlled, we cannot depend on their neatly fitting into the requirements of the present legal/regulatory system. The creation of proper mechanisms will take time, however, and once a technical–political issue has been identified the public must deal with it in some fashion. An appropriate legal/regulatory response will have four interdependent characteristics: (1) participation/education; (2) flexibility; (3) detail; and (4) comprehensiveness.

Participation/Education

In order to have productive public participation, a number of questions must be answered, including: What is adequate participation? In what form should that participation should take place? And who should participate? The quality of involvement is dependent to a large degree on the education of the participants. "Education" is not intended to refer to academic degrees or levels of intelligence but to familiarity with the basics of the technological issue under examination.

Much of the early debate over recombinant DNA is characterized by a lack of adequate public participation via elected representatives or public members on deliberative/advisory bodies. Determining in advance what is adequate participation so that it can be built into a policy mechanism depends on what aims public involvement is designed to fulfill, as Richard Trumbull observes:

If we are primarily providing an opportunity "to be heard" or "to feel as if they are participating," that is one thing. If we are merely interested in how the proposed program policy affects or appears to them, that is another.[28]

Ideally, the degree of public participation is sufficient to ensure that the decision is representative of the values of those affected by it. Such a result is not necessarily dependent on the quality of the public input. Success might be evidenced by a lack of opposition among an informed public. "The degree of participation in the issue," a member of a review board has said, "should depend on the political context of the technology itself. If the public wants entry, that should be sufficient reason to democratize the issue."[29]

Who is part of the relevant public and what format should be used to arrive at its decisions are further complicating problems. Competence is one suggested criterion for determining who should participate in the decision-making process. People with scientific and technical competence would be included, along with

---

[27]"Recombinant DNA and Technology Assessment," *Georgia Law Review,* Summer 1977, p. 798.

[28]Richard Trumball, in *Science Policy Implications,* p. 685.

[29]Sheldon Krimsky, Interview, Medford, Massachusetts, February 1–2, 1979.

those with "the ability to sense and express the fears or hopes and the confusions of lay persons, all of which are held to be data as cogent for decision making as technical facts."[30] Another criterion could be the nature of the technology—that is, whether it affects everyone in a society equally (should the United States build a neutron bomb?) or differentially (should the Concord be allowed to land at New York airports?). This distinction is not always an easy one. With recombinant DNA, many scientists argued that any dangers would be national in scope, while others contended that those closest to the laboratories were obviously more endangered.

The proper format for participation likewise depends on the issue. New England town meetings and public referenda are useful mainly when the questions can be narrowly defined. But there are hazards in reaching such decisions through the normal mechanisms of public participation. As James Carroll points out, this process

can degenerate into forums for the exercise of obstructionist veto-power techniques and paralyze public action. It can generate an overload of demands that agencies are not equipped to handle. It can be used as an instrument by an aggressive minority to capture decision-making processes and to impose minority views on a larger community. Finally it can lead to the dominance of technological know-nothings over the judgements of qualified individuals....[31]

An important problem with "participatory technology" is the difficulty it creates in arriving at national or even state policy with respect to technologies that pay no heed to man's political boundaries. Participatory, constitutive methods require a reduced scale of operation in order to be meaningful for the participants. Formulating policy for large areas involves putting distance between the decision-making process and the public.

It must be recognized that in many (perhaps most) cases one cannot speak of the "public" in an all-inclusive sense unless an issue is of overwhelmingly immediate consequence. In fact, it is not uncommon (for any number of reasons) for the public to take little or no interest in the issues associated with a new technology. Thus, one is faced with the task of building a mechanism aimed either at increasing the participation of a significant sample of the public or at elevating the quality of somewhat limited participation. These are not mutually exclusive, and aspects of both can be worked into the same process.

The committee format seems to be one of the best ways to serve the aim of participation. In contrast to an individual's making a decision on the basis of input from various sources, it allows for adequate participation in the actual decision, thereby increasing the chances the decision will be accepted as being correct. It is also applicable at almost every level, is easily modified to fulfill various aims, and is not dependent on widespread public interest.

[30]Donald Michael, in *Science Policy Implications,* p. 816.

[31]James Carroll, "Participatory Technology," *Science,* February 19, 1971, p. 652.

Productive public involvement is dependent on a process of education. It does not make sense to argue that meaningful, legitimate decisions can be made about technical issues without some attempt to learn about the technology in question. This does not mean that degrees and college courses in biochemistry are necessary to debate recombinant DNA any more than that a degree in physics is needed to talk about the dangers of nuclear power plants. But an understanding of the basic points of the technology allows the participants to recognize where technical questions end and political ones begin.

The desirability of education puts a time constraint on the breadth of public participation. Not everyone will be willing to devote the time needed to study the issue. But it does not impose any constraint of intelligence or training. The broad representation on the review boards in Cambridge and Princeton is particularly strong evidence that anyone can understand the issues and ask difficult, pointed questions.

Just as adequate public participation does not guarantee peaceable, workable solutions to technical policy problems, neither does education. "There is...a tendency to think that education only works if it resolves conflicts and people end up the educational process by agreeing."[32] But the issues involve political and moral choices, and individual decisions will reflect a wide range of opinion and belief. As George Kieffer has observed, we have yet to "solve" a single biomedical problem in terms of its moral implications.

Flexibility

This is an important characteristic when dealing with new technologies, because everything about them is undergoing constant change. If the control process is to retain any validity, it must keep up with these changes. Several kinds of flexibility are called for: the creation of bodies that are able to act as well as study, a willingness to transcend traditional intellectual and political boundaries, and the creation of regulation that is amenable to ready change.

The easiest path for an official to take when confronted with a controversial issue is to appoint a committee to study it and report back with advice. These committees, if properly constituted and organized, are excellent vehicles for promoting participation, education, and dialogue.

If the committee-study approach is to prove effective in the control of new technologies and not simply lead to an endless proliferation of unread reports, it must be not only a thinking group but one that can act or ensure that action will be taken. If the issues concerned are to be addressed responsibly, the group must be either highly visible, as was the Cambridge Review Board, or capable to some degree of enacting its recommendations, as legislative committees are.

Because technologies do not recognize divisions among intellectual fields or political boundaries, these bodies should be not only interdisciplinary or

[32]Alan McGowan, *Science Policy Implications,* p. 725.

interagency but capable of redrawing boundaries. For example, effective means of redefining the distinction between academic and industrial research are needed, something NIH has not succeeded in doing.

Simply as a matter of efficiency and efficacy, these groups should be designed to be willing and able to put themselves out of business. Technologies go through a developmental process, and it would be foolish to assume that the same format of control would be applicable at every stage. The group should be flexible enough so that it can intentionally change its purpose and its form as its perceptions of the various issues change. At some point it may be necessary to create whole bureaucracies, like FAA or the FCC. In other cases, the committee might have done its job so well or the technology developed in such a fashion that no further formal oversight is needed. A criterion of flexibility does not rule out legislation or any other form of rule-making. Regulation can be as flexible as the English language. It does mean that statutes with sunset clauses and provisions for simple revisions are preferable to laws that are put on the books until repealed and involve protracted amendment proceedings.

Detail

Any regulatory response should be sufficiently detailed so that it applies only to the technology in question in the specific circumstances where it is judged to need control. This is the best way to balance the concern that scientific inquiry might be unduly restricted with a healthy caution about where such inquiry might lead. The objections to the use of Section 361 of the Public Health Service Act to regulate recombinant research and the veto of a 1977 New York bill were both based on the perception that loose and generalized language in a statute would lead to excessive contraints on research and development that were not justified by the rationale for the original regulation.

Even if one did not need details to bring the issues to light in the first place, the need for them requires the use of experts. If policy decisions are to make any sense, their presence and informed advice are an absolute necessity. It will at times be necessary to restrain the experts in their pursuit of detail so as to avoid an exercise in competitive numerology. While various illustrious microbiologists debated with NIH over the proper values to assign to the likelihood of bacterial transformation inside the gut, the public was left with the impression of a decision that was "elitist, opaque and incomprehensible" and concluded "that their own democratic contribution had been bypassed."[33]

Evaluating the work of the Cambridge group, member Sheldon Krimsky wrote:

I don't believe it should be the function of the Board to review "hard" or "soft" scientific data. The Board should rather be looking at the data as interpreted by experts... The

[33] Amory Lovins, in *The Economic and Social Costs of Coal and Nuclear Electric Generation,* Stephen Barrager, Bruce Judd, and Warner North, eds., Washington, D.C., National Science Foundation, 1976, p. 107.

function of this review should be, therefore, to try to understand where the locus of disagreement lies, whether on an issue of scientific merit or on a value laden issue...[34]

The problems come in when one is attempting to use technical expertise while not simultaneously abdicating the ability to make decisions. Krimsky advises against having scientists and nonscientists on the same committee. "There is a strong tendency for the scientists to play a dominant and elitist role that intimidates lay persons."[35] Whether or not this actually takes place, a mechanism that inhibits nonscientists from making contributions is self-defeating. What is needed is a mechanism that fosters dialogue and forces people to confront each other's ideas. An example would be the use of an adversary arrangement where scientists on opposing sides are able to question each other. The creation of a dialogue between experts and nonexperts contributes substantially to the educational process in both groups and thereby improves the quality of participation.

Comprehensiveness

Three forms of comprehensiveness are necessary for appropriate legal/regulatory action: rules that apply to everyone doing the same thing; analysis and consideration of all the structural issues associated with a new technology; and equally rigorous examinations for all the unknowns in an area to the extent that this is feasible.

One of the great faults with the NIH guidelines and the whole self-regulatory process was that they applied only to those who chose to comply. Industry, although it claims to abide by the rules, is exempt from the sanctions of the enforcement system, as is anyone else who chooses to perform the research without government funding.

While scientists plunged headlong into recombinant DNA research, the overwhelming majority of their projects were aimed at realizing the potential benefits instead of evaluating the potential risks. The need for experiments to assess the level of risk and to determine whether the research was safe and under what conditions was obvious at the outset. The failure of the scientific community to begin assessment experiments of all kinds immediately, coupled with its reliance on the assurances of various committees, fostered the belief among nonscientists that the degree of potential risk was much greater than was being admitted.

Technical debates over potential risks versus potential benefits are inherently imbalanced. It is human nature to want the benefits intensely while regarding the risks as speculative. Therefore, methods must be found to provide a constituency for the risks to assure them equal consideration. One way to promote this kind of

---

[34]Sheldon Krimsky, "A Citizen Court in the Recombinant DNA Debate," *Bulletin of the Atomic Scientists,* October 1978, p. 38.

[35]*Ibid.,* p. 42.

analysis would be to require agencies like the NIH to allocate a fixed percentage of its funds for risk assessment experiments on the work supported by the rest of the money.

Focusing the discussion of the technique on issues of safety had a number of implications for the regulatory response. Risk and safety were something people knew how to talk about. In this regard, there was little difference between the scientists and the nonscientists. Public representatives were familiar with the terms, and the goals were obvious—minimize risk, maximize safety.

The emphasis on this approach furthered the impression that once risk assessment experiments were performed the matter would be settled, that it "had been a 'technical decision' from the outset, solely involving matters of laboratory technique and good scientific judgement."[36] The other major structural issues, those associated with ethical and social attitudes toward the eventual capabilities of the research, were definitely shortchanged in terms of the amount of consideration they received. Unlike work on the atom bomb, which was protected from such examination by the overriding desire to win the war, recombinant DNA research had no such shield. Thus, it is truly remarkable how little discussion of these issues took place during the course of the debate. Time and again, a group would recognize their importance and sidestep them. The Princeton community group, for example, simply listed some of the questions but made no attempt to reach a consensus or incorporate their concerns into recommendations.

And yet, these decisions are being made, perhaps by default. As recombinant techniques advance and are used in more and more situations, our ethical and social systems will adapt to them incrementally. To avoid discussion, to avoid confrontation, is to succumb to the fabled technological imperative and seriously undercut all the efforts to date to control the future of recombinant DNA.

[36]Michael Rogers, *op. cit.*, p. 208.

17. The Regulation of Recombinant DNA Research: A Defense

James M. Murtagh

It is probably impossible for me to convince everyone who reads this that research with recombinant DNA is a relatively safe and inexpensive means of exploring basic cell processes, and that, since recombinant research has many possible medical applications for the relief of untold human suffering, this research should be vigorously promoted. Six years of debate over recombinant DNA has left much of the public with the idea that this research is dangerous; it will be difficult to change this perception, even though we now have information showing that most of the earlier fears were groundless.

Moreover, I am well aware that there is a growing disenchantment with science and technology in general spreading throughout our society, and that any new breakthrough in technology, let alone recombinant DNA, would likely be viewed with suspicion in the current atmosphere of distrust that is beginning to surround scientific enterprise. Fed by the disaster at Three Mile Island, the problems of air pollution and nuclear proliferation, and the failure of science to end the hunger and population crises, the general disillusionment with science has grown in some quarters into a downright reaction against anything new. Faced with this increasing distrust of science, I realize that the task of defending recombinant DNA research has been greatly compounded. The prospective defender is called on to argue not only the specific merits of recombinant research alone, but also the global issues of man's involvement in science. It will be difficult to mount a defense of recombinant DNA research that will satisfy everyone, no matter how safe and beneficial I could show this research to be, but it is certainly worth the effort.

My defense will be a plain one, and I shall try to avoid the confusing technical jargon that has unfortunately dominated the current controversy and clouded some very thorny philosophic issues that need to be discussed. My remarks are divided into two parts: First, I shall speak to the public safety and epidemiological aspects of recombinant DNA, and show how recent laboratory findings make the research seem far safer than was previously thought. Second, I shall discuss some of the "higher" ethical concerns, beyond those of safety, that have been raised during the current debate. In each part, I shall concentrate on the principles under discussion, and show how the different premises held by advocates and opponents of recombinant DNA research have led to conflict. For the current controversy over recombinant DNA is more than a technical debate about the powers and limits of this research: it is also a debate about values. It is my hope that a close examination of the values involved will persuade the reader that research with recombinant DNA is both safe and ethical, and that this research should not be banned.

The Safety of Recombinant DNA

Probably the most troublesome uncertainty raised by the new recombinant technology, and certainly the one that dominated public concern during the early stages of the recombinant controversy, was the fear that some novel organism would be created inside a recombinant laboratory that might escape accidentally and cause incalculable harm. Because researchers splice together genetic material from widely disparate sources, it was inititally feared that a new bacterium might be created that would combine the pathogenic properties of several species, or would perhaps even possess entirely new properties, such as the ability to cause cancer. Fears were increased by the fact that most recombinant experiments are carried out in *Escherichia coli*, a normal inhabitant of the human gut. It was impossible during the early days of recombinant technology to predict what the effects would be of transplanting bits of DNA from higher animals, such as mammals, into *E. coli*, and therefore it was widely speculated that recombinant DNA could constitute a public health hazard.

Since the time these fears were first voiced, however, a number of events have occurred that make the possibility of a recombinant "superepidemic" seem exceedingly remote. First, elaborate guidelines prepared by the National Institute of Health have been adopted, which make it exceptionally difficult for the bacteria used during recombinant experiments to escape the laboratory, regardless of whether these bacteria are dangerous or not. These guidelines specify that recombinant experiments can take place only in specially constructed laboratories, using the methods of physical containment developed by health researchers to work with deadly organisms such as botulism and the plague. In addition, strains of bacteria have been developed that can thrive only in a special artificial medium, and that would self-destruct if they were exposed to the environment. These bacteria are so fragile that, even if they were directly ingested by humans, it has been estimated that less than one bacteria in a thousand billion would survive more than 24 hours in the human gut.[1] And it would have to be an exceedingly sloppy experiment in which even a fraction of that number of bacteria was swallowed by a technician. The result is that it is almost unthinkable that any bacterium, much less a pathogenic one, could escape a "high-risk" recombinant experiment and infect someone in the general population.

[1]B. Davis, "Evolution, Epidemiology and Recombinant DNA," in *The Recombinant DNA Debate,* D.A. Jackson and Stephen Stich, eds., Englewood Cliffs, New Jersey, Prentice Hall, 1979, p. 146.

But that is not all. There has been an even more important event, in addition to the development of NIH guidelines and self-destructing bacteria, that has rendered the possibility of creating a recombinant superbug far less likely than anyone previously had thought. During 1977, a National Cancer Institute team headed by Philip Leder[2] discovered that the genetic material of eukaryotes (that is, the higher animals and plants) is processed in a fundamentally different way from the genetic material of prokaryotes like bacteria, and that *E. coli* is totally unable to express any mammalian genome that it may have gained through recombinant research. Whereas bacteria express their genes in straightforward fashion, with every triplet of nucleotides corresponding to an amino acid in a protein chain, higher animals are much more complex. Their genetic information must be edited in a very roundabout fashion before it can be expressed meaningfully. Since *E. coli* lack the machinery to do this editing, eukaryotic genes are nothing but a bowlful of alphabet soup to these bacteria. It is harder for bacteria to read information from a mammal than for a six-year-old English schoolboy to read Swahili, according to Leder. To expect *E. coli* to pick up harmful characteristics by introducing random chunks of eukaryotic genes is ridiculous.

Bacteria are able to assimilate and replicate the DNA from higher animals when it is introduced through recombinant technology; however, the bacteria are not able to use it. Why, then, are researchers so anxious to put these seemingly inert specks of DNA into *E. coli*? For one reason, the very fact that *E. coli* replicates the foreign DNA is in itself an advantage. It is very desirable for any number of experiments to have large amounts of purified eukaryotic DNA. With recombinant technolgy, it is a simple matter to take a gene from a mouse, let us say, insert the gene into *E. coli*, grow the bacteria in perhaps a dollar's worth of medium, and then harvest all the purified DNA desired. Without recombinant technology, a researcher would need as many as 10,000 mice and an army of technicians to gather a comparable amount of material. Even then, the yield would not be as good.

And so, recombinant technology has made possible a myriad of experiments involving the sequencing and structuring of purified genetic material. In addition, the technique is enabling scientists to get around the coding barrier discovered by Leder, and to synthesize artificial genes that can be bred by *E. coli* to make useful mammalian protein. To do this, the scientists have to bypass the normal editing function of the mammalian gene that they are trying to mimic; they must take the gene from the mammal, figure out how the gene would read in a language understandable by *E. coli*, and then laboriously synthesize the gene step by step and insert it in its readable form into *E. coli*. It is an incredibly arduous task, but one that can have gratifying results. Already, recombinant geneticists have a bacteria that can make human insulin,and another that makes growth hormone. The medical implications of such manufacture are astounding, and it is certain that recombinant DNA will provide a cheap and high-quality source of these and other drugs if the research is continued.

Note, however, that the successful use of recombinant bacteria to produce

[2]Philip Leder, "Transcription of Mouse B- Gene," *Proceedings of the National Academy of Science,* **75**, No. 3, 1978, p. 1309.

mammalian protein does *not* involve the insertion of actual mammalian genes into bacteria. It is a completely synthetic string of nucleotides, and it contains no instructions except the ones that the scientists put there step by step. Bacteria cannot pick up characteristics randomly from higher animals; it is a tremendous achievement when a researcher forces an *E. coli* to pick up even the simplest of his artificial programs. We have no surprises to fear from this kind of process, for it is calculated effort and not blind chance that determines our result.

It is, of course, possible that a molecular biologist would someday use these techniques deliberately to create a pathogen, and it is this possibility that should alarm us more than the idea that something dangerous might slip out of the laboratory of a well-meaning researcher. There have been visions of terrorists and evil governments using recombinant superbugs as the ultimate weapon, and I suppose that these fears are not to be lightly dismissed. However, even this does not seem likely to me. It would take an incredible effort to convert *E. coli* into a pathogen; in fact, if people are bent on having biological warfare, they would probably get a more deadly result if they used techniques that are already available. It is hard to see why a terrorist would be interested in an *E. coli* strain containing, say, a gene for botulism toxin, when that gene is already housed in the naturally occurring *Clostridium botulinun*—a readily available, easily cultivated bacterium. With that organism, the terrorist could manufacture, at the cost of a few dollars, enough botulism toxin to wipe out an entire city.

In any event, it is certainly clear that research with recombinant DNA is far safer than anyone ever realized, even two years ago. Our new experiences with handling recombinant bacteria, and especially the finding that *E. coli* cannot process raw mammalian genes, should lead us to a new foundation of confidence in the safety of genetic research. Certainly, we still have much to learn, and there may be dangers in these experiments that we do not suspect. But we should meet any safety problems as they appear. There is no reason to let fear rob us of the benefits of the new DNA research.

The Ethics of Recombinant Research

Questions about public safety are not the only types of concerns voiced during the controversy over recombinant DNA. There is also the more troubling anxiety that recombinant DNA violates some sort of higher ethic, regardless of whether or not it is dangerous, and many critics are asking whether this is not a Promethian adventure that our society will regret. Is it moral to interfere in the course of natural evolution? Is it acceptable for humans to probe the code of life itself, to decipher the secret of our creation? I submit that these criticisms of the research flow more from vague and unformulated fears that recombinant DNA might be a form of human presumption, a playing of God, than from any rational examination of the research itself. To be sure, it is natural that the discovery of such important and powerful techniques would arouse these feelings. Every major advance in human history, from the discovery of fire to the invention of the airplane, has probably evoked the cry that man was about to obtain a power he was preordained not to have, and fears of the wrath of heaven. But, as I shall show, to use such fears to justify a ban on recombinant DNA is indefensible. I shall turn, then, to the three most common "nonsafety" issues.

George Wald's "Natural Barriers" Argument

Perhaps the most popular argument currently employed by opponents of recombinant DNA research is the theory of "natural barriers" first championed by George Wald in an issue of *The Sciences,* and subsequently taken up by both Robert Sinsheimer and Erwin Chargaff, in various forms. The argument goes like this: Recombinant DNA methods enable scientists to move genes back and forth across natural barriers, "particularly the most fundamental such barrier, that which divides prokaryotes from eukaryotes."[3] This power is so fundamental and so awesome, Wald concludes, that it ought not be used, *regardless* of whether or not recombining of genes is actually dangerous. Man was not meant to meddle with the powers of evolution, according to Wald, and since recombinant DNA amounts to the creation of "essentially new organisms," it should be expressly forbidden.

This argument, at first glance, does seem attractive, and it may well evoke widespread public sympathy. No one would like to be the first Frankenstein and assert that it is right to create "essentially new" creatures. However, when one looks more closely at the principle underlying this position, it becomes apparent that Wald's logic is flawed. To argue that it is simply wrong to breach natural barriers is to condemn man to the Stone Age. *All* activities designed to improve the condition of man, from the building of dikes and railroads, to the establishment of schools, were made possible by opposing some sort of natural barrier. Moreover, men have been creating "essentially new organisms" since the dawn of history: The breeding of cattle, the domestication of house pets, and the hybridization of food crops are all activities that fit into this category. The history of human betterment *is* a history of deliberate intervention into evolutionary processes; and no man who sits with a full belly at a desk with an electric light ought to argue that such "breaching of natural barriers" is necessarily evil.

It may be argued that breaching the division between prokaryote and eukaryote is a transgression of a different kind and that breaching any of the other barriers in nature. But short of some theological argument, it is hard to see what would make this "barrier of nature" different from any of the others that men have already toppled. Indeed, at least one philosopher has observed that Wald's argument has a theologic ring to it; for if a person were to argue that it is *intrinsically* wrong to cross a natual boundary, he would have to condemn the creation of a new *E. coli* that produced insulin, *even if* the act of creation generated *no unwelcome side effects.*[4] Unless a person were to claim a direct commandment

[3] George Wald, "The Case against Genetic Engineering," *The Sciences,* September 1976.

[4] See the argument of Carl Cohen, "When May Research be Stopped?" *New England Journal of Medicine,* **296**, No. 21, 1977, p. 1206.

from God, it is hard to imagine why he would want to defend this point of view.

In fact, we should realize that the charge that geneticists are breaching forbidden natural barriers rests on the sort of assumptions that men have used throughout the ages to resist every major scientific advance. It is the same type of reasoning that underlies the statement that "if man were meant to fly he would have been born with wings." It is indeed ironic that Wald and other critics of DNA research fail to notice the echoes of earlier ages in their all-too-familiar arguments against the new technology. The charge that we are breaching forbidden barriers shows that there are some aspects of human nature that remain constant, and the fear of new and unknown forms of science is one of them.

The "Slippery-Slope" Argument

The second "nonsafety" issue to be dealt with is an argument made through analogy to the example of someone sliding down an icy mountain slope. This argument asserts that, even though research with recombinant DNA is not in itself wrong, it should still be banned, since the research may make other things possible that are wrong. Critics who use the "slippery-slope" argument maintain that, unless DNA research is stopped now, genetic engineering will inevitably be applied to humans sometime in the future. Some suggest that a tyrant could seize the new genetic powers to create a Brave New World. Research, they argue, takes on a momentum of its own, once it starts down a certain path; if we foresee consequences at the end of a series of investigations, we should stop the research dead in its tracks at the beginning, lest it proceed at an uncontrollable rate and until it becomes impossible even to stop.

How to reply to the "slippery-slope" argument? I share the critics' concerns about possible future misuse of genetic knowledge. A Brave New World would indeed be awful, and we would do well to remain vigilant against this possibility. However, to adopt the "slippery-slope" principle, and assert that we should ban research simply because that research might be grievously misused in the future, would be a tragic mistake. For recombinant DNA is hardly alone in potentially leading to knowledge that might be disastrously misused. In fact, it is hard to think of an area of scientific research that could *not* lead to the discovery of potentially dangerous knowledge. If this principle were adopted, almost all scientific investigation would be either severely restricted or entirely abandoned.

Moreover, even if the "slippery-slope" argument were an acceptable moral principle, it is not clear where the slope would begin in the present example. The discovery of recombinant DNA technology is no more radical a move toward genetic engineering in man than are many other steps, which have aroused no public terror. These steps include the purification of restriction enzymes, the isolation of the gene, the description of the genetic code, and the development of pure bacterial strains. Even the discovery of the laws of gene segregation in pea plants by Mendel would have to be counted in the chain of events that may someday lead to human genetic engineering. If one is to condemn recombinant DNA through the "slippery-slope" argument, then one must also hold that all these other events are equally evil. This is an absurd result, but it clarifies the central illogic of the "slippery-slope" position.

A more rational alternative to the problem of future misuse is a more flexible approach. Instead of indiscriminately banning all research that *might* lead to harmful results, we should stop only those applications that are plainly morally abhorrent, but at the same time continue to allow work to go forward on those projects designed to improve the human condition. We cannot allow vague and ill-defined fears of what someone else might do in the future to deter us from pursuing, in the most prudent fashion possible, the welfare of human beings. Certainly we have a responsibility to recognize the hazards and costs of research as they become visible, and vigilance is to be greatly appreciated. But vigilance must be directed at specific, definable actions, not at what may lie at the bottom of the icy mountain. The argument from the "slippery slope" fails utterly.

The Forbidden Knowledge Argument

The final criticism I wish to address in this essay is the one that has probably worked most persuasively on the lay imagination. It combines elements of all the previous arguments, and I have saved it for last because it is the one with which I disagree most strongly. Throughout the controversy over recombinant DNA, there have been repeated suggestions that it is the *knowledge* that might be gained, and not the actual process, or the dangers, or the possible misuses of this research, that makes it wrong. Because the knowledge gleaned from recombinant research is of such an awesome nature, and has such potential to change man's sense of self, it has been argued that this knowledge is intrinsically unfit for human acquisition. To probe too deeply into man's genetic heritage, the argument continues, is to reduce man to a mechanism and to rob him of his humanity. Thus, people who hold this view argue that we would be better off not knowing how to fit pieces of DNA together, and would suppress further exploration in this area.

The principles that underly these suggestions of "forbidden knowledge" are very questionable. I submit that there is *no* area of knowledge, and *no* item of knowledge, that is intrinsically wrong to possess. All great discoveries, including those of Copernicus, Darwin, Freud, and Einstein, have involved great challenges to the conventional wisdom of their times, and have certainly forced tremendous changes in man's sense of self. To argue that man is best left in a state of self-ignorance is intolerably repressive and ignores the central reality of our existence: that man is by nature inquisitive, and has a need to know. Certainly, principles of privacy may render certain sorts of knowledge not proper for public scrutiny, but there is a great difference between the claim that some knowledge is not properly public and that some knowledge is inherently immoral. Persons may limit their own pursuit of knowledge if they honestly hold beliefs of the second sort, but it would be a tragedy if a restriction were to be imposed on scientists who pursue knowledge in good faith using moral means.

The philosopher Carl Cohen explains the dangers of the "forbidden knowledge" argument in much more eloquent terms than I can muster:

If one did believe that there are domains in which human knowing is taboo, molecular genetics might indeed be one of them. Inquiry into nuclear fusion or celestial exploration might then equally be taboo, as might be the study of relativity, or the development of contraceptive techniques. The penetration of every intellectual frontier threatens deeply held convictions. Every striking advance in human prowess frightens many, horrifies some and appears to a few as the profane invasion of the holy of holies. The difficulty lies not in discriminating between the real holy of holies and those only mistakenly supposed; it lies in the unwarranted assumption that there are any spheres of knowledge to which ingress is forbidden.[5]

What more can be said? Either we must reject the argument of forbidden knowledge, or abandon our claims to scientific objectivity.

At the beginning of this paper, I made the observation that the controversy over recombinant DNA involves a debate over values, a debate as old as the activity of science itself. The issues of natural barriers, "slippery slopes," and forbidden knowledge did not originate with recombinant DNA. These issues have played major roles in many past scientific revolutions, such as those of Copernicus and Einstein, and will probably also apply to other revolutions in the future. All recombinant DNA research has done is to focus public attention on these criticisms in a new light. And this is good; science must remain a problematic activity, for, as Hans Jones has noted, "The very process of attaining knowledge leads to manipulation of things to be known." One must weigh one's values to see if knowledge is worth the price of this kind of manipulation, which is the heart of the scientific method.

Certainly, there are times in which everyone accepts restrictions on research. When research involves human subjects, or entails substantial risk to the public, to the environment, or to private property, there are demands of justice that can be satisfied only by substantial restriction or control. However, in view of recent experiences that show that recombinant DNA is not a danger to public safety, and with the adoption of the NIH guidelines, we should be assured that this research does not fit into the category that needs to be further restricted. The fact that the research has passed the public safety test, and passed it so well, should be the basis for the settlement of this controversy, not the three other questionable principles espoused by critics of the research.

In closing, I should like to recall the words of Søren Kierkegaard who said, "To venture causes anxiety, but not to venture is to lose oneself." With the finding that the present research is not as dangerous as we once thought, and in light of the

[5]Carl Cohen, *ibid.*

tremendous benefits that the new DNA research may make possible, the time has come to venture. Research with recombinant DNA is both safe and ethical, and it could provide the means to ease a great deal of human suffering. It is an activity that deserves widespread public support.

18. Genetic and Epigenetic Intervention: Fanciful and Realistic Prospects

Clifford Grobstein

I shall begin by clarifying several terms in my title to give better definition to my subject. First, I am going to talk about prospects; hence, I am presuming to prognosticate. I hasten to say that my time frame is of the order of a decade. I regard this as a useful task rather than an exercise in ingenious speculation. Forecasting in longer time frames—of the order of a century—is possible only for phenomena that are cyclical (planetary rotation) or have other stable long-term trends. Research is not such a phenomenon. Scientific knowledge tends to grow in fits and starts, like water seeping through a partly compacted and partly fragmented rock stratum. Targets of opportunity, as well as serendipity, play large roles along with social trends in determining the flow of discovery.

Second, I have made a distinction between fanciful and realistic prospects. By fanciful, I mean anything that can be imagined as a projection from the present, leaving aside practical feasibility and real time. By realistic, on the other hand, I mean not only what can be imagined but what safely can be regarded as imminent. Realistic prospects are important to establish because they often need to be prepared for; they are the grist for anticipatory assessment and policy-formation. Fanciful prospects, in contrast, are materials for science fiction as well as for the spinning of startling scenarios, whether for entertainment or to promote individuals, causes, or ideologies. Such scenarios feed on uncertainty and usually are based on best or worst-case possibilities, depending on the purposes to be served.

My effort, therefore, is directed toward realistic prognostication of what may be expected from genetic and epigenetic intervention. Genetic, of course, refers to hereditary transmission between generations. On the basis of advances of the past century, today we define genetic technically as biologically significant information that is contained in replicable nucleotide sequences—most of them DNA.

Epigenetic is a less widely used term but increasingly a useful one. It includes mechanisms for the transcription and translation of replicative genetic messages, the well-known DNA to RNA to protein triad. Epigenetics extends beyond, however, to include the synthesis of basic molecules other than proteins that depend upon the genetically established enzymatic properties of proteins. Also included are the molecular interactions within this mélange that assemble and elaborate into the microstructure of cells and eventually into multicellular macrostructure and function. In short, epigenetic covers all the genetically dependent events that radiate outward from the replicative genome—events that dominate early development. There is, moreover, a continuingly important epigenetic component in all development, on into the adult and, very likely, even expressed in the timing of death.

To complete the clarification of my title and subject, I emphasize my deliberate use of the term "intervention." Genetic and epigenetic processes can be influenced in a number of relatively imprecise ways, including some that reflect human purpose. These latter certainly represent a form of intervention. But here I mean to emphasize planned, precise, and direct manipulation of genetic and epigenetic processes to yield specific, preconceived ends. This kind of genetic intervention has been called genetic engineering, and its counterpart would be epigenetic engineering. I prefer not to use the term engineering because it has unintended connotations that stem from its customary usage in relation to mechanical, physical, and chemical systems. Nonetheless, biological intervention shares with engineering the use of techniques precisely and effectively aimed at achieving human purpose.

So much for definitions. I shall comment first on genetic intervention, which is not only imminent but actually under way, after a very long history of less precise human manipulation of hereditary properties in organisms of many kinds. Thus, selective breeding probably began unconsciously not long after early domestication of plants and animals. Experience undoubtedly brought greater sophistication and eventually scientific artificial selection. Combined with means to increase mutation frequency, artificial selection in modern times became a potent intervenor and manipulator of hereditary properties underlying much of current agriculture. The same means laid the foundation for classical chromosomal genetics, which, together with biochemistry and biophysics, spawned molecular genetics. With the advent of the latter, truly precise genetic intervention became imminent. In the last few years, recombinant DNA techniques brought such intervention to fruition at so precise a level and with such facility as to constitute, in practical terms, a novel modality for genetic intervention.

Since this subject has been widely discussed in the last few years, I shall be brief in dealing with the scientific basis. Available techniques now make theoretically possible transfer of any nucleotide sequence and, therefore, of any genetic message of higher organisms into bacteria. The sequences can be replicated along with the bacterial DNA and, under appropriate conditions, can also be transcribed and translated. In effect, bacteria can thus be genetically programmed to produce highly specific products (for example, protein hormones) normally produced only by higher organisms. Under laboratory conditions, these techniques have been utilized successfully to yield human insulin.

The prospects that can be imagined for this form of genetic intervention are very wide indeed. The fanciful worst-case suggestions include the production and either accidental or deliberate release of new health-threatening and environment-despoiling organisms, the capability to modify human heredity according to authoritarian whim, and the pollution of supposedly pristine evolutionary lines by wholesale miscegenation among widely disparate life forms. The "culture shock" produced several years ago by these worst-case scenarios has now subsided, at least overtly. Among the fanciful best-case prospects are solutions to food, energy and environmental problems, correction of human hereditary defects, and the birth of babies tailored to parental desires rather than by the meiotic wheel of chance. Sober assessment suggests that few, if any, of these fanciful prospects are likely to be realized in the decade just ahead.

The realistic prospects, however, are still a very impressive list. Following the period of considerable uncertainty and controversy about possible hazards, recombinant DNA techniques are today being used for an increasing variety of purposes, under guidelines promulgated by the National Institutes of Health. Although there are persistent objectors, the general consensus is that such recombinant DNA techniques *can* be used safely and with little hazard to investigators or community. There does remain, however, residual uncertainty and concern as to whether the techniques can also be used *unsafely*, particularly if hosts other than those so far assessed are employed to receive the recombinant DNA.

Setting such considerations of potential hazard aside (as is currently the trend), the prospects are for rapid advance in at least two directions through this powerful new form of genetic intervention. The first is in application of the technique to the study of genetic and epigenetic mechanisms in complex organisms, including humans. The obstacles presented by the large size and complexity of the genomes of higher organisms, as well as by the long generation times of these organisms, are being circumvented, at least in part, by fractionation and replication of specific portions of the genomes in bacterial hosts. Although formidable problems remain, we may expect considerably greater understanding of complex genetic systems by the end of the decade ahead. These advances will, in turn, open new opportunities in medical research, not the least of which will be a heightened probability of finding out what happens when normal cells transform to cancer cells. Practical benefits in improved rationales for cancer prevention are likely to result near the end of the coming decade, considerably before any general nostrum is compounded for those actually afflicted by cancer. In that sense, genetic intervention is likely to contribute to the shift of emphasis in cancer research from therapy to prevention.

The second prospect for advance through genetic intervention lies in intensification of the earlier-mentioned successful efforts to program bacteria to produce useful products they themselves never learned to make. The potential of this advance, at this time, justifies characterization as spectacular. Bacteria are fantastically effective in making more of themselves at a very high rate—two generations an hour are readily achievable under favorable conditions. A hand calculator quickly shows that, at least mathematically, one bacterium can thus multiply to produce more than 10^{14} descendants in one day. If each descendant cell has been programmed by recombinant techniques so as to produce 100 molecules of a desired material per generation, the yield in a week can be 10^{17} molecules, orders of magnitude more than could be obtained from the natural species by older methods of extraction or chemical synthesis. These theoretical estimates are likely, of course, to be scaled down as actual practical conditions are tested.

The techniques already successfully tested are versatile enough to produce any natural polypeptide and to create desired modifications or even entirely new substances by design. A whole vista has been created, not only for pharmacology but conceivably for new and cheaper processes of nutrient production in fermentation vats rather than by the hallowed processes of agriculture. The next decade will not realize these longer-term (more fanciful) prospects but will provide extensive testing and assessment of this truly new scale and direction of potential biotechnology. Parenthetically, the decade will test our social wisdom in dealing with the fruits of fundamental science. The decade just past has brought the possibility of large-scale economic payoff for several decades of public support of the fundamental science of molecular genetics. The same decade has been one of tightening budgets for all fundamental research. The goose that laid the golden egg is not starving, but it is no longer thriving either. The general tightening of public budgeting is cutting back the long-term sources of new productivity. The social challenge lies in finding mechanisms to ensure a fractional direct return of the economic value added by such research as molecular genetics to pierce the fiscal cap that has been placed on the ultimate resource of fundamental science.

I shall mention only one other realistic prospect stemming from molecular genetics. Central to the whole new molecular concept of heredity and its expression is the complementarity principle, which generates affinity between large molecules. Based on conformance of molecular shapes and distribution of electrical charge, complementarity leads to highly specific bonding between molecules, as earlier recognized in immunologic bonding between antigen and antibody. Immunologic specificity has been put to use not only in diagnostic tests for infectious disease but to determine the precise location within cells of particular molecular species. It has similarly proved possible to exploit the complementarity of genetic nucleotide sequences as "probes" to search for and measure amounts of molecules of similar affinity. This is likely to lead to new powerful diagnostic tests for both genetic and epigenetic pathology. The coming decade will see an expansion of such diagnosis, heightening the effectiveness of genetic counseling, and conceivably initiating a path toward genetic and epigenetic therapy.

In turning from genetic to epigenetic prospects for intervention, we enter murkier waters. The epigenetic pathways from gene to realization are still to be disclosed in complex organisms. Nonetheless, epigenetic intervention, like genetic, is not a new phenomenon. Although still relatively imprecise, the prospect for the next decade is for a gradually rising capability. Investigation may move in a number of directions. One example may suffice to illustrate the potential. Certainly the first recorded birth of a human infant based upon external or *in vitro* fertilization was publicly among the most spectacular biomedical events of the seventies. Strictly speaking, this was not epigenetic intervention. Rather, it represented surgical intervention to circumvent mechanical blockage of the normal passage of eggs and sperm through the fallopian tubes or oviducts. Nonetheless, the procedure has significant consequences for epigenetic intervention, as we shall see.

Natural fertilization involves fusion of egg and sperm cells specialized for this particular function. Cell fusion leads to genetic mixture and joint genetic influence of parents on the succeeding generation. Genetic information from both parents is represented in the newly established chromosomal genome. Natural fertilization, however, is not the only way in which cell fusion followed by joint genetic influence can be brought about. During the past decade, extensive work has been done in cell culture on experimentally fused cells derived from both the same and different species. For example, fusion of human and nonhuman cells yields composite genomes with varying combinations of chromosomes from from the two species. Correlation of the nature of the mixed chromosomal complement with the species-specific characteristics of the cells provides precise mapping of the location of genes on human chromosomes. Such mapping is essential not only for understanding the organization of the human genome but for interpreting the abnormal chromosomal complements associated with human genetic disease.

Understanding of the process of genetic fusion, whether in natural fertilization or in artificial cell fusion, therefore is an entrée to epigenetic mechanisms of expression. A number of variants of fertilization are being studied in nonhuman species with this same objective in mind. For example, activation of eggs without sperm entry (parthenogenesis) results in a developmental course influenced genetically only by the female parent. Conversely, removal of the egg nucleus prior to fertilization by sperm results in a developmental course influenced genetically only by the male parent. Such enucleated eggs can also be provided through microinjection with a diploid nucleus from a nongerminal source—the technique by which the first amphibian so-called clones were produced some three decades ago. Such eggs would be expected to be equivalent in their genetic background to the nuclear donor source—comparable to identical twins, but in two succeeding generations.

Not only cell fusion but also embryo fusion has been demonstrated in nonhuman mammals. Instead of yielding genetic fusion within a single cell, this technique produces a genetic mosaic, an individual consisting of cells with different genetic constitutions. In effect, these developing embryos have more than two "parents"; mouse embryos with as many as six genetic parents have been carried to term and the multiparental origin confirmed by their expression of the appropriate mixture of hereditary characteristics. Thus, embryo fusion, cell fusion, and the several variants of fertilization have different genetic and epigenetic consequences that provide toeholds for clarifying the still only dimly understood mechanisms involved in complex organisms. The prospect is for steady progress in these lines of investigation in the decade ahead.

The novelty of such genetic and epigenetic manipulations, now being carried out on nonhuman mammals, is exciting and encouraging for new understandings about mechanisms of mammalian and human development. Practical benefits can be foreseen also in animal production for food and other purposes. It is also clear, however, that human eggs are biologically similar to those of other mammals and, logistic and ethical issues aside, are likely to be accessible to the same manipulations. Discussion of the implications of such intervention in human development has already begun and may be expected to intensify in the next decade. Once again, it is important to distinguish the fanciful from the feasible. The net of questions raised by what clearly is feasible is complex and difficult enough without confusing our thinking by simultaneously addressing the fanciful.

What we know to be feasible, because it already has been accomplished, is to exteriorize a mature human egg, to fertilize it externally with the husband's sperm, to have it develop for at least several days in a laboratory dish, and to reimplant it into the physiologically receptive uterus of the ovum donor to continue development to term. What also seems certainly feasible, although not demonstrated in humans, is to fertilize the exteriorized egg with sperm other than that from the husband or to reimplant the egg into a uterus other than that of the egg-donor.

The chief scientific question being debated about external fertilization by the husband's sperm is whether externally fertilized eggs have any greater risk of developmental abnormality than naturally fertilized eggs. We can expect in the next decade to clarify this matter, since it seems likely that the desire to relieve infertility will lead to continued clinical trials, whether in the United States or in other parts of the world. On the basis of what is currently known about humans and animals, one can anticipate reports of occasional unfortunate effects but a high enough frequency of successes to ensure that external human fertilization will become an accepted medical practice in many places by the end of the decade. The questions of nonspousal fertilization (which biologically is little different in consequence from nonspousal artificial insemination) and of so-called surrogate motherhood will continue to be debated during the decade. Although these issues do not differ scientifically from the spousal case, they obviously do differ from social and ethical perspectives. From the history of artificial fertilization and abortion, it seems reasonable to forecast that both nonspousal external fertilization and surrogate motherhood will be reported by the end of the decade. Neither, however, seems likely to become frequent or to become accepted medical practice in that time frame.

More difficult to forecast and to resolve are the prospects that lie just on the boundary of the realistic and fanciful. These arise from the obvious fact that procedures for external fertilization have opened a window, whether for observation or for intervention, on previously inaccessible, very early stages of human development. To accomplish reimplantation successfully, it is desirable for the egg to undergo at least a brief period of external development. Without such a period, one would be less certain that fertilization had occurred or that the product was not grossly abnormal. Moreover, if the conceptus were immediately returned to the uterus, it would be arriving prematurely, since its transit through the fallopian tube normally requires several days. Whatever the consequences or judgments about this brief external development of the human egg, it has so far been accepted as prudent to observe the developing egg externally for at least several days.

Other mammalian eggs have been maintained externally under similar conditions and for longer periods. There reportedly is significant species variation in requirements for external maintenance. Although there are serious technical obstacles, no theoretical limit can currently be set on the length of time mammalian embryos might be maintained in culture. This has given rise to the fanciful prospect of completely external mammalian development, possibly extended to humans as well. Such a prospect seems quite unlikely, even if it were to be judged desirable, in the coming decade.

What certainly is feasible, and likely, is increased experience and enhanced understanding of early mammalian development as it occurs externally. During the decade ahead, the growth of knowledge about nonhuman species in preimplantation and implantation stages will make it increasingly attractive to apply that knowledge to human development. Attention is likely to focus particularly on such phenomena as twinning and cloning, the high incidence of loss of preimplantation human embryos, the genesis of certain dangerous tumors derived from early embryonic cells, and the effects of toxic substances on embryos in early stages of pregnancy. This trend will add to existing tension surrounding the question of the social, ethical, and legal status of the early human embryo.

What is the prospect for useful consensus by the end of the decade on this extraordinarily complex and difficult matter? One asks the question because advancing knowledge is likely a decade hence to have turned some now-fanciful prospects into feasible ones. If social policy is to lead rather than to be tortured by advancing knowlede, it must anticipate emerging issues. Certainly the next century will record advancing *capability* to intervene genetically and epigenetically in human affairs. It is not too early to begin seriously facing the issues, both within the scientific community and between it and the general community that will be affected.

By the end of the 1980s, let us hope that the scientific background that even now must be taken into account in defining the status of the early human embryo will be more widely understood. The definition cannot be based on scientific facts alone; it is too deepy entwined with other perspectives and deeply held convictions. The definition, however, can neither ignore nor deny well-established scientific understandings that bear upon the issue. And these understandings have been growing rapidly in the past several decades.

The essential matters that must be kept in mind are these. What we designate scientifically as human life does not have a beginning at any sharply defined stage of development. Life is continuous between generations, and humanness is a characteristic of our species at all stages of life. Both prefertilization and postfertilization eggs are human and alive and so, too, are previous and subsequent stages—up to death. Beyond death, the remains also are biologically human as long as any vestige of genetic identity persists. If we thus define human life purely biologically, every shed cell, every drop of blood, every amputated organ must be treated with the same respect as would be given a fully functioning human being. And if we limit our definition only to identifiable humanness, both the quick and the dead, and all their parts, have the same status.

Both scientifically and in common sense, however, we know that there is something about mature, fully functioning human organisms that is not displayed by any molecule or even by any single cell, whether ovum, sperm, or neurone. It is that complex and still-unfathomed something that has traditionally been accorded respect in social, ethical, and legal terms. There is a precursor to the unique something in the egg, at least much of it is the genome. We also know, however, that the precursor is very different from the entity traditionally identified socially, ethically, and legally. The entire complex process of epigenetic and developmental change lies between the precursor and the entity. The subject of social, ethical, and legal concern gradually appears or emerges in the course of those complex changes. To say at what point the critical entity is present requires that we decide what aspect of the entity is crucial. That is a matter for judgment, taking fully into account the scientific facts but also the purposes we have in mind in making the judgment.

As examples, important aspects of the protected entity are general awareness, self-awareness, and the capability to make choices. Stated more generally, what concerns us in assigning human status is capacity for individual experience, whether of joy or pain. To the degree that this is crucial, we must ask what we know of these properties in early embryos. Concerning human embryos, we know very little directly and nothing for certain. Concerning other species, however, there is relevant information. The fertilized egg is not yet a single entity or individual in terms of what it is capable of forming. As the egg divides to form two, four, eight, and then more cells, each cell of the earliest stages can give rise to a complete individual, even though it normally gives rise only to part of one. Thus, the fertilized egg will usually give rise to one individual but can give rise to more—in fact, in some species it generally does. This, of course, occurs in humans as identical twinning. Moreover, a four-cell stage that normally will produce a whole individual will produce only half of one if embryo fusion is performed with another four-cell stage. Clearly, individuality is not fixed at fertilization and for at least several cell divisions thereafter. Individuality develops as the fertilized egg becomes an embryo; it is not present initially, even though fertilization establishes a new genome. Individuality in the sense of entity arises epigenetically.

Similarly, what we call experience also arises eipigentically and much later. Everything we know scientifically ties human experience to the nervous system and the complex circuitry of its neurons. The earliest recognizable precursor of the nervous system does not arise until well after implantation of the developing embryo in the uterine wall; there is no sign of it during the stages involved in external fertilization and reimplantation. Moreover, the first cells recognizable as neurons do not appear until several weeks after implantation. Complex circuitry of the brain is only beginning to be established as late as six months into pregnancy. Thus, speaking biologically, the essential objective substrate for the individual experience we so highly prize and jutifiably protect gradually is established epigenetically as the embryo matures. The process is continuous; hence one must be quite specific in defining what part-aspects of individual experience are crucial in order to estimate when each appears individually or in ensemble.

These are deep and difficult problems. We shall do well in the coming decade if we succeed in clarifying them and developing a consensus on the agenda for policy-making. The task cannot be expected to be carried out dispassionately, as the debate over recombinant DNA in the past five years has indicated. Earlier papers in this section have made clear how complex the background for decision-making in biomedical public policy can be. Values, aspirations, and fears are as determinative as scientific facts.

One hopes that certain principles will be held to. Mutual reliance among adversaries must be able to be placed in scientific testimony, because scientific knowledge is a common stock for all. Scientists, despite their own predilections, must seek to assure equal access for every point of view to the information that scientists can best gather and interpret. If this is not the case, passion may totally override rationality.

New social and political mechanisms, as noted in the earlier papers in this section, may be required to accomplish this. If we succeed in establishing such mechanisms, the prospect will be good that we can profit from the vast new interventive options that now confront us. If we fail, we may fall prey to either fancied or realistic worst-case scenarios. Such risk accompanies all innovative but untried options. The management of this kind of uncertainty is the test of each new human cultural stage, as well as between sets of contending cultures. To avoid the effort to handle these uncertainties is to be less than humans might and should be.

IV. Scientists and Political Issues

Primack, Stone, and Chalk discuss the ways scientists and technologists can play an active role in politics, and both the risks and the rewards entailed. Chyba compares the effort of Leo Szilard and other atomic scientists to take responsible action by curbing publication of their research results with the efforts of self-regulation by recombinant DNA researchers. Biren examines efforts to control the harmful effects of hazardous substances, and Price considers the social consequences of the mechanization of agriculture. Revelle examines the energy needs of developing countries, using Asia as the case in point, and the potential role of outside technical assistance and capital. Brown argues that arms control is one of a number of mutually related priorities that must be accepted in the new "era of limits."

19.   Scientists and Political Activity: Insiders, Outsiders, and the Need for a New Ideology

Joel Primack

When I was starting out as a postdoctoral fellow, I went off to Harvard and spent some time at MIT. I was taught by Victor Weisskopf, a famous physicist and professor of physics at MIT, that one of the most important things a physicist has to do is be able to give a good talk. Vicki said that a good rule is to open with general comments, out of respect for the general audience. Then give a lot of detail, out of respect for expertise. Conclude by speculating and talking about things you do not understand very well, out of respect for physics! That is what I shall do, but my subject will be science and political activity. The first part of my presentation will deal with the advisory system and other formal ways that scientists participate in making decisions in our society. Then I shall discuss the role of the scientist as activist, the new organizations that are needed, and similar subjects. Finally, I shall speculate about science, technology, and society in the 1980s.

When I was a graduate student, the traditional role of those few scientists who were involved in policy discussion, as it is now called, was to participate in science advisory committees for the federal government. In fact, I became interested in the question of science and policy through my Ph.D advisor, Sidney Drell, who went to Washington approximately once a week the entire four years I was his student. He was a member of the President's Science Advisory Committee and was the chairman of its strategic weapons panel for a couple of those years. I would ask him, when he got back, what he did there. (This was the Vietnam War period, and I would also ask him why the war was still going on!) He would tell me that he could not tell me. That made me curious as to whether science advising did any good.

I spoke to a number of science advisors and read a lot of reports from science advisory committees, and I taught a course on the subject at Stanford. The conclusion I came to was that, yes, science advising does do some good. One of the best things it does is bypass bureaucratic channels. Scientists learn early in their education that good ideas often come from outside the establishment. Senior scientists would be very foolish to ignore ideas just because some unknown student is responsible for them. Government bureaucrats often do not understand that. One of the best functions of science advising is to let good ideas and good criticism rise quickly to the top. The problem is that the science advisory system generally is not capable of stopping a bureaucracy from doing something it has already decided to do. What science advising can do is tell a bureaucracy how to do better what it already wants to do. If that is what is needed, then science advising is appreciated.

There are a number of problems, however. One is that the mere existence of the system implicitly supports the status quo. Suppose a science advisor is consulted on a government policy question—for example, the supersonic transport. What can be done, and was done in that case, is that the people responsible for selling the idea to Congress simply say, "We have consulted with the greatest scientists, and here is our conclusion." They do not, however, say what the scientists said. Unless the scientists come out and specifically say, "They consulted with us, and we told them this program isn't worth anything," then, of course, the natural assumption is that the scientists agreed with them. Consequently, it is very important that those scientists who were part of the process, and who find that they are in sharp disagreement with the conclusions, speak out—particularly if they have the impression that important information has been withheld needlessly. That is an ethical obligation.

An inherent problem with science advising is that scientists with strong views are automatically excluded from advisory committees. The rule is that a committee is supposed to achieve a consensus. A committee that does not reach consensus only confuses the policy maker whom the committee is supposed to be advising. The result is that some major concern may be raised by a provocative scientist or other spokesman, but when the committee investigates it nobody can understand what the problem is because that person was never asked to join the committee, since he might not be willing to compromise.

A third problem is that debates, such as those before congressional committees, are often not very useful because the scientists who participate talk past each other. Jeremy Stone described the nature of such debates with his example of the hawks and the doves, who are actually worried about very different kinds of concerns: The hawks are worried about the terrible Russians, and the doves are more worried about the terrible nuclear weapons. The difficulty is that, since the scientists are talking jargon, the average congressman or president cannot figure out what the argument is really about, and thus the general practice is to ignore all those issues they do not understand and make a decision on the basis of how many senators they owe favors to, or some other consideration often irrelevant to the nature of the problem but politically expedient.

There are solutions to these problems, but unfortunately I cannot discuss them in detail in a short presentation. I shall discuss one very nice solution, developed by Nancy Abrams and Steve Berry, called Scientific Mediation. In this approach, instead of having public debate or appointing a committee that is supposed to reach a consensus, the government agency responsible for making a technical decision selects a scientist to represent each of the different points of view. People with strong points of view are included, not excluded. These two (or three) scientists work together with the assistance of a mediator to explain in a single document what it is that they agree and disagree about. For each element of disagreement, they have to explain to the other's satisfaction, and in the face of the other's questioning, what their point is and the basis for it, including value assumptions if need be. They are forced by the process to speak precisely to the points the other is raising. That is almost never done elsewhere. The mediator makes sure that they are not too polite to ask each other tough questions and also that their joint paper is presented in intelligible English. The result is a document co-authored by leading experts in the field that really addresses the crucial questions in language understandable to the politicians, the media, and the public. This method has been tried only once, in Sweden, in connection with the debate over what to do with radioactive waste from nuclear reactors. The results were very good, and a tremendous number of issues were brought out very effectively. That is one example of how the advisory system can be improved.

There is a final problem with the system, however, that I think is intractable: The process is controlled by the officials, often for political purposes. For example, one of the main reasons advisory committees are appointed is as a backstop. A decision has already been reached, but whoever is in charge wants a committee to lean back on in case he is criticized. We can never be sure that an advisory committee will be called into existence when it is actually needed. Unless someone from outside raises an issue, it may never be addressed by the executive branch of government or by any other large organization. This is the basic reason why it is necessary to have another avenue for experts of all kinds to become involved in the political process. (These outsiders play a role different from that of the insiders, but you should not imagine that they are necessarily different people. In a given week, a person like Jeremy Stone or me may serve on a government committee and also criticize some government action—not necessarily by the same organization that hired the committee, but sometimes.)

Congress and the public need independent sources of technical advice, because it is dangerous to rely exclusively on the people who invented or are pushing a particular technology. Providing this independent expertise is the rationale for public interest science. It is very hard to find in the nuclear establishment, among the people who design, build, own, or operate reactors, anyone who will say anything negative about nuclear power, for example. Apparently there are not many people in the business who have negative feelings about it, and those who do, tend to keep them to themselves. Good decision-making requires input from people who know all the issues involved and who are both capable of articulating the problems and willing to do so.

It is important to be realistic about the influence of technical information and advice on political decisions. For one thing, many public officials do not pay much attention to the information they get from their own internal sources. They pay more attention to what they read in the newspapers. This is ironic but true. If you have a bright idea or an important piece of information that you must communicate to the people in charge, then, of course, tell it to them. But also tell the press. This is illustrated very well in a story that Saul Alinsky tells in his book, *Rules for Radicals*. It seems that in the 1930s labor leaders had a long talk in the White House with President Franklin Roosevelt about a labor issue. Finally, Roosevelt said, "All right, you've convinced me. Now go out and bring pressure on me." That is politics. Presidents do not make decisions in a political vacuum—nobody does.

Raising issues for the public through the news media is not easy. If you raise an issue at the wrong time, when people are not ready for it, when you do not have allies, when the press will not pick it up, you have lost. Also, you cannot raise an issue in the public arena and expect people to be interested unless they become convinced that they are personally endangered. It is one of the ironies of the weapons race that most people do not feel very threatened by the missiles on the Russian side or all those on our side that could be triggered into a nuclear war. Yet, when President Johnson decided to place antiballistic missiles in major cities all over the country, people suddenly became very concerned. They could see a hole being dug right outside Chicago, right in a park in the middle of Seattle, right outside Boston and Los Angeles—and the problem became tangible. They knew that, if those bombs went off, it would be the end of those cities, and somehow that was enough to convince people that they should be worried. Congressmen got letters from their constituents, and the program was killed very quickly. Then Nixon came into office, and to save the ABM program he placed it farther out, guarding Minuteman missiles far from population centers. But, even though there were no consitituents out there, the issue had already been raised to the level of public consciousness represented by headlines in newspapers, and the newspapers did not drop it. Congress could no longer decide in the usual "to whom do I owe favors?" way. Instead, the issue was debated on its merits, and its merits were nonexistent. It made no sense to defend hardened Minuteman silos with very soft ABM radar. The result was that so much resistence developed in Congress that the program could never be expanded, and the armed forces were willing to trade it away in the SALT I negotiations.

The defeat of the SST and ABM, the success of Rachel Carson in raising the issue of DDT, of the Environmental Defense fund in seeing the issue through, and of Matt Meselson and others in fighting for an end to chemical and biological warfare—these examples show that, although you cannot always fight city hall and win, it is easier than people might think. A few scientists, a few lawyers—such people working on the outside—can often do a great deal more than all the prestigious Nobel Prize winners and National Academy of Science committees working through channels. We need both approaches.

There are some problems with this outsider approach, however. Outsiders are often second-hand experts. The first-hand experts are usually the people who have created the technology, and they are afraid to come out into the open for reasons that are sometimes poignant—reasons that are expressed clearly by Rosemary Chalk. We must try to gain recognition of the need for public interest science and whistle blowing, although it will never be easy. We should recognize that one of the reasons people choose to be scientists is that they do not like public disputes. People who enjoy public disputes go into law, not science. Scientists like to sit quietly and calculate. They work with computers that do not talk back. We should appreciate that scientists on the whole do not have the desire or the fortitude necessary for public action. The number of scientists who become involved will always be a small fraction, but it is an important fraction. The environment of science should be such that there are "ecological niches" for scientists who want to become involved, either part-time or full-time. I am a part-timer, and I think that being in a small town on the West Coast gives me a certain perspective on problems that is not often available to people who are constantly putting out brush fires in Washington. Both perspectives are needed.

One of the ways my friends and I have tried to create niches for public interest scientists is to get scientific societies involved as sponsors and therefore legitimizers of various desirable activities. For example, discussion of issues like the arms race or nuclear reactor safety at American Physical Society meetings is now legitimate, and we have a whole section, the Forum on Physics and Society, devoted to such matters. The American Physical Society has also organized a number of summer studies that have prepared critical reports on such subjects as nuclear reactor safety and the physics of energy conservation. In addition to illuminating neglected issues, these studies have educated and given credentials to new groups of independent scientists.

The Congressional Science Fellowship Program is an effort to bring more scientists into direct contact with the political process in Washington. Those of you who have been congressional interns realize that you could never have imagined how this country is governed without seeing it firsthand. Nothing could prepare you for it—not any number of courses in politics. Nobody is in charge. The place is a madhouse. Everybody is working at cross-purposes. That is how the United States is governed. Many scientists have now gone through the Congressional Science Fellowship Program, over fifteen each year, and many have stayed on and played important roles. Jessica Tuchman Matthews, for example, started as a Congressional Science Fellow with Congressman Morris Udall, and then helped to run his 1976 presidential campaign. She later became one of the top staff people in President Carter's National Security Council, playing a tremendously important role in developing the nonproliferation policy of the Carter administration as well as his human rights policy. She finished a Ph.D. in microbiology just a few years ago, and she has already had a remarkable career in government, from which she has now resigned to join the editorial staff of the *Washington Post*. Her career reflects some of the possibilities for independent scientists in public service.

The Science for Citizens Program of the National Science Foundation represents an attempt to legitimize and fund political activity by scientists working with citizen groups. Intervenor funding, before Nuclear Regulatory Commission Proceedings, for example, is another way of supporting and legitimizing participation of independent technical specialists. Public interest science as a profession needs a great deal more development. Money and opportunities are available, and there is a lot of room for creativity. The Congressional Science Fellowship Program did not exist until it was created; somebody had to think of it and carry it out. Is it enough then, to have on the one hand the insider advisory system, and on the other hand a group of outsider scientist–activists? Will our problems be solved if we civilize the dissenting process? Let us consider the history of the development of nuclear power as a public issue. After World War II, there was the atomic scientists' movement, and the big issue was, should there be civilian or military control of atomic energy? The scientists converged on Washington, had a good fight, and won! We then had civilian control of atomic energy, and everybody went to sleep until a decade later, when the test ban became the big issue. Nuclear testing was being carried out in the atmosphere, and dangerous radioactivity was being spewed out by those bombs. Linus Pauling won the Nobel Peace Prize for his efforts to stop atmospheric testing. After tremendous activity, the test ban came along. Then came the new problem of low-level emissions by reactors, and then the issue of reactor safety. More recently people have become increasingly concerned about nuclear waste and proliferation.

What do we learn from this review? First, we see that these public issues were mostly raised by independent scientists. But there is a second, more important lesson: Only recently have scientists—I included—begun to realize that earlier, when we were fighting all those individual little fights, we somehow lost our perspective. What we should have been doing was developing the technology for more efficient energy use and for generation of solar energy, but those of us who were involved had been trained that the way to succeed in science is to identify manageable problems and attack them. The manageable problems that confronted us were such things as figuring out just how much danger could be caused by radioactive accidents of various kinds. Well, I could handle a problem like that. I persuaded the American Physical Society to organize a study of reactor safety. But I did not take a step back and realize what I have realized only in recent years—that we also needed to address the larger question of how energy fits into society as a whole. Thus, in answer to the question of whether it is enough to have scientist–activists counterbalancing the science advisory system, I would have to conclude that it is not enough. Scientists should be aware that their tendency to attack "manageable problems" also requires that they collaborate with people of other backgrounds and perspectives to assure that, in framing the questions and choices they present to society, they have not focused on details and missed the more important issues.

Let me emphasize the importance of this conclusion by approaching it from a different direction. There are two different attitudes that people may have toward their family physician. Some people go to the doctor, tell him what is wrong, and ask him to prescribe. Others go to the doctor and say, "Tell me what is wrong, and give me my options. I want to make the decision." People have the same dual attitudes toward experts as they do toward doctors. Some believe that the public is incompetent in such matters and should leave the important decisions to experts; other people think that in a free nation the public must decide important issues.

Anne Cahn teaches a crucial lesson in her Ph.D. thesis, *Eggheads and Warheads*. She discovered that almost all the scientists who were in favor of the ABM were of the view that it was up to the experts to decide; those who were opposed were of the view that the public should decide. There was a clear-cut division. If you knew how scientists felt about the ABM, you could predict very accurately their position on whether the public was competent to decide or not.

This ambivalence about expert advice reflects a general debate that has been going on in the political science community, and in society generally over whether the United States is moving toward technocracy or not. There are those who think that ideology is basically not useful and that the world is facing such complicated and difficult problems that it is up to the experts to decide. The economists should decide what to do about inflation, the nuclear scientists should decide what to do about nuclear power, and so on. Then there are people like me, who think that that is not wise. I believe that most social problems are like an undetermined system of equations. All of us learn in mathematics that if you have N equations, but $N + 5$ unknowns, then you have a five-dimensional space in which the solution lies. The solution is not uniquely determined. Practically all social problems are like that. In some cases it makes little difference what choice is made. The choice is not predetermined by technical considerations. In the arms race, for example, there was no reason known only to experts why the United States had to build 1000 Minuteman missiles. It was a choice pulled straight out of the air. That is what Herbert York showed in his book *Race to Oblivion*. He was the one who made the decision, but he had no good reason for it, and he regretted it later. The "missile gap" had been created by President Kennedy as a campaign slogan, and the "missile gap" required some missiles to fill it.

It is typical of social problems that they do not have unique solutions. However, in order to mobilize people toward a common end, it is politically necessary to convince them that this is The Solution our society must strive for. In fact, however, there really is no completely determined solution, and you must base your choice on some other considerations—considerations that are moral, ideological, etc. The truth is that in reality such considerations are required whether they are made explicit or not. You cannot decide most issues on purely technical grounds. It is very important for scientists to recognize this. It is equally true, I think, when a person goes to his doctor. It is true when we discuss any of the issues that I am talking about. You have to bring considerations to your decision.

I feel that scientists have been politically naive. This is the one-hundredth anniversary of Albert Einstein's birth. Einstein is widely regarded as a brilliant physicist, possibly the most brilliant who ever lived. He also thought very deeply about public affairs. After World War II, Einstein realized that world government was the only path that could ultimately prevent nuclear war. He fuly understood the meaning of world government—that the central task of modern government, which is redistribution of resources, would have to extend to the whole world and not, as it does now, only to individual countries. Einstein was regarded by his fellow scientists as being politically naïve. In fact, it was the scientists who were naïve and who did not understand that Einstein was not merely visionary but fundamentally right, and that as usual he saw much further into the future than they did.

It is necessary that scientists, as part of their involvement in public debates, learn to present the issues with the ideological component articulated. When I say ideological, many of you may think of Marxism or traditional liberal American-ism, but that is not what I have in mind. The "ideology business" requires tremendous creativity. People are going to have to create new ideology, new concepts of intellectual frameworks, new visions of the future in which it is possible to articulate such issues as the role of energy conservation as part of a larger picture of society, not just how much energy can be conserved with new technology. I am not quite sure how to bring this about, but one thing is clear— scientists cannot do it alone. They need many expert and non-expert people, in some kind of political association or maybe several kinds. The task of such an association would be to develop an ideology that puts technology in its political place, rather than encouraging the technologizing of politics itself, which happens when a force as powerful as technology is not understood politically. Maybe what we need is a new, green version of Americans for Democratic Action. I think that this is going to be the main problem that the next decade faces, and I hope that we can all work together on it.

20. Scientists as Lobbyists

Jeremy J. Stone

After World War II, the building of the bomb gave rise to four organizations, of which Pugwash is one. The first organization formed, I believe, was the one I represent, the Federation of Atomic Scientists, as it was then called, or the Federation of American Scientists or FAS, as it is now known. It was set up as a civic organization and lobby, and that is what we do today in Washington—lobby on issues involving science and world affairs. The *Bulletin of the Atomic Scientists* was started as an educational arm of the scientists' movement in Chicago, and subsequently Pugwash was set up as the international arm of this movement relating to scientists' responsibility. Still later, Leo Szilard had a clever idea for the election of senators: he organized the Council for a Livable World, which encourages people to send donations to specific states in which the senatorial contest is between a very good candidate and a very bad one, and where the state is small enough so that the donations can have a real effect. This organization is the electioneering arm. So, these four political organizations of scientists were established, three just after World War II, and one in 1962.

The reason these four arms are not better manned than they are is that the scientific community has taken two approaches to scientific responsibility. I did not fully realize this until about a year ago, but now I believe that I understand it quite well. The critical question you have to ask yourself is "scientific responsibility to whom?" And when you ask yourself that question, you discover that the traditional scientific view was that a scientist was responsible to the scientific community, whereas the new view is that the scientist is responsible to the public. In the old view a scientist was not supposed to demean his profession by acting in an unscientific manner. He was not supposed to speak about things he did not fully understand. He was supposed to have his papers refereed; he was supposed to continue in this traditional fashion.

Of course, if you are going to work in the arena of public policy, you have a serious problem with these restrictions, because you can not be absolutely sure of most of the things people need to know about. You can say what you know, and you can distinguish values from facts, but you cannot wait until your papers are refereed. In fact, by the time everyone is in agreement, on every last detail, the debate would be over. The public policy debates are about uncertainties. The values are about as important as the facts, and they have to be mixed together in any kind of analysis. One of the problems we face is that the older scientific tradition falls like a shroud over most scientists and inhibits them from risking the hazards of the public policy arena—as, for example, when John Chancellor says, "Well, Dr. Smith, we have thirty seconds left on NBC, tell us now, are reactors safe or unsafe?" The traditional scientist usually shrinks from such exposure, since nothing he can say is really "right." That situation I would shrink from too.

Also, the traditional scientist is often misquoted, and the press seldom gets things right. That is why there is such a small number of scientists whose names are constantly appearing in the press. Some scientists become accustomed to the hazards of exposing their heads above the trenches. They often are relaying information from their colleagues as they understand it. It would be better, of course, if a much larger group of scientists entered the fray. For those who are considering getting involved, I can tell you it is a very pleasant occupation. There is a camaraderie among the people engaged in it. We see each other only too often, because there are so few of us. As a result, because we are working in what we feel is the public interest, there is a special feeling among us. There is a minimum of slandering in this community—less than there would be in others, because we do have a common purpose.

If you want to function in this business, you have to combine a certain pragmatism with a certain idealism. These are the two main ingredients. You must be idealistic enough to realize what should be sought after, but you must also be crazy like a fox in some ways to try to achieve it. I do not know quite how to explain what I do except by giving a few examples of capers that were successful and some that were not. From these examples you can get at least some sense of what I do. Two of these examples have to do with new approaches to the arms race.

At one time, I considered the question that, if aerosols were supposed to be doing something dangerous to the ecology, would not a 10,000-megaton war do so also? Such a war must do something unexpectedly serious too! And if it were known that it would do so, people might be inhibited from launching an attack even under desperate circumstances, because they would know that it would destroy the planet. Eventually, the Arms Control and Disarmament Agency gave the National Academy of Sciences $55,000 for a study on the effects of nuclear war on the ozone, the climate, the ecology, and so on. I thought that the Academy would come up with something. Study groups were brought together, and they did discover a lot of uncertainties about the post-attack food chain, the post-attack climate, and so on. Unfortunately, the summary letter sent out by the president of the Academy did not fully express the purpose of the study. The gist of the letter was that the Academy had established the fact that, even in a 10,000-

megaton war, it is not certain that every person on earth would be killed. This statement was exactly the opposite of what one would have wanted. A reporter called me the morning of its impending release and said, "Did you see what they were about to say?" I said "No." That morning I put out a quick press release, and I had a man standing at the Academy door handing them out so that at least the newspaper could report that the FAS felt that the study had gone astray. The moral of that tale is that the best-laid plans often do go wrong.

Another example is that of the B-1 bomber. This was a case where, in desperation, trying to forestall the purchase of the bomber, we persuaded specialists on the arms race to sign a petition. The petition was reduced to one sentence stating, in effect, that the B-1 bomber was not worth the money. We thought that this statement would get the widest possible support. The senior specialists were led by Clark Clifford. But to our embarrassment, no newspaper picked up the press release. One of our members was going to be on the Today Show, and I asked him to mention it, but he did not.

Then Marilyn Berger, working for NBC, called up and said, "I hear you have a press release; has anyone picked it up?" I said, "No," and she said, "Well, if no one else has picked it up, then according to the NBC rules, I can use it." I said, "Great." She came out with a camera crew and they filmed Clark Clifford saying that the B-1 bomber was not worth the money. Then they showed me at FAS typing up the petition, and then on NBC news they showed Rockwell International refusing to let the camera crews in. Afterward Marilyn Berger was saying, "Well, here's this petition and here's Rockwell, but here's the prestigious opposition that seems to be winning."

Senate vote on the bomber was going to be held in three days. I talked to Marilyn Berger the next day and said, "Marilyn, do you realize that, in the entire history of the cold war, no strategic weapons program that has reached the floor of the Senate has ever been defeated? Your prediction is somewhat risky." She had just gotten the job at NBC, and she suggested that I had better work hard lining up votes the next week.

To my amazement, the Senate did vote to put the B-1 bomber decision off until after Carter's election—which was the vote at issue—and Carter then killed the B-1 bomber. So this was a very important vote.

Afterward, an experienced high-level participant in Washington said to Marilyn, "You know, Marilyn, from my experience in Washington, you probably decided that vote, because senators like to be on the winning side." When a statement is made that it looks as if one side is winning, and if it takes only two or three votes to change the whole situation, this can happen. So here we have an example of both self-fulfilling prophecy and media confusion. This episode went from a press release that was not picked up at all, paradoxically, to something that may have changed the vote and done in the B-1 bomber.

I have also had an experience with an earthquake—an experience that reveals the problems of lonely responsibility. It involves two senior scientists in America who got in touch with me and told me that there was an earthquake coming on the East Coast. Then a third scientist, even more famous, called me and said he wanted me to "handle" the situation. In effect, what he said was, "these two guys think an earthquake is coming, Jeremy, figure out what they should do." I proceeded to talk to them, and I realized that they were going to hold a press conference if I wanted one, but not otherwise. It was somehow left up to me, although the two of them were quite convinced. They had reason to think an earthquake was coming because of some mysterious explosions off the East Coast, which were identified by the Navy as the work of a fighter aircraft. The two scientists thought that they were methane explosions, which were about to cause earthquakes. They had heard stories about bottom-dwelling fish found on the surface, and other unusual occurrences. I spent a week looking into such things as whether the animals of the zoo were restive—calling all the zoos on the East Coast.

Meanwhile, upper levels of government were not worrying too much. I contacted the senator from California, Alan Cranston, thinking that at least one senior senator ought to know about it. I greatly admire Senator Cranston, and I knew he had a special interest in the situation because he had an earthquake bill in the Senate. I spent a week talking to seismologists, and I even went to the Chinese Embassy and asked them for help from their experience. (They sent a telegram to Peking, and the Chinese went through their entire earthquake register and produced two relevant examples of earthquakes. These involved examples of bottom-dwelling fish surfacing and gaseous explosions.)

I talked to seismologists who had been to China, and I tried to track down the rumors that the American scientists had reported about sightings of bottom-dwelling fish. Well, the alarm was a lot thinner than one would believe. After a week of chasing around, I decided to look at the data myself, and I realized that most of the Canadian data were associated with the Concord supersonic aircraft. By this time, however, the earthquake rumors had reached Canada, and the *Canadians* were worrying about it. Ironically, one arm of the Canadian government knew that the Concord was causing these booms, while another arm was preparing for earthquakes.

Throughout this experience, I was sure that if I called a press conference there would not be an earthquake, whereas if I did *not* call a press conference there *would* be an earthquake. This is a problem of social responsibility. It is not true, as this shows graphically, that every alarm should be sounded.

A last example: I did have one good idea of how to end the arms race that goes beyond the studies for the Academy. I suddenly realized (and I did a poll to check it) that the vast majority of senators—and almost all congressmen—had never been to Russia and did not really know what it was that they were voting defense money for. They had never seen the object of their anxiety. I knew also that most Soviet Politburo members had never been in the United States. In due course, a senator introduced a bill saying that every senator who had not been to Moscow should have the right to go with his spouse for two weeks. This bill included congressmen.

It would, after all, cost a fraction of the cost of a missile to arrange for these people to see their adversary. It is clear, as may be found in books that span a century—the reports of De Custine in 1830s and of Andre Gide in 1930—that the response of Western exposure to Russia, including the Soviet Union, has always been the same. Hawkish visitors are always tranquilized, and dovish visitors are always disillusioned. This has almost the certainty of a theory of physics. The hawks see that the Russians are much poorer and much more afraid of war than they had realized, and no one seems visibly to be in chains. The doves are disillusioned by the suffocating intellectual atmosphere. They see that there is no popular belief in the pretentious slogans that are strewn around. In the end, Americans fear the Russians less, but they trust them less too—a quite healthy combination of effects. Unfortunately, the Nixon administration killed the bill after the Senate had passed it.

We polled the senators again and discovered that 50% have now visited Russia—an increase resulting from two congressional delegations. In the House the number is still 20%. And thirty years into the cold war, only 40% of the Politburo has come here to see what they are talking about.

In short, one very important way to resolve much of the problem of the arms race is to persuade the leading figures of each country to visit the other country. One of the ways by which I induced some of the senators to go, and the Soviets to receive them, was to take a poll of how many senators were going to China—as opposed to Russia—and I discovered that many more were going to China. The State Department used the results of this poll effectively with the Russians in an effort to stimulate such exchanges.

I am supposed to comment not only on ways but also on means of achieving our goals, but I am a lot weaker in that area. There are so many public interest groups now that, in some sense, they are strangling each other. So many groups are asking for money that it is very difficult to get any. So if large groups of students arrived on my doorstep and said they wanted to work, I could not employ them. That is a source of great regret to me. One of the things that needs to be figured out, as a part of the whole problem of science and public policy, is how to fund a large public interest sector. I think the Federation may spend a good part of next year just determining how we might expand the scientific sector, to see if we cannot bring more professionals into the trenches. There are, after all, many scientists who work on these issues part-time, but the problems have become too big for occasional jaunts from Cambridge to Washington. You really have to be in Washington most of the time, because much of what happens there happens quite randomly.

21.   Scientists as Whistle Blowers

Rosemary Chalk

> Corporate employees are among the first to know about industrial dumping of mercury or fluoride sludge into waterways, defectively designed automobiles, or undisclosed adverse effects of prescription drugs and pesticides. They are the first to grasp the technical capabilities to prevent existing product or pollution hazards. But they are very often the last to speak out, much less to refuse to be recruited for acts of corporate or governmental negligence or predation. Staying silent in the face of a professional duty has direct impact on the level of consumer and environmental hazards. But this awareness has done little to upset the slavish adherence to "following company orders."
>
> From *Whistle Blowing*,
> edited by Ralph Nader *et al.* (1972)[1]

"Whistle blowers" is a modern term coined in the 1970s to describe employees who speak out about wrongdoing by their bosses. Although the word is fairly modern, the issues it represents—conflicts between freedom of speech and "following company orders," the responsibility of scientists to protect the public interest and also to be loyal to their employers, etc.—are classic in form. A brief review of the dilemmas associated with the decision of whether or not to speak out about wrongdoing in the use of science and technology will illustrate the real-life difficulties of exercising social responsibility in science today.

Science and Bureaucracy

To understand the significance of scientists as whistle blowers, it is necessary to review two major phenomena that characterize our postindustrial society: (a) the impact of science and technology in producing the basic goods and services used

[1] Ralph Nader, Peter Petkas, and Kate Blackwell, eds., *Whistle Blowing*, New York, Grossman, 1972, p. 4.

by our society, and (b) the role of bureaucracy in developing and applying scientific and technical knowledge. These two forces create the social environment in which scientists must decide when and how to speak out.

Bureaucratic organizations—public and private—are important and perhaps essential to maintain the modern way of life. These structures organize and allocate the resources necessary for producing our food, energy, health services, and transportation among other functions. Because of the contributions that science and technology make in producing these goods and services, scientists and engineers play an important and unique role within all bureaucratic organizations. They bring to the organization the skilled expertise and information needed to develop, regulate, and control science and technology in accord with private and public interests. In addition to their skilled knowledge, scientists and engineers also bring a tradition of professional independence, embedded in the independent academic environment in which they were first trained. As these individuals move from the classroom to the organization, however, conflicts surface between the academic concept of independent inquiry and judgment in science, and the structure of authority and delegated decision-making that characterize modern organizations. These conflicts in some cases clearly illustrate the political uses of scientific and technical information as a tool of authority and power. It is not the scientists alone who decide how the organization's resources, including their own knowledge, are to be used, or who will have access to their information. By the nature of their employment, professionals may bind themselves unknowingly into a master/servant relationship, where their information and knowledge is to be allocated only with the consent of those who hired them.

As was noted in the introduction, a whistle blower is broadly defined as any employee who discloses information about wrongdoing. In many cases the disclosure is intended to reveal acts by others within the organization that are contrary to the public interest. Whistle blowers have revealed fraud or criminal activity or other illegal actions, but they may also disclose actions that are not illegal. The whistle blower may speak out about mismanagement or capricious decisions, or may identify potential risks or hazards affecting the public health and safety that have been ignored in the development of new goods and services. It is in this area, the disclosure of potential risk, where scientists are most apt to "blow the whistle."

The classic case of a whistle blower in a government bureaucracy is that of Ernest Fitzgerald, who disclosed a $2 billion cost overrun of the C-5A aircraft in testimony before a congressional committee in 1969.[2] Fitzgerald, a cost analyst employed by the Pentagon, documented and disclosed to the Congress—without authorization from his employers—significant cost increases associated with the development of a controversial new military project that previously had never been reported. After he gave his testimony, Fitzgerald's job was abolished by the Pentagon.

[2] U.S. Joint Economic Committee, "The Dismissal of A. Ernest Fitzgerald by Department of Defense," Hearings, November 17–18, 1969. Summarized in Nader *et. al., op. cit.*

Fitzgerald's disclosure took the form of documented cost figures. The majority of cases that involve scientists or engineers as whistle blowers, however, do not involve questions of illegality or mismanagement. Instead, these critics are more likely to point to areas of possible harm to public safety or health. Their disclosures are based on individual judgments of the risk presented by a specific scientific or technical product or procedure. In most cases these risks are potential and uncertain, and many factors will enter into the collective decision as to whether the risk is a significant threat to the public interest.

The wrongdoing that a scientist may disclose as a whistle blower may not be an illegal action, but may instead be the decision by an organization to ignore an area of potential risk to the public. In many cases the whistle blower is a professional employee responsible for assessing potential risk from a technical product, as in the nuclear or drug industry. The scientist's perception of the significance of a potential risk may be shaped by judgments not shared by others within the organization. If the scientist is alone in expressing concern about a particular product, the organization's management may decide to ignore the warning until evidence of harm is visible.

We should recognize that the scientist's loyalty to the employer, to the profession, and to the public are often harmonious and complementary. But in cases where judgments based on assumption of risk, differ, the individual scientist must determine how to resolve these differences. In some cases the scientist will decide that, although he is employed by a particular organization, his paramount responsibility is to use his professional knowledge in the public's best interest. If the scientist disagrees with his employer's decision, therefore, he may feel professionally and morally obligated to take his information directly to the public. In some cases, he may also be legally obligated to do so.

Determining whether a potential risk is harmful or not is a difficult social process. In many cases it is not possible to know if the whistle blower is right or not. If a scientist is working in areas of high uncertainty or great disagreement as to the extent and significance of a particular risk, it may not be possible to validate the whistle blower's claim. What is important is to know whether the whistle blower is calling attention to an issue of significant uncertainty or an area of reasonable disagreement that should be taken into consideration in the private or public policy decision-making associated with a particular technology. The fact that there is disagreement or uncertainty among those directly familiar with the technology under review is the important factor, and it is that fact of dissent or uncertainty that should be communicated outside the organization.

A review of whistle-blowing cases indicates that the way bureaucracies often respond to dissent by a member within the organization is to get rid of the dissenter by dismissal or transfer. If a bureaucracy has already decided that an issue of concern to one of its employees is not important, the individual who consistently raises that issue is interfering with the decision-making rules of the organization. In many cases the dissenter is branded as a troublemaker, and the personality of the dissenter may become more significant in his performance review than his technical expertise.

Two Examples of Whistle-Blowing Scientists

To illustrate how scientists may exercise their responsibility to the public as whistle blowers, I shall briefly review two cases. The first involves three engineers employed by a mass transit system; the second involves a single chemist employed by a municipal water system.

The BART Engineers

In the early 1970s, the San Francisco region decided to develop a mass transit system for its commuters. The Bay Area Rapid Transit (BART) District was established to monitor and review technical plans of private contractors who were building the transit system. Three engineers (Max Blankenzee, Robert Bruder, and Holger Hjortsvang) were employed by BART to review the contractors' performance through a special consortium arrangement that administered the contracts between BART and the individual private firms.[31]

During their review of the consortium's designs, the three BART engineers became concerned about potential safety hazards associated with BART's braking system as well as more general management difficulties involved in the BART/consortium/contractor arrangement. The engineers believed that this arrangement resulted in a lack of adequate testing for certain design components within the transit technology itself. They discussed their concerns with their management supervisors, who did not agree with their viewpoint. The discussions between the engineers and their supervisors focused not so much on the technical analysis, however, as on the right of the engineers to be concerned with such matters as monitoring the consortium's management performance.

After experiencing great frustration in trying to work through their own system, the engineers privately contacted a member of BART's Board of Directors in early 1972 and discussed the safety and management issues with him. This director then raised these issues during a board meeting. The subsequent discussion became a general critique of the overall quality of BART's management performance. The BART general manager aggressively defended his contractors and the consortium arrangement, and the result was a vote of confidence by the board in BART's management. Following this meeting, the general manager identified the engineers as the source of the leak of criticism to the board members. The engineers were called in to account for their action, and their immediate dismissal was ordered with no written explanation. In the months following their firing, the engineers pursued their concerns with their professional engineering society, the California State Legislature, and the press. They

---

[3]The BART case is thoroughly described in a research study by Robert M. Anderson *et al., Divided Loyalties: A Case of Whistle-Blowing,* Purdue University, Science, Technology and Society Press (in print). The case is also discussed in the *Newsletter on Technology and Society* (Issues No. 4 and No. 12) of the Committee on Social Implications of Technology, Institute of Electrical and Electronics Engineers, New York, 1973 and 1975.

also initiated a law suit againt BART based on the lack of written notice in their dismissal, and they were supported in their litigation by an *amicus curiae* brief submitted by the Institute of Electrical and Electronics Engineers, a national professional society.

Further investigation of the engineers' safety concerns probably would not have been undertaken seriously except for a dramatic illustration of the deficiencies in the BART system. In October 1972, eight months after the board's meeting, a BART train failed to stop at a station platform, ran off the track, and crashed into a commuter parking lot. The National Transportation Safety Board immediately began an investigation of this accident and eventually identified a defective design component as the cause. This investigation also criticized the consortium arrangement and indicated that this arrangement prevented direct technical review by BART of the contractor's design. In early January 1973 the engineers made a financial settlement with BART out of court. Each of them eventually found employment elsewhere. In 1978, the three engineers received the first Public Service Award presented by the Institute of Electrical and Electronics Engineers.

Glenn Greenwald

The second case of a whistle-blowing scientist is not as well known as the BART incident, but it reflects the same divided loyalties. It involved a young chemist named Glen Greenwald, who was an employee of the municipal water system of the City of North Miami Beach in the mid-1970s.[4] He had a B.A. in chemistry, and his job responsibilities included testing water in household residences. In early 1977, Greenwald collected some water samples from a relatively affluent residential neighborhood and became concerned about the color, taste, and odor of one sample, which he believed contained some bacterial contamination. He began chemical tests on the water and the next day discovered that some contamination was present, although the level was uncertain. He looked for his supervisor, who was not immediately available, and then requested his department head to order a daytime flushing of the area, which required the opening of water hydrants in the area to allow the water in the system to clean itself out. Daytime flushings are not common, and they may cause concern and alarm in the neighborhood. The department head, trusting Greenwald's judgment, ordered the flushing.

When the supervisor learned of the action, however, he became very upset, because he considered it to be premature and unnecessary. He told Greenwald to continue testing the water from the affected household, but not to discuss the issue of possible contamination with any resident in order to avoid further

---

[4]Rosemary Chalk, and Frank von Hippel, "Due Process for Dissenting Whistle-Blowers," *Technology Review,* June/July 1979, pp. 49–55. The Greenwald case summary is based on material related to Mr. Greenwald's appeal to the Department of Labor under the employee protection section of the Safe Drinking Water Act of 1974.

speculation of risk or hazard. Greenwald went to a household to collect a sample and was asked by a resident why so much testing was required. The chemist responded that there was a possible contamination problem with the water and further advised the resident not to drink the water until he had been advised that there was not a health hazard. Greenwald returned to his laboratory and reported the conversation to his supervisor. He was immediately fired for insubordination, and despite several appeals, the firing was upheld.

Heroes or Villains?

If an individual scientist truly believes that a risk to public health and safety is being ignored, what are the appropriate actions to take from the viewpoint of ethical responsibility? This question has many answers, derived from various approaches to defining the primary role of scientists in society. Whistle-blowing scientists believe that their primary responsibility is to inform the public about areas of potential risk. As indicated earlier, however, the issues of concern to these scientists involve judgments that are not based solely on scientific evidence. In making a determination that an area of potential risk poses significant harm to the public, the scientist steps beyond the traditional limits of the scientific method.

Over thirty years ago, Robert Merton identified a set of norms as the "ethos of science."[5] The Mertonian definition included disinterestedness as one of the four elements in the ethos of science, derived from his definition of "the extension of certified knowledge" as the institutional goal of science. In taking their arguments to the public, whistle blowers are often acting on the basis of a new institutional goal of science, which Jeremy J. Stone has called "public interest science."[6] The public interest scientist realizes that his knowledge has not been certified by the traditional means of peer review, but he believes that, when science presents a significant risk to the public, the public has a right to be directly informed about it. Furthermore, the whistle blower may believe that the absence or presence of this risk cannot be confirmed by the traditional peer review procedure in science. In making this choice of public disclosure over peer review, the scientist is violating the canon of "disinterestedness" as presented by Merton. This action may lead the scientific establishment to brand a whistle blower as "professionally irresponsible."

Many professional societies, however, are expressing concern that the traditional peer review process and "disinterestedness" principle are not adequate to a modern definition of social responsibility in science. These societies are revising

[5]Robert Merton, "The Normative Structure of Science," in *The Sociology of Science,* Chicago, University of Chicago Press, 1973.

[6]"To Whom Are Public Interest Scientists Responsible?" *FAS Public Interest Report,* Washington, D.C., Federation of American Scientists, December 1976.

their codes of ethics to accommodate new statements of the public rights and responsibilities of scientists, such as:

Engineers shall hold paramount the safety, health and welfare of the public in the performance of their professional duties (Engineers Council for Professional Development, Code of Ethics of Engineers).

[Psychologists] accept responsibility for the consequences of their work and make every effort to insure that their services are used appropriately. (American Psychological Association, Ethical Standards of Psychologists).

In research, an anthropologist's paramount responsibility is to those he studies. When there is a conflict of interest, these individuals must come first. The anthropologist must do everything within his power to protect their physical, social and psychological welfare and to honor their dignity and privacy (American Anthropological Association, Principles of Professional Responsibility).

In some cases, whistle-blowing scientists have indicated that they were guided in their disclosure actions by principles embedded in their professional codes of ethics. As a result, many professional societies have taken an interest in developing ways to support their members in conflict with employers involving questions of professional judgment. The American Association for the Advancement of Science (AAAS) has been one of the leading professional groups urging professional society involvement in supporting members who experience difficulty as a result of an "ethical" action. The AAAS published a report in 1975 titled *Scientific Freedom and Responsibility*, which examined several whistle-blowing incidents and concluded:

How active can and should professional societies be in actively fighting on behalf of their members who are attempting to defend the public interest? ... In matters directly related to the professional competence of members of the society, where the public interest is clearly involved, we believe that the societies can and should play a much more active role than in the past.[7]

The AAAS interest in the whistle-blowing issue and other questions related to the professional rights and duties of scientists resulted in the formation of the AAAS Committee on Scientific Freedom and Responsibility, currently chaired by John T. Edsall of Harvard University, who is also the primary author of the 1975 AAAS report cited above. The committee is urging the scientific and technical

[7]John T. Edsall, *Scientific Freedom and Responsibility,* A report of the Committee on Scientific Freedom and Responsibility, Washington, D.C. American Association for the Advancement of Science, 1975, p. 39. (The report was summarized in *Science,* May 16, 1975, pp. 687–693.)

societies who are affiliated members of the AAAS to examine ways in which they can defend and support members who raise ethical issues relevant to their professional competence.

This committee surveyed the professional societies affiliated with the AAAS and asked them to describe important issues of social responsibility in science today.[8] The responses varied greatly, but they shared a common assumption that scientists have a responsibility to be aware of the social implications of their research. They emphasized the responsibility of scientists to establish zones of certainty and uncertainty within perceived risk or benefit areas arising from the social implications of their research. The respondents indicated that scientists have a responsibility to communicate to the public these perceptions of risks and benefit, and to participate in decision-making processes that involve science and technology.

Professional society involvement in developing ethical guidelines for science today is clearly moving beyond the "disinterestedness" ethos.

Is Whistle Blowing Enough?

Perhaps it is time to build collective standards for professional responsibility that recognize that the scientist today works as part of a collective enterprise. The individual scientist working in a laboratory or on a project from the very beginning of a concept to its technical and commercial application is a very rare phenomenon. Scientists work in groups, and the ethical principles that govern their social behavior should address this modern phenomenon of collective work and shared decision-making.

To be of real benefit, model ethical standards and guidelines must provide the concerned scientists with an environment in which they will be encouraged to act in the public interest and at the same time enabled to pursue their traditional goal of advancing knowledge. Whistle blowing is part of a much larger concern with the situation of the professional scientist or engineer within bureaucracy. Scientists and society need mechanisms that will encourage a broad public review of potential problems, but bureaucracies are not designed to deal with dissent, uncertainty, or disagreement, particularly after decisions have been made with respect to the development or regulation of science and technology. New ways need to be designed that will allow dissent and disagreement to become part of the style of decision-making within a bureaucracy. Due process procedures are one model for dealing with dissent; social impact statements are another way in which scientists could be urged to consider the social implications of their work and to register those concerns as part of their professional responsibilities. Such procedures would not unnecessarily delay the development of science or technology, since they would be information and communications mechanisms rather than a part of the decision-making process itself. Due process for dissenters

[8]1978 Annual Report, AAAS Committee on Scientific Freedom and Responsibility, Washington, D.C. pp. 24–35.

would assist the public in understanding the preliminary and uncertain nature of concerns about risk that might initially appear extremely threatening. The procedures would also provide a forum in which colleagues could disagree and review the basic assumptions that create their disagreement.

Some organizations have begun to insitutute due process procedures within their organizations. The Nuclear Regulatory Commission has solicited comment on proposed procedures for handling dissenting professional opinions within the organization. Xerox and IBM both have "open-door" and ombudsman procedures that permit employees openly to register dissent or disagreement with upper levels of management.[9] The newly revised Civil Service Commission now has an Office of the Special Counsel, which is authorized to initiate inquiries into any adverse personnel action affecting a whistle blower within the federal government. Most important, a number of "employee protection" sections exist in various environmental legislation, including the Occupational Safety and Health Act, the Toxic Substances Act, and the Federal Water Pollution Control Act Amendments.[10] These protections provide a direct means of appeal through the office of the Secretary of Labor for any employee in a private organization who has been discriminated against as a result of an attempt to carry out the regulatory intent of the legislation. There are some problems with these employee protection sections, such as the short (30-day) time limit to the appeal procedure, but these actions represent first steps in opening up private and public bureaucracies and making them more responsive to social concerns. Finally, beyond the development of due process for dissenters, we need to recognize that risk assessment is a social rather than a scientific process. In some cases, the whistle-blowing scientist has a general and ambiguous concern about the social impact of a technical product that is not easily defined. The scientist may want to delay work until the risks are shown to be small; others may recommend that development be continued until the risks are proved to be large. No due process procedure will resolve this disagreement, yet these basic assumptions should be made explicit to members of the public, who ultimately bear the social impact of advances in science and technology.

[9]David W. Ewing, *Freedom inside the Organization: Bringing Civil Liberties to the Workplace,* New York, Dutton, 1977.

[10]These protections were listed in a press release by the AAAS Committee on Scientific Freedom and Responsibility, March 28, 1979. See also Chalk and von Hippel, *op. cit.*

22.　The Recombinant DNA Debate and the Precedent of Leo Szilard[*]

Christopher Chyba

In 1975, the American Association for the Advancement of Science released its report, "Scientific Freedom and Responsibility."[1] In a section titled "Should There Be Forbidden Areas in the Realm of Basic Research?" the report spoke dramatically of the recent debate over recombinant DNA research:

> Recently in a statement probably unprecedented in the history of science, an eminent group of researchers, headed by Paul Berg of Stanford University, has deliberately renounced, for the time being, the performing of certain experiments of probably great scientific interest, because of potential though unproven hazards to human health.[2] The proposed experiments would involve the use of some newly discovered enzymes, which serve to introduce genetic material of other species into bacteria and other living cells...
>
> Clearly this declaration represents a landmark in the assumption of scientific responsibility by scientists themselves for the possible dangerous consequences of their work.[3]

Yet history provides one obvious precedent to the recombinant DNA debate: Before the Second World War, a group of atomic scientists, led by the Hungarian-born physicist Leo Szilard, attempted to effect a ban on the publication of research into splitting the atom. These physicists, well aware of the "possible dangerous consequences of their work," feared that each publication of an experiment in nuclear fission moved Nazi Germany one step closer to an atomic

[*] I would like to thank Joel Primack, Clifford Grobstein, and other members of the first U.S. Student Pugwash Conference for valuable discussions on the topic of this paper. Of course, only I should be held accountable for the final product.

[1] "Scientific Freedom and Responsibility: A Report of the AAAS Committee on Scientific Freedom and Responsibility," prepared by John T. Edsall, Washington, American Association for the Advancement of Scince, 1975. (The Edsall Report)

[2] Paul Berg *et al.*, "Potential Biohazards of Recombinant DNA Molecules," *Science*, July 26, 1974, p. 303.

[3] The Edsall Report, pp. 13-14.

bomb. Hence, there were extensive and dedicated efforts to prevent the results of fission research from entering the public domain.

This precedent from the 1930s for the DNA controversy of the 1970s is widely noted in the popular literature on recombinant DNA. References are often made to the similar, far-reaching consequences of atomic and genetic science,[4] to the parallels between the physicists' and biologists' attempts to control potentially hazardous results stemming from their basic research,[5] and to the effects, in one author's words, of being "seared by the nuclear flame" upon "confronting the more subtle implications of the innermost language of life...."[6]

A comparison between the roles of the physicists in the 1930s and the biologist in the 1970s should not be made too glibly, however. The differences are substantial. Most obviously, the physicists were attempting to halt *publication* of certain results of fission experiments, whereas the biologists who concerned themselves with the implications of recombinant DNA sought restrictions on certain types of research itself. Nevertheless, the two examples probably represent the strongest and most important attempts at exerting control at the level of basic research that scientists have ever made. A careful analysis of these two examples should help to illuminate the strengths and weaknesses of both "movements," as well as provide a more precise context for drawing comparisons between them.

The attempts of some atomic scientists to control publication of fission experiments originated with the Hungarian physicist Leo Szilard. It was in September 1933 that Szilard, having fled to England to escape the Nazi persecution of the Jews, first conceived of the possibility of using neutrons to create a nuclear chain reaction. Shortly thereafter, Szilard learned of the discovery of artificial radioactivity by Irène Curie and Frédéric Joliot at the Laboratoire de Chimie Nucléaire of the Collège de France. He quickly realized that his conception of a chain reaction might now be tested experimentally and discussed the subject with several other physicists. He was unable to evoke any enthusiasm, however. When he brought the subject to the attention of Lord Rutherford, the famous experimental physicist at Cambridge, the reaction was not favorable: "I was thrown out of Rutherford's office," Szilard later told the physicist Edward Teller.[7]

By this time, Szilard had also foreseen the possibility of the creation of an atomic bomb. As his fears began to grow, he worried that, "unfortunately, it will appear to many people premature to take some action until it will be too late to take any action."[8]

---

[4]"We are on the brink of scientific breakthroughs that make the atomic bomb seem tame (!)," reads the back cover of *Bio-Revolution: DNA and the Ethics of Man-Made Life,* by Richard Hutton, New York, New American Library, 1978.

[5]See, for example, Nicholas Wade, *The Ultimate Experiment: Man-Made Evolution,* New York, Walker and Co., 1977, pp.36-37.

[6]Clifford Grobstein, "The Recombinant-DNA Debate," *Scientific American,* July 1977, p. 28.

[7]Stanley A. Blumberg and Gwinn Owens, *Energy & Conflict: The Life and Times of Edward Teller,* New York, G. P. Putnam's Sons, 1976, p. 86.

[8]Spencer R. Weart, "Scientists with a Secret," *Physics Today,* February 1976, p. 23.

Szilard considered several elements as possible candidates for the chain reaction he envisioned. In 1934, he ruled out his initial guess, that of beryllium. Despite his inability to discover an appropriate element, however, Szilard filed a patent containing the words "chain reaction" in the spring of 1934. Not wishing the patent to become public, he assigned it to the British Admiralty.

Szilard saw such patents as one method of maintaining secrecy in atomic research. In a letter to F.A. Lindemann, director of the Clarendon Laboratory at Oxford, Szilard addressed another method as well. Noting that there was ample reason to be "deeply concerned about what will happen if certain features" of atomic physics "become universally known," he wrote:

... In the circumstances, I believe an attempt, whatever small chance of success it may have, ought to be made to control this development as long as possible.

There are two ways in which this can be attempted. The more important one is secrecy, if necessary, attained by agreement among all those concerned that another form of publication should be used as far as the dangerous zone is concerned, which would make experimental results available to all those who work in the nuclear field in England, America and perhaps in one or two other countries, but otherwise keeping the results quiet...

The other way, the less important one, is to take out patents.... Obviously it would be misplaced to consider patents in this field private property....Also one has to avoid applying for patents wherever secrecy is endangered or in countries which are likely to misuse them; so far I have carefully observed this point.[9]

Szilard then turned to indium in his search for an element in which a chain reaction would occur. In the spring of 1936, however, in collaboration with several other physicists, he determined that indium was not such an element.[10] Szilard had hopes of systematically examining the entire period chart, but he was unable to obtain the funding required for such a series of experiments. After expressing his concerns once more to Lord Rutherford and the physicist John Cockcroft, Szilard's fears lay largely dormant until January 1939.

It was late in 1938 that Otto Hahn and Fritz Strassman, working in Berlin at the Kaiser Wilhelm Institute for Chemistry, detected a radioactive barium isotope among the by-products of a uranium target they had bombarded with neutrons. Hahn communicated his results to his colleague Lise Meitner, who concluded that the barium indicated that a new process, atomic fission, had taken place.

Szilard learned of this discovery through Eugene Wigner at Princeton. He realized that, if the fission of uranium released neutrons, the possibility of a chain reaction was moved dramatically closer to reality. If so, he wanted to keep this knowledge from the Germans. He therefore tried to contact Enrico Fermi and Frédéric Joliot, the two physicists he thought most likely to determine if neutrons were, in fact, released.

[9]Spencer R. Weart and Gertrude Weiss Szilard, eds., *Leo Szilard: His Version of the Facts,* Cambridge, Massachusetts, The MIT Press, 1978, pp. 41-42.

[10]These results were eventually published by M. Goldhaber, R.D. Hill, and L. Szilard in *Physical Review,* 55, 1939, pp. 47-49.

On January 26, 1939, Fermi and Niels Bohr had publicly presented the discovery of fission at the Fifth Washington Conference on Theoretical Physics. Through I.I. Rabi, professor of physics at Columbia, Szilard contacted Fermi to speak with him about the implications of this discovery. Fermi's reaction disappointed Szilard:

> From the very beginning the line was drawn; the difference between Fermi's position throughout this and mine was marked on the first day we talked about it. We both wanted to be conservative, but Fermi thought that the conservative thing was to play down the possibility that this [a chain reaction, and hence (perhaps) an atomic bomb] may happen, and I thought the conservative thing was to assume that it would happen and take all the necessary precautions.[11]

In February 1939, Szilard expressed his concerns in a letter to Joliot. "The only reason for my writing you this letter to-day is the remote possibility that I shall have to send you a cable in some weeks," Szilard began. Nevertheless, he made his position clear. Szilard pointed out that, if more than one neutron were liberated in uranium fission, "a sort of chain reaction would be possible. In certain circumstances this might then lead to the construction of bombs which would be extremely dangerous in general and particularly in the hands of certain governments." Szilard informed Joliot that whether censorship on publication should therefore be imposed was being discussed among physicists in the United States. Fermi was conducting experiments at Columbia to determine if there was indeed neutron emission in uranium fission. Had Joliot begun such experiments? Finally, Szilard wrote: "Should you come to the conclusion that publication of certain matters should be prevented, your opinion will certainly be given very serious consideration in this country."[12]

Neither Joliot nor his co-workers, Hans von Halban and Lew Kowarski, responded immediately to Szilard's letter. However, shortly thereafter, they published in *Nature* the results of an experiment that showed that neutrons were indeed released during uranium fission.[13]

Evidently Szilard's letter made little impression upon the French. The letter seemed to express the concern of only a single physicist. A note from Enrico Fermi a few days later, informing Joliot that Fermi was working on uranium fission, but making no mention of abstaining from publication, lent credence to this impression. Furthermore, as time passed and Szilard's promised cable did not appear, the French, in Kowarski's words, "considered that probably the whole idea was abandoned. We simply published."[14]

[11]Weart and Szilard, *op. cit.*, p. 54.

[12]*Ibid.*, p. 69.

[13]H. von Halban, F. Joliot, L. Kowarski, *Nature,* **143** 1939, pp. 470-472.

[14]Testimony of L. Kowarski before the U. S. Atomic Energy Commission's Patent Compensation Board, Docket 18, 16 March 1967, Energy Research and Development Administration, Germantown, Maryland. Cited in Weart, *op. cit.*, p. 24.

Apart from problems over Szilard's letter itself, Joliot's biographer Pierre Biquard believes that Joliot published on the basis of scientific principle:

In principle the scientist is hostile to any kind of secrecy with regard to fundamental research. International scientific cooperation is an essential condition of scientific progress and cannot be reconciled with secrecy. Thus Joliot, in disagreement with Szilard, continued to publish.[15]

Others take a less idealistic view of Joliot's motivations. Robert Jungk, in his history of the atomic scientists, offers a particularly negative perspective. He writes that Joliot

... was just on the point of experimental realization of the chain reaction to which Szilard's anxious communication had referred. He was determined not to be deprived, under any circumstances, of the credit for being first with this discovery. When the experiment succeeded ... he did not entrust the account of it, as in the case of all his previous work, to a French periodical. He sent his report to the British magazine *Nature* because it usually published the work sent in to it more quickly than any other journal concerned with natural science ... Kowarski traveled, on March 8, to the airport of Le Bourget, only an hour's journey from the center of Paris, and personally supervised the document's deposit in the London mailbag. To such a race, for the sake of a few days, had atomic research already degenerated by the spring of 1939. A wholly new spirit of keen international competition had now arisen.[16]

Before hearing of Joliot's results, both Szilard and Fermi had performed experiments that independently showed that neutrons were emitted in uranium fission. Soon after this discovery, Szilard, Fermi, and Wigner met in the office of George Pegram, Chairman of the Physics Department and Dean of the Graduate Faculties at Columbia. Wigner expressed the opinion that neutron emission was too important for the scientists to keep to themselves; the government must be informed. On March 16, Pegram wrote a letter to the Navy in which he stated that uranium might "liberate a million times as much energy per pound as any known explosive."[17] Fermi, however, was dubious about such prospects. In any event, nothing came of this initial contact with the government.

The American physicists who were involved in the experiments that showed the emission of neutrons entered into intensive discussions over whether to publish the results. Szilard and Walter Zinn (the physicist at Columbia with whom Szilard had performed the neutron experiments), and Fermi and Herbert

[15] Pierre Biquard, *Frederic Joliot-Curie: The Man and His Theories*, New York, Paul S. Eriksson, 1966, p. 45.

[16] Robert Jungk, *Brighter Than a Thousand Suns: A Personal History of the Atomic Scientists,* translated by James Cleugh, New York, Harcourt Brace Jovanich, 1958, p. 76.

[17] R. G. Hewlett and O. E. Anderson, *The New World, 1939/1946,* Vol. 1, A History of the United States Atomic Energy Commission, University Park, Pennsylvania, Pennsylvania State University Press, 1962, p. 15. Cited in Daniel S. Greenberg, *The Politics of Pure Science,* New York,, The New American Library, 1967, p. 73.

Anderson (a graduate student working with Fermi), each sent a paper to *Physical Review,* but with a request to delay publication.

Szilard, Fermi,and Teller met in Washington to discuss whether or not these results should be published. Szilard recalls:

Both Teller and I thought that they should not. Fermi thought that they should. But after a long discussion, Fermi took the position that after all this was a democracy; if the majority was against publication, he would abide by the wish of the majority . . . [18]

It was while they were still in Washington that these physicists learned of Joliot's publication in *Nature.* Szilard again relates:

At this point Fermi said that in this case we were going to publish now everything. I was not willing to do that . . . However, from that moment on Fermi was adamant that withholding publication made no sense. I still did not want to yield and so we agreed that we would put up this matter for a discussion to the head of the physics department, Professor Pegram. [19]

While Pegram considered his decision, the physicists in favor of withholding publication intensified their efforts. The editor of *Physical Review* was approached with the request that authors who submit manuscripts dealing with certain areas of fission be asked to contact Szilard's group. Attempts were also made to bring the English into the self-censorship agreement. Victor Weisskopf, a physicist from the University of Rochester visiting Princeton, cabled P.M.S. Blackett, a physicist at Victoria University, Manchester, with the suggestion that papers dealing with nuclear fission be sent to periodicals as usual, but with the request that they not be published until further notice. Nevertheless, Weisskopf suggested, experimental results could still be circulated privately between U.S., English, French, and Danish laboratories. Weisskopf sent a similar telegram to his friend Hans von Halban, asking for Joliot's reaction.

At the same time, Wigner wrote to P.A.M. Dirac, the well-known physicist at Cambridge. Enclosing a copy of Szilard's letter to Joliot, Wigner noted that Fermi's and Szilard's neutron experiments did not dispel the fears of a possibly dangerous application of fission. Requesting Dirac to contact Blackett, Wigner pointed out that self-censorship in the United States alone could not succeed. About a week later, Weisskopf was informed by Blackett that the collaboration of *Nature* and the Royal Society could be expected.

The agreement of Niels Bohr, then visiting the United States, was also obtained. Although he was skeptical of the chances of success, Bohr drafted the following letter to his institute:

The Columbia group is busy organizing cooperation among all the physics laboratories outside the dictatorship countries, to keep possible results from being used in a

[18]Weart and Szilard, *op. cit.,* p. 56.
[19]*Ibid.*

catastrophic way in a war situation, and I must therefore ask you, if work along these lines is going on in Copenhagen, to wait before you publish anything until you have cabled me about the results and received an answer.[20]

The French, however, declined to join in the self-censorship. Joliot's group sent the following telegram:

SZILARD LETTER RECEIVED BUT NOT PROMISED CABLE STOP PROPOSITION OF MARCH 31 VERY REASONABLE BUT COMES TOO LATE STOP LEARNED LAST WEEK THAT SCIENCE SERVICE HAD INFORMED AMERICAN PRESS FEBRUARY 24 ABOUT ROBERT'S WORK STOP LETTER FOLLOWS
                            JOLIOT HALBAN KOWARSKI[21]

Szilard immediately replied, noting that Robert's work, as reported by Science Service, concerned only a type of neutron emission that was not dangerous, and furthermore, that his group had already been approached and had agreed to cooperate. Nevertheless, Joliot replied:

QUESTION STUDIED MY OPINION IS TO PUBLISH NOW REGARDS JOLIOT[21]

Shortly after this answer, Pegram decided that it was hopeless to attempt a censorship. Szilard's colleagues at Columbia agreed with Pegram's conclusion, and it was decided to publish.

Thus the effort at self-censorship failed. Ultimately, of course, secrecy was imposed. Indeed, in June 1940, Gregory Breit, who was familiar with Szilard and Wigner's arguments, was named to the National Academy of Sciences in the Division of Physical Sciences of the Academy's National Research Council, where he argued in favor of censorship. Eventually, both the scientific journals and the scientists themselves agreed. But, as Breit wrote the physicist Ernest O. Lawrence, "As recently as six months ago, I should have been opposed to any such procedure, but I feel now that we are in many respects essentially on a war basis."[23]

There are important differences between the atomic scientists' attempt at self-censorship just described and that of the biologists in their quest to restrict DNA research. These differences become apparent in an examination of the early stages of the recombinant DNA debate.

This debate had as its focus the safety of the public and of the researchers themselves, rather than fear of possible dangerous applications of the research. The debate really began in the summer of 1971, when Robert Pollack, a cancer researcher lecturing at the Cold Spring Harbor Laboratory, learned of an

[20]Bohr Scientific Correspondence, cited in Weart, *op. cit.*, p. 26.
[21]Weart and Szilard, *op. cit.*, p. 73.
[22]*Ibid.*, p. 74.
[23]Lawrence Papers, Bancroft Library, Berkeley, California, cited in Weart, *op. cit.*, p.30.

experiment planned by Professor Paul Berg, of the Stanford University Medical School. Berg intended to insert the DNA of an animal tumor virus (SV40) into a bacteriophage and then recombine this DNA in the host bacterium *Escherichia coli*. Pollack was alarmed by the proposed experiment, both because SV40 was known to cause cancer in mice (although it seemed harmless to man), and because *E. coli* commonly resided in the human gut. Pollack feared that the SV40 genetic material might somehow cause its host bacterium to activate cancer, which— since *E. coli* flourishes in man—could have grave consequences should any of the bacteria escape Berg's laboratory. He therefore called Berg to express his concern over the proposed experiment.

Berg was sufficiently impressed with the potential hazards of his experiment that he then began to question other biologists about its safety. At MIT, David Baltimore, later to be a Nobel laureate in medicine, voted against Berg's proposed experiment. Maxine Singer, at the National Cancer Institute (NCI), expressed similar criticism.

Berg also talked with Andrew Lewis, a virologist at the National Institute of Allergy and Infectious Diseases (NIAID),who had once made the distribution of a hybrid SV40 virus to other scientists conditional upon the recipients' promises to obey certain safety precautions. At the time, several prominent biologists— including Berg, had initially refused to agree to such conditions. Now, Berg and Lewis found that they had mutual concerns.

As news of Berg's proposed experiment spread, other scientists expressed their fears. Wallace Rowe, also of NIAID, commented that "the Berg experiment scares the pants off a lot of people, including him."[24] George J. Todaro, of NCI, felt simply that Berg's experiment "is one of those which I think just shouldn't be done."[25]

Six months after Pollack's initial telephone call, Berg informed Pollack that the experiment was postponed indefinitely.

By the summer of 1973, the power of recombinant DNA techniques had increased greatly, largely through the work of Herbert Boyer and Robert Helling, of the Department of Microbiology at the University of California at San Francisco, and of Stanley Cohen and Annie Chang, of the Stanford University School of Medicine. Using these new techniques, Cohen and Chang found that they could introduce genes that had provided resistance to penicillin for their original parent bacteria into *E. coli*. It was found that the *E. coli* then also became resistant to penicillin.

In July 1973, Cohen and Chang performed an even more dramatic experiment. They found that genes from *Xenopus llevis*, a South African toad, would be reproduced after insertion into bacteria. This experiment represented the crossing of tremendous evolutionary distances.

---

[24]Nicholas Wade, "Microbiology: Hazardous Profession Faces New Uncertainties," *Science,* November 9, 1973, p. 567.

[25]*Ibid.*

Both these experiments depended upon a particular cloning vehicle, pSC101, which had been first obtained in Cohen's laboratory. Cohen and Chang soon found themselves inundated with requests from other scientists for pSC101. Before providing samples of pSC101, however, Cohen and Chang asked that it not be used for certain potentially hazardous experiments. Moreover, in order to maintain some control over the spread of pSC101, they requested that those receiving samples not pass pSC101 on to other laboratories.

More formal attempts at exerting control in recombinant research were initiated in the summer of 1973. In June, at the Gordon Research Conference on Nucleic Acids, Herbert Boyer presented a paper that discussed the techniques he and Cohen had developed for recombining DNA. In particular, he described the experiment in which the gene for penicillin resistance had been inserted into *E. coli*. The issue of possible biohazards arose, and Maxine Singer and Dieter Söll, cochairpersons of the session, agreed to consider the safety issues raised by the new technique. In a fifteen-minute session on the last day of the conference, a majority of the remaining participants agreed to send an open letter to the National Academy of Sciences (NAS) and the National Institute of Medicine. The letter, which was published in the September 21 issue of *Science,* read in part:

We are writing to you, on behalf of a number of scientists, to communicate a matter of deep concern. Several of the scientific reports presented at this year's Gordon Research Conference of Nucleic Acids... indicated that we presently have the technical ability to join together, covalently, DNA molecules from diverse sources... This technique could be used, for example, to combine DNA from animal viruses with bacterial DNA, or DNA's of different viral origin might be so joined. In this way new kinds of hybrid plasmids or viruses, with biological activity of unpredictable nature, may eventually be created. These experiments offer exciting and interesting potential both for advancing knowledge of fundamental biological processes and for alleviation of human health problems.

Certain such hybrid molecules may prove hazardous to laboratory workers and to the public. Although no hazard has yet been established, prudence suggests that the potential hazard be seriously considered.

A majority of those attending the Conference voted to communicate their concern in this matter to you and to the President of the Institute of Medicine (to whom this letter is also being sent). The conferees suggested that the Academies establish a study committee to consider this problem and to recommend specific actions or guidelines, should that seem appropriate...[26]

After the receipt of this letter, an official of the NAS contacted Singer, who suggested that the NAS speak to Paul Berg. Berg, in turn, contacted James Watson, Nobel laureate and director of the Cold Spring Harbor Laboratory. Together they decided to call an international conference to examine the recombination experiments being performed and to consider appropriate safety

[26]Maxine Singer and Dieter Soll, "Guidelines for DNA Hybrid Molecules," *Science*, September 21, 1973, p. 1114.

precautions. In preparation for this conference, they held a preliminary meeting of eight concerned scientists[27] at MIT in April 1974.

By July, this group had reached several conclusions. Most important, it was decided to call for a moratorium on certain types of recombinant research believed to be the most hazardous. Furthermore, the group proposed that the National Institutes of Health (NIH) establish guidelines for scientists working in this area of research. Finally, the scientists officially called for an international meeting of involved researchers to discuss ways to deal with the possible hazards posed by recombination.

The group made its conclusions known in two ways. On July 18, it took the biohazard problem directly to the public in the form of a press conference. Later that month, the group, now with the title of Committee on Recombinant DNA Molecules of the National Academy of Sciences, published its concerns in the form of a letter in *Science* magazine. For this letter, the group had also obtained the signatures of several well-known West Coast biologists, including Stanley Cohen and Herbert Boyer. The letter thus represented the opinion of the most prominent recombinant DNA researchers on both coasts and invoked the prestige of the Academy. The letter read:

Recent advances in techniques for the isolation and rejoining of segments of DNA now permit construction of biologically active recombinant DNA molecules *in vitro* ...

Several groups of scientists are now planning to use this technology to create recombinant DNA's from a variety of viral, animal, and bacterial sources. Although such experiments are likely to facilitate the solution of important theoretical and practical biological problems, they would also result in the creation of novel types of infectious DNA elements whose biological properties cannot be completely predicted in advance.

There is serious concern that some of these artificial recombinant DNA molecules could prove biologically hazardous. One potential hazard in current experiments derives from the need to use a bacterium like *E. coli* to clone the recombinant DNA molecules and to amplify their number. Strains of *E. coli* commonly reside in the human intestinal tract, and they are capable of exchanging genetic information with other types of bacteria, some of which are pathogenic to man. Thus, new DNA elements introduced into *E. coli* might possibly become widely disseminated among human, bacterial, plant, or animal populations with unpredictable effects.

... The undersigned members of a committee, acting on behalf of and with the endorsement of the Assembly of Life Sciences of the National Research Council [of the NAS] on this matter, propose the following recommendations.

First, and most important, that until the potential hazards of such recombinant DNA molecules have been better evaluated or until adequate methods are developed for preventing their spread, scientists throughout the world join with the members of this committee in voluntarily deferring the following types of experiments.

Type 1: Construction of new, autonomously replicating bacterial plasmids that might result in the introduction of genetic determinants for antibiotic resistance or bacterial toxin formation into bacterial strains that do not at present carry such determinants. ...

[27] Paul Berg (chairman), David Baltimore, Herman Lewis, Daniel Nathans, Richard Roblin, James Watson, Sherman Weissman, and Norton Zinder.

Type 2: Linkage of all or segments of the DNA's from oncogenic [tumor-causing] or other animal viruses to autonomously replicating DNA elements such as bacterial plasmids or other viral DNA's. Such recombinant DNA molecules might be more easily disseminated to bacterial populations in humans and other species, and thus possibly increase the incidence of cancer or other diseases.

Second, plans to link fragments of animal DNA's to bacterial plasmid DNA ... in light of the fact that many types of animal cell DNA's contain sequences common to RNA tumor viruses ... should not be undertaken lightly.

Third, the director of the National Institutes of Health is requested to give immediate consideration to establishing an advisory committee charged with (i) overseeing an experimental program to evaluate the potential biological and ecological hazards of the above types of recombinant DNA molecules; (ii) developing procedures which will minimize the spread of such molecules within human and other populations; and (iii) devising guidelines to be followed by investigators working with potentially hazardous recombinant DNA molecules.

Fourth, an international meeting of involved scientists from all over the world should be convened early in the coming year to review scientific progress in this area and to further discuss appropriate ways to deal with the potential biohazards of recombinant DNA molecules.

The above recommendations are made with the realization (i) that our concern is based on judgements of potential rather than demonstrated risk since there are few available experimental data on the hazards of such DNA molecules and (ii) that adherence to our major recommendations will entail postponement or possibly abandonment of certain types of scientifically worthwhile experiments. Moreover, we are aware of many theoretical and practical difficulties involved in evaluating the human hazards of such recombinant DNA molecules. Nonetheless, our concern for the possible unfortunate consequences of indiscriminate application of these techniques motivates us to urge all scientists working in this area to join us in agreeing not to initiate experiments of types 1 and 2 above until attempts have been made to evaluate the hazards and some resolution of the outstanding questions has been achieved.[28]

As several members of the Committee on Recombinant DNA made preparations for the proposed international meeting, the moratorium they had thus invoked was enjoying the observation of biologists worldwide. Nicholas Wade writes:

As it turned out, the moratorium was scrupulously observed throughout its requested duration, from July 1974 until the convening of the conference seven months later. As far as is known, it was heeded by scientists in Europe and the Soviet Union as well as by those in the United States. In England the Medical Research Council (the equivalent of the National Institutes of Health) ordered its scientists not to undertake any of the experiments in the Academy group's letter.[29]

The International Conference on Recombinant DNA Molecules was held at the Asilomar Conference Center in February 1975. The conference included

[28] Berg *et al., op. cit.*, p. 303.

[29] Wade, *The Ultimate Experiment*, p. 39.

ninety American and fifty foreign scientists, as well as representatives of the press and several lawyers concerned with the public policy aspects of science. The conference all but unanimously agreed to several levels of physical and biological precautions in DNA experiments. Moreover, the conference passed a motion calling for the banning of the most dangerous types of experiments. In June 1976, the NIH finally presented detailed and specific guidelines for recombinant DNA research, based largely upon the conclusions drawn at Asilomar.

Thus, it can be said that the biologists attempting to restrict research in recombinant DNA succeeded in their aims. Although the NIH guidelines by no means satisfied everyone involved in the debate, they did represent the culmination of a collective attempt by the scientists concerned to inject caution into their own area of resarch.

While the recombinant DNA debate shared with Szilard's movement such a collective attempt, it clearly differed both in motivation and, indeed, in its level of success. The biologists were concerned with specific health hazards arising from their work, and, despite efforts by some scientists to expand the debate,[30] on the whole it maintained this narrow focus. The controversy in nuclear fission, on the other hand, grappled with the broad questions of secrecy in scientific research in the light of international politics. Such questions would seem less easily addressed.

Thus, secrecy was the essential concern of Szilard, whereas such an issue never entered the DNA controversy. Szilard and the other concerned physicists wanted to prevent dangerous applications of their research, as opposed to incidental hazards encountered during the research itself.

One severe problem both the biologists and the physicists clearly shared was that of the intense competition at the frontiers of scientific research. We have already noted Jungk's analysis of Joliot's motives for continuing to publish: Joliot "was determined not to be deprived, under any circumstances, of the credit for being first with this discovery," for "a wholly new spirit of keen international competition" had arisen. An analogous problem presented itself to those attemtping to restrict research in recombinant DNA. One observer has written:

One reason why safety was being neglected is the high-pressure atmosphere and intense rivalry of modern science, particularly in such fast-moving fields as molecular biology. The fierce pace of competition, though highly efficient at getting results, does not encourage researchers to handicap themselves with excessively rigorous safety precautions.[31]

[30]For example, "there arises a general problem of the greatest significance, namely, the awesome irreversibility of what is being contemplated. You can stop splitting the atom; you can stop visiting the moon; you can stop using aerosols; you may even decide not to kill entire populations by the use of a few bombs. But you cannot recall a new form of life. Once you have constructed a viable *E. coli* cell carrying a plasmid DNA into which a piece of eukaryotic DNA has been spliced, it will survive you and your children and your children's children. An irreversible attack on the biosphere is something so unheard-of, so unthinkable to previous generations, that I could only wish that mine had not been guilty of it. The hybridization of Prometheus with Herostratus is bound to give evil results." See Erwin Chargaff, "On the Dangers of Genetic Meddling." *Science*, June 4, 1976, pp. 938–940.

[31]Wade, *The Ultimate Experiment,* pp. 30-31.

Indeed, when the eight-member Committee on Recombinant DNA met at MIT in 1974, it received a letter from the virologist Andrew Lewis, who observed:

It is unlikely that in the competitive atmosphere in which science functions that broad unenforceable requests for voluntary restraint will contain the potentially hazardous replicating agents which arise from the widespread application of the plasmid recombinant technology.[32]

Yet the moratorium proposed by this committee did gain widespread, even universal, acceptance. Why should this have been the case? More important, why did the biologists succeed in this attempt at "requests for voluntary restraint" while Szilard's group failed?

The answer would seem to lie in the broader-based origin of the recombinant DNA debate. Nicholas Wade emphasizes the importance of the role of a prestigious leader, in this case Paul Berg.

There is a direct line of descent from Berg's first scruples to the decision reached by the Asilomar conference...but the sequence of events was by no means a foregone conclusion. Probably few other people could have asked for a moratorium, got it to stick worldwide, and then handled the issue with the openness and disinterest that disarmed resentment and led the world's scientific community to a notable and generally harmonious consensus.[33]

Yet this analysis is clearly insufficient. Berg was only one force in the debate. The letters of both the Gordon Conferences on Nucleic Acids and the Committee on Recombinant DNA indicated broad support among the most prominent American biologists for placing controls on the research. Not only were many individual biologists represented, but the letter from the Committee on Recombinant DNA indicated the support of the National Academy of Sciences as well.

As the techniques involved in recombinant DNA research originated in the United States, this represented strong restraint on the part of those most able to benefit from the techniques in question. Yet these biologists did not hesitate to expand the debate to an even broader base in the form of the international Asilomar Conference.

Szilard's group suffered from the absence of precisely these strengths. The atomic scientists' efforts were undeniably primarily the work of one Hungarian physicist. While Szilard was successful in drawing more scientists into the debate, his success was never sufficiently communicated to the French. One author has written that the French

could scarcely believe that everyone would adhere to an unprecedented pact, a pact pushed forward, so far as they knew, only by two Central European refugees on the

---

[32]Lewis to the Committee on Recombinant DNA, Natural Academy of Sciences, November, 29, 1974, cited in Wade, *The Ultimate Experiment,* p. 32.

[33]Nicholas Wade, "Genetics: Conference Sets Strict Controls To Replace Moratorium," *Science,* March 14, 1976, p. 935.

outskirts of the Columbia scientific community. (Had Fermi, Bohr or a leading American scientist written them about the scheme, the French might have found it more plausible.)[34]

Szilard's group never published an impressive letter in a leading scientific journal in the manner of the letters published by the biologists in the 1970's. Of course, it is highly unlikely—given the necessity for secrecy—that such a letter would ever have been possible. Similarly, the atomic scientists did not enlist the official endorsement of a government agency. Jungk writes:

In the first place he (Joliot) had not taken Szilard's letter seriously simply because he had supposed it to be a solo performance by his Hungarian colleague. Weisskopf's telegram... had strengthened this impression that the proposal had been made unofficially by a minority of scientists. So important a matter, in the opinion of the formally minded French, should have been broached by the American Academy of Sciences instead of being raised by a few "individualists" and "outsiders."[35]

Thus, the very nature of the restraints Szilard wished to impose limited his effectiveness. Ultimately, the biologists, while overcoming problems similar to those faced by Szilard, had the distinct advantages of a broader, more visible base, official sanction, and, perhaps, a different perspective among themselves than that held by most scientists of Szilard's day. This new perspective was addressed by Sidney Brenner (of the Medical Research Council laboratory of Molecular Biology in Cambridge, England) at the Asilomar Conference:

The issue that I believe is central is a political issue. It is this: we live at a time where I think there is a great anti-science attitude developing in society, well developed in some societies, and developing in government, and this is something we have to take into consideration.... Who really believes that natural science will increase your GNP? Maybe this is the end of this era. It is very hard to tell in history where you really are.... I think people have got to realize there is no easy way out of this situation: we have not only to say we are going to act but we must be seen to be acting.[36]

Ultimately, any "anti-science attitude" among the public must owe much to the onset of nuclear weaponry. Thus, Szilard's failure may ultimately have increased the likelihood of the biologists' success.

---

[34]Weart, *op. cit.,* p. 28.

[35]Jungk, *op. cit.,* p. 77.

[36]Wade, "Genetics: Conference Sets Strict Controls To Replace Moratorium," p. 935.

23. Controlling Science and Technology: The Limitations of Health and Environmental Laws

Curt Biren

Over the past few centuries, science and technology have been commonly thought to be the engines of progress, the means by which our needs and desires will be satisfied.[1] In a speech given just fifteen years ago, the editor of *Science* magazine, Philip Abelson, enthusiasticaly extolled the contributions of research to human health and welfare:

> Our increasing abundance almost universally rests on science.... Let me remind you of just a few of the developments of the last twenty-five years: the wonder drugs; tranquilizers; polio vaccine; synthetic rubber; foamed plastics; new fibers; synthetic adhesives; synthetic gems; paints and corrosion-resistant coatings; high-fidelity sound reproduction; television; new methods of printing and reproduction such as Xerox; color photography and almost instantaneous photography; transistors; magnetic tape; new materials like pure titanium, tantalum, and zirconium; nuclear reactors; nuclear-powered submarines; jet aircraft; artificial earth satellites; the electronic computer.[2]

Although the benefits from science and technology have clearly been enormous, it cannot be denied that disastrous health and environmental consequences have accompanied these benefits. The familiar threats to our health and environment are numerous. Workers are threatened by exposure to asbestos, vinyl chloride, coke oven emissions, benzene, arsenic, cotton dust, lead, coal tar pitch volatiles, and beryllium.

Drugs, food additives, and even clothing made with chemical products can be very hazardous. The long list of hazardous or potentially hazardous substances includes the cancer drug laetrile, saccharin, "Tris" (the flame retardant formerly

[1]See various contributions in Sanford A. Lakoff, ed., *Knowledge and Power: Essays on Science and Government,* The Free Press, New York, 1966.

[2]Philip H. Abelson, "Science in the Service of Society," in Lakoff, *op. cit.,* p. 471.

used in children's sleepwear), and methapyrilene, an antihistamine used in many commerical sleep-inducing compounds.

Pesticides such as DDT, kepone, 2,4,5-T used in the defoliant Agent Orange, the fire ant retardant Mirex, and polybrominated biphenyl (PBB) all pose significant hazards. Other environmental contaminants include carbon tetrachloride, polychlorinated biphenyl (PCB), fluorocarbon, chloroform, mercury, cadmium, barium, dibromochloropropane (DBCP), sulfur emissions from coal plants, and hydrocarbons, carbon monoxide, and nitrogen oxide from automobile emissions.

A total of 40,000 metric tons of toxic waste is believed to be generated each year. According to some estimates, 90% of this waste is disposed of improperly, resulting in imminent threats to human health and the environment at hundreds, perhaps thousands, of dump sites around the country.[3]

The hazards of nuclear power are well-known. Serious nuclear accidents are real possibilities. Long-term exposure to low-level radiation may cause cancer. Further threats to human health and the environment include uranium mill tailings, the radioactive dirt produced from the milling of uranium, of which more than 125 million tons were left in various parts of the West, virtually unnoticed and sometimes ungaurded for years.[4]

This is just a sampling of some of the bitter fruits of our progress. Of the more than 30,000 chemical substances in commercial use today, with 1000 new ones introduced each year, over 2000 are suspected by the National Institute of Occupational Safety and Health to be carcinogens. Only 20 of these substances, incidentally, are now being controlled under full standards issued by the Occupational Safety and Health Administration.[5]

Furthermore many of these hazards are not even known today; of those that are known or suspected, the effects may not be observed for decades. It was only recently that Vietnam veterans learned that exposure to Agent Orange, widely used in the Vietnam war, could cause genetic mutation, cancer, and a variety of other ailments.[6] It took over twenty years for us to realize that fallout from nuclear testing in the Utah area in the 1950s could increase the incidence of childhood leukemia.[7]

[3]United States Congress, *Oversight—Resource Conservation and Recovery Act,* Hearing before the Subcommittee on Oversight and Investigations of the Committee on Interstate and Foreign Commerce, U.S, House of Representatives, October 30, 1978, Washington, D.C., U. S. Government Printing Office, 1979.

[4]See "Radiation: Our Chemical Chickens Have Come Home To Roost," a special report in the *Chicago Tribune,* April 1-5, 1979, p. 17.

[5]Library of Congress, Congressional Research Service, "OSHA's Regulations of Possible Carcinogens: Marshall v. American Petroleum Institute," Issue Brief Number IB 79223, May 25, 1979.

[6]See United States Congress, "Involuntary Exposure to Toxic Herbicide and Pesticide Products," Hearings before the Subcommittee on Oversight and Investigations of the Committee on Interstate and Foreign Commerce, U.S. House of Representatives, June 26, 1979.

[7]Joseph L. Lyon, *et. al.,* "Childhood Leukemias Associated with Fallout from Nuclear Testing," *New England Journal of Medicine,* February 22, 1979.

To stem the tide of hazardous substances into our food, our drugs, and our environment, the government has enacted numerous laws designed to control the use of these substances. The Federal Food, Drug and Cosmetic Act, enacted in 1938, and the Federal Insecticide, Fungicide and Rodenticide Act, enacted in 1948, were among the first. They were followed by a multitude of other laws enacted in the late 1960s and 1970s, including the Federal Hazardous Substances Act, the National Environmental Policy Act, the Occupational Safety and Health Act, the Hazardous Materials Transportation Act, the Safe Drinking Water Act, the Clean Air Act, the Federal Water Pollution Control Act, the Consumer Product Safety Act, the Solid Waste Disposal Act, the Toxic Substances Control Act, and the Resource Conservation and Recovery Act.

Although the length of this list of protective laws is indeed impressive, the effectiveness of the legislation is another matter. First, the laws are, in many cases, very complicated and difficult to administer. Second, the legislative intent of some is extremely vague, often leading to endless legal challenges to the accompanying regulations. Third, the scientific basis for determining that a substance is hazardous (or safe) is, in many instances, quite suspect. Finally, many of these laws require administrators to consider the costs (or risks) versus the benefits, which can be exceedingly difficult, if not virtually impossible, to carry out.

The most comprehensive law ever enacted dealing with the regulation of chemical substances is probably the Toxic Substances Control Act (TSCA). Passed in 1976, it gives the Environmental Protection Agency (EPA) the authority to regulate the manufacturing, processing, and commercial distribution of any chemical substance if EPA finds that

there is a reasonable basis to conclude that the manufacture, processing, distribution in commerce, use, or disposal of a chemical substance or mixture, or that any combination of such activities, presents, or will present an unreasonable risk of injury to health or the environment.[8]

(Certain products currently regulated under other laws, such as pesticides, tobacco, nuclear material, firearms, ammunition, food, food additives, drugs, and cosmetics, are exempt from regulation under TSCA.) Under this law, EPA can impose an outright ban on the substance or can direct the manufacturer merely to provide a warning on the use of the substance.

The magnitude of EPA's task is staggering. First, EPA must publish an inventory list of all existing chemicals manufactured or processed in the United States. Second, it must assess the safety of all new chemical substances, on the basis of test data required from the manufacturer; if there is inadequate information on a chemical substance or evidence that indicates toxicity, EPA may impose limits on the use of the substance or ban it altogether. Third, EPA may require a manufacturer or processor of potentially harmful chemicals to conduct

[8]15 U.S.C. 2605.

and pay for tests on a chemical if EPA can demonstrate that (a) the chemical may present "an unreasonable risk to health or the environment"; (b) there are insufficient data and experience upon which the effects of the substance on health or the environment "can reasonably be determined or predicted"; or (c) testing of the substance "is necessary to develop such data."[9] In applying this authorization, EPA must "consider the environmental, economic, and social impact of any action" it takes or proposes to take.[10]

Brief mention of some of the difficulties in administering this complex law will suffice to indicate the immensity of EPA's task. In preparing its initial inventory of some 44,000 existing chemical substances, [11] EPA has been wrestling with a number of imponderable questions concerning exactly what counts as a chemical substances.[12] It has had difficulty determining how to obtain sufficient data to approve with confidence a new substance for commercial use without requiring so much information that the costs of testing will inhibit technological innovation.[13] EPA must also deal with a number of issues pertaining to its guidelines for testing, including the availability of testing resources, testing priorities, interagency coordination, generation and interpretation of test data, and quality assurance of test data.[14] Finally, EPA must somehow balance its responsibility to protect public health and the environment with its mandate not to place an unfair burden upon industry.

The Occupational Safety and Health Act, enacted in 1970, presents the government with another arduous task. The act grants the Occupational Safety and Health Administration (OSHA) the authority to regulate the occupational exposure to certain hazardous substances.

The foremost difficulty in implementing this act is, perhaps, its vague wording. The act states that in each case a standard shall be set

which most adequately assures, to the extent feasible, on the basis of the best available evidence, that no employee will suffer material impairment of health or functional capacity even if such employee has regular exposure to the hazard... for the period of his working life.[15]

The standards must be "reasonably necessary or appropriate."[16]

[9] 15 U.S.C. 2603.

[10] 15 U.S.C. 2601.

[11] Environmental Protection Agency, "Toxic Substances Control: Initial and Revised Inventories; Premanufacturing Notification Requirements and Review Procedures," *Federal Register*, May 15, 1979, p. 28559.

[12] John B. Ritch, Jr., "Toxic Substances Control Act: A Massive Responsibility, An Arduous Task," *ASTM Standardization News*, March 1979, p. 12.

[13] *Ibid.*, p. 14.

[14] Library of Congress, Congressional Research Service, "The Toxic Substances Control Act: Testing Issues," Issue Brief Number IB78050, May 23, 1979, p. 1.

[15] 29 U.S.C. 655.

[16] 29 U.S.C. 652.

Such vague language inevitably leads to costly and protracted legal challenges to the regulations, which delay the implementation of the legislation. The interpretation of the OSHA language cited above is at the heart of two major pending lawsuits. The first is a suit brought against OSHA by the American Petroleum Institute (API), the petroleum industry group, in regard to OSHA's regulations for benzene. The API contends that OSHA, in failing to perform a cost–benefit analysis of the regulations, has violated the law. According to the API, the best available evidence does not demonstrate that the new exposure limitation is reasonably necessary or appropriate. OSHA, on the other hand, contends that it is required to consider only whether the cost of the regulations is affordable.[17]

The case, which is now pending before the Supreme Court, will have far-reaching implications for OSHA's ability to regulate benzene as well as other hazardous substances. If the Supreme Court rules in favor of the API, OSHA will be forced to conduct extensive, costly, and extremely difficult cost–benefit studies before promulgating any regulations that limit the exposure to hazardous substances.

A second suit brought against OSHA involves its regulations for lead. The final regulations, which were issued in November of last year, were challenged by the Lead Industries Association (LIA). The LIA has charged that the lead standard "pays no heed to economic or technical reality or to the inflationary impact it will have...[It] will impose enormous costs upon the lead industry, without providing health benefits that would not be available under less expensive alternate proposals."[18] The case, which is now before the District of Columbia Court of Appeals, will undoubtedly also be important for the development of subsequent regulations by OSHA.

Another difficulty in implementing these laws is that of trying to prove scientifically the danger (or safety) of a particular substance. This problem is exemplified by the efforts of the Food and Drug Administration in its implementation of the Federal Food, Drug and Cosmetic Act. This act, which has been amended five times since its enactment in 1938, provides for the prohibition or limitation of the commerical use of any food "if it bears or contains any poisonous or deleterious substance which may render it injurious to health,"[19] any food additive "if it is found to induce cancer in man or animal, or if it is found, after tests which are appropriate for the evaluation of the safety of food additives, to induce cancer in man or animal,"[20] any drug that has not been shown to be safe,[21] and any cosmetic "if it bears or contains any poisonous or

---

[17]Library of Congress, Congressional Research Service, "OSHA's Regulations of Possible Carcinogens; Marshall v. American Petroleum Institute," *op. cit.*

[18]Paul G. Engel, "Controversy over the Lead Standard Moves Into Federal Courts," *Occupational Hazards*, April.

[19]21 U.S.C. 342.

[20]21 U.S.C. 348.

[21]21 U.S.C. 355.

deleterious substance which may render it injurious to users."[22]

Although the authority granted to the FDA under this law is fairly clear and unambiguous, the procedures used to identify potentially hazardous substances are not. For example, identifying substances that pose significant cancer risks can be extremely difficult. Whether researchers use animal tests, short-term tests for mutagenicity, or epidemiological studies of humans, the assessments of the effects on humans are usually very uncertain.

Animal tests, for instance, which are currently considered to be the best methods for predicting the carcinogenic effect of substances in humans, do not give reliable estimates of the numbers or locations of cancers that might occur in humans.[23] In commenting on this problem in a memorandum to the Commissioner of the Food and Drug Administration, Arthur Upton, the Director of the National Cancer Institute, stated:

A given exposure to a carcinogen may cause a very low incidence of tumors in one species, whereas the identical exposure may cause a very high incidence in another species. An estimated risk of 4.2 cancers, for example, per 220 million people, as calculated by extrapolation from mouse or rat data, might turn out in reality to be as low as no human cancer, or as high as 420,000 cancers. Although the occurrence of very large errors should be rare, each such error could be a catastrophe. One would not know such errors had occurred until many years after human exposure.[24]

Short-term tests and epidemiological studies give much less reliable assessments. In short-term tests, such as the Ames test, several specially constructed strains of bacteria are used to detect mutageenic changes resulting from exposure to certain chemicals. Although it can be reasonably assumed from a positive result of a well-validated test that the particular chemical substance being tested will probably be carcinogenic in animals, negative results are merely suggestive and in no way definite.[25]

In epidemiological studies, researchers attempt to ascertain if there is a positive causal relationship between exposure to a particular substance and the occurrence of some disease in humans. Again, although positive findings sometimes provide strong evidence of a causal relationship, negative results are extremely difficult to interpret. The data on exposure are almost always very uncertain and, unless the data were obtained from large numbers of people and over long periods of time, the findings may be so tentative as to be of little use, if any.[26]

[22] 21 U.S.C. 361.

[23] See United States Congress, Office of Technology Assessment, *Cancer Testing Technology and Saccharin*, U.S. Government Printing Office, Washington, D.C., 1977, pp. 11-12.

[24] Quoted in "How to Assess Cancer Risks," *Science,* May 1979, p. 813.

[25] See United States Congress, Office of Technology Assessment, *op. cit.,* pp. 12–14.

[26] *Ibid.,* p. 14.

Finally, the consideration of cost versus benefits or of risk versus benefits is increasingly being required in administrators' decisions on regulations. Risk–benefit analyses or the consideration of factors other than safety are explicitly required under the Toxic Substances Control Act, the drug provisions of the Federal Food, Drug and Cosmetic Act, and the Consumer Product Safety Act.[27] OSHA may be required to do cost–benefit analyses depending on the outcome of the API case discussed above. Efforts are under way to incorporate risk– benefit considerations under the food additive provisions of the Federal Food, Drug and Cosmetic Act, largely as a result of the proposed ban on saccharin.

These analyses, however, present administrators with a multitude of problems. OSHA, for example, would have enormous difficulty if it had to translate animal studies into exact numbers of human lives saved, assign some dollar value to lives saved or health protected as a result of implementing a new standard, and attempt to quantify secondary and tertiary benefits to be derived from a standard, including effects on productivity, worker's compensation, the welfare system, health care systems, and income taxes.[28]

Furthermore, even when a cost–benefit or risk–benefit analysis is feasible, it remains to be seen how an analysis should be used as part of the decision-making process. Reliance on these studies for decision rules may yield the most economically efficient regulations; however, relying on such analyses may distort the thrust of the legislation itself, which is to provide protection for those who may suffer from exposure to the particular hazard.[29]

All these particular problems in implementing our health and environmental laws, outlined above, epitomize the difficulty of the task we face in protecting ourselves from our own technological inventions. Given these problems, we can by no means rest assured, especially in light of further advances in science and technology, that the public health and environment will be adequately protected.

What this suggests is that more fundamental changes may be needed in order to bring under control our uses of science and technology. One commentator has concluded that our current malaise is attributable to the deeply rooted view of science and technology as the means by which to control nature for our own needs and desires.[30] Perhaps this common ideology now needs to be reinterpreted in a different light—to mean, for example, the control over the relationship between nature and humanity. Perhaps only through such a change in our basic views can we honestly expect to put an end to the perilous uses of science and technology.

---

[27] *Ibid.,* p. 16.

[28] Library of Congress, Congressional Research Service, *op. cit.,* p. 2.

[29] See Burke K. Zimmerman, "Risk–Benefit Analysis: the Cop-Out of Governmental Regulation," *Trial,* February 1978.

[30] William Leiss, *The Domination of Nature,* New York, G. Braziller, 1972.

24.   The Mechanization of Agriculture: Costs and Benefits

Barry L. Price

The mechanization of agriculture is presently one of the most controversial issues on the California political agenda. At issue is whether or not the state should continue to fund the University of California to undertake research on new agricultural technologies known to have significant social costs. On one side of this issue are manufacturers of farm machinery, large farmers, and many of the university's agricultural scientists. These groups argue that research into new farm technologies is worthy of continued public support because the total social benefits generated by these agricultural innovations far outweigh whatever social costs they may entail. On the other side of this issue are farmworkers and small farmers. These groups argue that public funds should not be used to support research on farm technologies that force small farmers off the land and leave large numbers of already poor farmworkers unemployed.

Although there are thoughtful proponents on both sides of this issue, many of those most active in the public debate have continued to make exceedingly broad, and often overly simplistic, generalizations regarding the costs and benefits associated with changing technologies in agriculture. In the hope of helping myself and others avoid the misconceptions that result from relying on such simplistic generalizations, I have reviewed a large number of existing studies focusing on the costs and benefits generated by different types of new technologies adopted in agriculture. Because this literature is so extensive, I have not attempted to discuss all of it. Instead, I have attempted to distill from these studies only the major generalizations and necessary qualifications regarding the costs and benefits generated by the principal kinds of agricultural technologies for the main groups in society affected by them.

Technological Change and the Consumer

Most analysts have concluded that in general consumers are major benefi-ciaries of technological advances in agriculture.[1] Technological innovations have generated dramatic increases in agricultural productivity. This is true whether one looks at the ratio of outputs to individual factors of production such as land and labor, or the ratio of outputs to the combined factors of production.[2] By stimulating increases in productivity, technological change has sharply expanded the total food supply. Zvi Griliches, for example, has concluded that almost all the observed growth in output in agriculture in the United States between 1949 and 1963 was due to technical change.[3] Technologically generated increases of the food supply in the face of relatively inelastic consumer demands have brought major consumer benefits in the form of lower prices and increased consumption.

Although most analysts have concluded that in general consumers are major beneficiaries of farm innovations, they acknowledge that the accuracy of this generalization rests heavily on the assumption that the prices of commodities affected by these innovations are determined largely by the forces of supply and demand operating in competitive markets. Most observers of American agricul-ture still consider this a realistic assumption in general.[4] But they are careful to acknowledge that this may not be the case for all farm commodities.[5] In these cases they have cautioned that potential consumer gains from technological change may be partly or completely captured by other groups in the agri-business.[6] The factors most commonly thought to negate the generalization that consumers are the major beneficiaries of technological change in agriculture are (a) the existence of economic organizations with sufficient market power to influence the price of commodities affected by technological change; and (b) the

[1]This generalization is supported by a wide range of economists. See, for example, Willard Cochrane, *Farm Prices: Myth and Reality*, Minneapolis, University of Minnesota Press, 1958, p. 85; Theodore W. Schultz, "A Policy To Redistribute Losses from Economic Progress," *Journal of Farm Economics*, 43 August 1963, p. 559; or Yujiro Hayami and Rober W. Herdt, "Market Price Effects of Technological Change on Income Distribution in Semisubsistence Agriculture," *American Journal of Agricultural Economics*, 59, May 1977, p. 245.

[2]Harold O. Carter and Warren E. Johnson, "Agricultural Productivity and Technical Change: Some Concepts, Measures and Implications," Paper presented at Workshop on Mechanization and Rural Employment sponsored by the Giannini Foundation of Agricultural Economics, Cooperative Extension and the Kellog Program, April 1978, pp. 15-22.

[3]David Sills, ed., *International Encyclopedia of the Social Sciences*, New York, Macmillan, 1968-79, s.v. "Productivity and Technology," by Zvi Griliches.

[4]See, for example, Daniel Suits, "Agriculture," in *The Structure of American Industry*, 5th ed., Walter Adams, ed., New York, Macmillan, 1977, p. 2.

[5]For a useful brief overview of the economic concentration levels existing in the principal commodity industries at the production level, see Federal Trade Commission, *Staff Report on Agricultural Cooperatives*, prepared by the Bureau of Competition, September 1975, pp. 108—118.

[6]Jurg Bieri, Alain DeJanvry, and Andrew Schmitz, "Agricultural Technology and the Distribution of Welfare Gains," *American Journal of Agricultural Economics*, 54, December 1972, p. 808.

existence of government price support programs for commodities affected by technological change; and (c) the heavy trading of commodities affected by technological change in international markets.

Economic concentration has been increasing at all levels of the food industry. Grower organizations, labor unions, and giant corporations at the producer, processor, and retail levels are acquiring sufficient market power to influence prices. Commenting on the existence of economic concentration at the producer level, Nick Kotz has noted, for example, that "twenty large corporations now control poultry production [in the United States]... United Brands, Purex, and Bud Antle, a company partly owned by Dow Chemical, dominate California lettuce production."[7]

Other analysts have made similar observations about the existence of market power at the processor and retail levels. A University of California economist, for example, recently told a State Senate hearing that Safeway Stores, Inc., and other giant supermarket chains use their enormous market power to cut the prices that cattlemen receive for their beef by 10%. At the same time, he claimed, they are able to inflate the prices consumers pay for this beef in the supermarket by another 10%.[8] These and numerous other observations of the exercise of market power by organizations at every level of the food industry have encouraged many analysts to challenge the assumption that agricultural prices are determined competitively, and with it the conclusion that the benefits of technological change inevitably accrue to consumers. Instead, they have suggested that the distribution of benefits generated by new agricultural technologies needs to be viewed more in terms of social and economic conflict between the major groups within the food production and distribution sectors. Accordingly to this perspective, the ultimate distribution of benefits generated by each type of technological change "will depend on how much of the economic surplus can be extracted by each group."[9]

Consumer gains from technological changes also may be seriously eroded if the prices of commodities affected by these changes are determined by government support programs, or in the international market. Under both of these conditions, increases in the aggregate supply of commodities affected by technological changes may not lower domestic prices or increase domestic consumption. In these cases the benefits generated by technological change will accrue primarily to producers rather than consumers.[10] Support for this conclusion was provided in part by an analysis of the welfare effects of new cotton varieties adopted in Brazil. In this analysis, Ayer and Schuh observed that the majority of benefits resulting from the improved varieties were captured by cotton producers in Sao Paulo rather than passed on to domestic consumers in the form of lower

[7] Nick Kotz, "Agribusiness," in *Radical Agriculture*, Richard Merril, ed., New York, New York University Press, 1976, p. 43.

[8] *The Sacramento Bee,* July 28, 1977.

[9] Bieri *et al., op, cit.,* p. 808.

[10] Hayami and Herdt, *op, cit.,* p. 246.

prices because an important part of that cotton was sold on the international market, leaving domestic supply and price relatively unaffected.[11]

Two other factors in addition to those already mentioned may affect the amount of benefits that consumers can expect from technological changes in agriculture. Several studies have demonstrated that, other things being equal, the less sensitive consumer demand is to price changes of a commodity, the greater the consumer benefits that may be generated by an innovation affecting that commodity.[12] When a technological advance occurs in a commodity industry, it stimulates an increase in the supply of that commodity on the market. This expanded supply of the commodity will bring about a decline in its market price. The extent to which the commodity's price declines depends primarily on how much of the new supply is taken up by additional purchases of the commodity at lower prices. The smaller the rate of increase in quantity demanded relative to the rate of price decline (the smaller the price elasticity of demand for the commodity), the lower the new equilibrium price will be, and the greater will be the benefits consumers derive from the new technology. Since the elasticity of demand varies greatly among agricultural commodities, the amount of consumer benefits that may be generated by innovations affecting them also will differ. In general, technological advances in the production of subsistence crops are thought to generate larger potential benefits for consumers because they evidence less elasticity of demand than many of the specialty commodities.[13]

A final consideration affecting the amount of consumer benefits generated by new farm technologies arises from the fact that different farm commodities are consumed by different segments of consumers. Commodities such as potatoes or onions are staples consumed in roughly equal amounts by consumers in all major income classes. Other commodities such as cream or certain fresh fruits and vegetables are largely luxury items disproportionately consumed by a much smaller number of more economically privileged shoppers. Finally, a few commodities such as cabbage or dried beans may be disproportionately consumed by poorer segments of the buying public.[14] Technological advances affecting those commodities that are most widely consumed are, of course, likely to generate larger consumer benefits than are innovations that affect commodities consumed largely by narrow segments of the public.[15]

---

[11]Harry Ayer and G. Edward Schuh, "Social Rates of Return and Other Aspects of Agricultural Research in Sao Paulo, Brazil," *American Journal of Agricultural Economics,* **54,** November 1972, pp. 559—566.

[12]Reed Hertford and Andrew Schmitz, "Measuring Economic Returns to Agricultural Research," in *Resource Allocation and Productivity in National and International Agricultural Research,* Thomas M. Arndt, Dana G. Dalrymple, and Vernon W. Ruttan, eds., Minneapolis, University of Minnesota Press, 1977, pp. 155—156.

[13]*Ibid.*

[14]Suits, *op, cit,*, p. 10.

[15]For a discussion and practical example of the importance of this factor, see Grant Scobie and Rafael Posada, "The Impact of Technical Change on Income Distribution; The Case of Rice in Colombia," *American Journal of Agricultural Economics,* **60,** February 1978, pp. 85—92.

Regardless of the value of the economic benefits that consumers ultimately reap from technological innovations in agriculture, their gains must be discounted by two important costs. The first and potentially most serious of these costs is an increased risk tht the foods eaten by consumers may be harmful to their health. This greater health risk results from a sharp jump in the use of new agricultural chemicals[16] and either the unwillingness or the inability of government regulatory agencies to institute fail-safe requirements for the testing of these new chemicals.[17] Even the staunchest supporters of the continued use of new agricultural chemicals concede that their use will increase the risks to human health. They argue only that the increased health risks caused by these chemicals are outweighed by their production benefits.[18]

The second kind of cost that needs to be discounted against consumer gains generated by new agricultural technologies is the constraint these innovations place on consumer choice. This narrowing of consumer choice stems from the fact that both mechanical and biochemical innovations encourage farmers to produce more uniform plant varieties.[19] If a farmer is attempting to increase his yields per acre, he will concentrate his efforts on a few particularly high-yielding plant varieties.[20] Similarly, if a farmer is attempting to substitute farm machines for farm labor, he will plant those relatively few plant varieties specifically developed for machine handling.[21] Most farmers, of course, are simultaneously attempting to increase their overall productivity by increasing yields per acre and substituting farm machines for farm labor. They therefore are encouraged to plant only those very few varieties that optimize both of these productivity objectives. The result is a relatively uniform crop production that limits the range of consumer choice.

Technological Change and the Farmer

Most analysts believe that farmers, at least in the aggregate, have been victims of technological change in agriculture.[22] Moreover, these analysts believe that the

---

[16]Carter and Johnson, op. cit., p. 22.

[17]Marilyn Chow and others. World Food Prospects and Agricultural Potential. New York, Praeger Publishers, 1977, p. 141.

[18]Bayard Webster, "Deciding What Is a Carcinogen," New York Times, April 13, 1976, p. 18, cited in Marilyn Chow and others, op. cit., pp. 146—147.

[19]This observation was made long ago by S. Giedion in his Mechanization Takes Command, New York, Oxford University Press, 1948, pp. 132—134.

[20]Michael Perelman, "The Green Revolution: American Agriculture in the Third World," Radical Agriculture, pp. 118—119.

[21]William Friedland and Amy Barton, Destalking the Wily Tomato, Research Monogram No. 15, Department of Applied Behavior Sciences, University of California, Davis, California, June 1975, pp. 23–24.

[22]See, for example, Earl O. Heady, "Basic Economic and Welfare Aspects of Farm Technological Advance," Journal of Farm Economics, 31, May 1949, p. 301; Schultz, op. cit., p. 562, and Scobie and Posada, op. cit., pp. 86–87.

losses suffered by farmers from technological changes are closely tied to the benefits these innovations have provided for consumers. This point of view was outlined by Willard Cochrane in 1958, when he emphasized that the continuing low incomes received by farmers of that period were due largely to low market prices resulting from the fact that technologically generated growth in aggregate farm supply had been outpacing aggregate demand. Although Cochrane and most other economists believe that farmers as a whole have been victims of technological change, they have been careful to point out that the losses suffered by farmers have not fallen uniformly on all segments of this population. Farmers who adopt new agricultural technologies early generally enjoy handsome profits, at least in the short run. They enjoy handsome profits because they are able to use the new technologies to reduce their per unit costs and expand their output in the face of constant prices. In the long run, however, other farmers also desirous of these profits adopt the new technologies. The additional output of these latter adopters ultimately will expand aggregate supply to a point where market prices decline, eliminating the opportunity for above-average profits. Thus, while early adopters make attractive short-run profits by adopting new farm technologies, the average farmer is thought to be on a "technological treadmill," because by the time he adopts a new technology market prices have already fallen to a point where his profit is no greater than it was prior to the advent of the technology.[23]

Generally, large farmers are the first to adopt new technologies and capture the large short-run profits they generate. Before farmers can adopt a new technology, they often must have access to large amount of capital, or other potentially costly prerequisites. Because larger farmers normally have greater access to capital and other technological prerequisites, they are more likely to adopt new technologies ahead of smaller farmers. This is particularly true with regard to large-scale mechanical innovations, which require large initial cash outlays.[24] The newest mechanical tomato harvesters, for example, cost $140,000. This amount is well beyond the means of most small farmers. Nor is the size of the initial outlay the only reason that large-scale mechanical innovations are beyond the means of most smaller growers. Because the use of such machines generally provides economies of scale, they are also more costly per unit of output for smaller farmers than for larger farmers. These higher costs per unit of output have often forced smaller growers to abandon production of commodities affected by such technologies. Adoption of the mechanical cotton harvester, for example, forced roughly 54% of the smaller growers out of that industry.[25] Similarly, adoption of the mechanical tomato harvester in California reduced grower ranks in that industry by approximately 85%. Most of those leaving were smaller growers having acreages far below the capacity of this new technology.[26]

[23]Cochrane, *op. cit.*, pp. 85–93.

[24]Bieri *et al.*, *op. cit.*, p. 807.

[25]Seymour Melman, "An Industrial Revolution in the Cotton South," *The Economic History Review*, 2, 1949, p. 67.

[26]Friedland and Barton, *op. cit.*, p. 54.

Elimination of the smaller grower through the process of technological advance provides an important benefit to larger growers adopting these innovations. It often facilitates the organization of remaining larger growers into effective grower organizations capable of exercising significant market power, or government influence. Adoption of the cotton harvester, for example, is thought to have contributed greatly to organizing remaining larger growers into a producer organization capable of lobbying effectively for cotton price supports.[27] Similarly, adoption of the mechanical tomato harvester is thought to have greatly facilitated organization of remaining growers for purposes of collective price bargaining with major food processors.[28] In each of these cases, elimination of numerous smaller growers through technological change facilitated the organization of the remaining larger growers. These organizational advantages were then translated into concrete monetary gains for those who remained in the industry in the machine era.

Many analysts believe that biological and chemical innovations are less likely than mechanical ones to bias the distribution of benefits in favor of the large farmer. They base this belief on the claim that biological and chemical innovations are more "scale-neutral" than large-scale mechanical technologies. This claim and the counterclaims it has generated are best illustrated in analyses of the Green Revolution now in full swing in the developing nations. Writing in the *Pakistan Development Review* in 1969, Hiromitsu Kaneda argued that

> biological and chemical innovations that have brought forth the "green revolution" are, by their very nature, neutral to the scale... Small-scale peasant farms can adopt these innovations with relatively minor adjustments in contrast to technical innovations involving tractors and combines.... The benefits of the "green revolution" have not been limited to a few large farmers.[29]

Writing in a subsequent issue of the same journal, Bose and Clark strongly seconded this opinion. According to these authors, "There are virtually no economies of scale associated with their use. New seeds and fertilizers are as productive on small holdings as on large.... They cannot benefit the small farmers as much as the large."[30]

In spite of such claims that the biological and chemical innovations of the Green Revolution are scale-neutral, many authorities on this question still feel that these innovations have benefited mainly larger farmers. Keith Griffin, for example, has argued that, because the new high-yielding varieties of rice and wheat introduced recently in many Third World nations grow best on irrigated

---

[27]Melman, *op. cit.*, pp. 68–70.

[28]Friedland and Barton, *op, cit.*, pp. 56–59.

[29]"Economic Implications of the 'Green Revolution' and the Strategy of Agricultural Development in West Pakistan," *Pakistan Development Review*, **9**, Summer 1969, p. 136.

[30]"Some Basic Considerations on Agricultural Mechanization in West Pakistan," *Pakistan Development Review*, **9**, Autumn 1969, p. 273.

land and when aided by heavy use of chemical fertilizers, large farmers have been the primary beneficiaries of these new technologies.[31] The economic logic underlying Griffin's argument was provided by Clifton Wharton, Jr., when he pointed out that rice farmers in the Phillippines using traditional seed and fertilizer practices faced investment costs of approximately $20 per hectare, whereas those farmers wishing to use the new high-yielding varieties of rice faced investment costs of approximately $220 per hectare.[32] The significantly higher investment costs associated with the high-yielding rice varieties strongly suggest that larger farmers are in a better position than their smaller counterparts to take advantage of this important new technology. Numerous empirical studies support this contention. In a study of small farmers in India, for example, Gyaneshwar Ojha found that the percentage of farmers using new hybrid varieties of maize was positively correlated with farm size. Sixty-one percent of the farmers growing between five and ten acres of maize used the new high-yielding varieties. But only 26% of the farmers growing less than five acres of maize used the high-yielding varieties. Ojha concluded that, in terms of returns per acre, adoption of the new high-yielding varieties of maize had further increased the disparity of income among Indian farmers.[33]

Apparently, differences in the distributive effects of biochemical innovations and mechanical innovations are more a matter of degree than kind. Regardless of where they are adopted, biological and chemical technologies generally require additional capital investment on the part of the farmers.[34] They also may require that adopting farmers possess irrigated land and above average farming "know-how." If larger growers have easier access to these technological prerequisites than smaller farmers, they have an advantage in the race to innovate. This bias in favor of the large farmer may be less in the case of biological and chemical technologies than it is in the case of mechanical innovations because the former are "scale-neutral," but it nonetheless does exist.

To this point I have focused on the role of technological change mainly as it affects farmer income by influencing prices they receive in the commodity markets. Herdt and Cochrane have reminded us, however, that farmers also may gain or lose from technological change in their role as land owners. In 1966, they argued that, under conditions of government-supported farm prices, retiring farmers and landowners who sell farmland have been the primary beneficiaries of technological change in agriculture. Under these conditions technological changes lower farmers' production costs and increase thier profits per acre. This encourages farmers to expand their acreage in order to increase their total profits. In pursuit of these increased profits, farmers bid up the price of land, distributing the gains of technological change to those selling farmland. These landowner gains

[31] Keith Griffin, *The Political Economy of Agrarian Change*, Cambridge, Harvard University Press, 1974, p. 56.

[32] "The Green Revolution: Cornucopia or Pandora's Box?" *Foreign Affairs*, **47**, April 1969, No. 3. p. 470.

[33] "Small Farmers and the HYV Programme," *Economic and Political Weekly*, April 4, 1970, pp. 603–605.

[34] Heady, *op. cit.*, p. 297.

may even eliminate the consumer gains thought to be generated by technological change under more competitive market conditions. This occurs if rising land costs push up production costs to a point where the costs saved by the new technology are completely canceled out. Price supports, of course, are crucial to this interpretation of winners and losers. When there are no price supports, lower per-unit costs of production lead to increases in supply, which in turn bring lower commodity prices. The lower commodity prices make land less attractive, reducing and perhaps even eliminating entirely the gains to landowners resulting from technological change.[35]

Technological Change and Farm Labor

Technological changes in agriculture have serious consequences for farm laborers as well as for farmers and consumers. As in the case of the latter groups, the consequences of agricultural innovations for farmworkers vary greatly, depending on the particular kind of innovation involved. Mechanical innovations have the most clear-cut effects of farm labor. In general this kind of technological change has (a) drastically reduced employment opportunities for farmworkers, and (b) significantly increased the productivity of those farmworkers who remain employed. Numerous accounts document the massive displacements of farm labor that have accompanied such major mechanical innovations as the tractor,[36] the mechanical cotton harvester,[37] and the mechanical tomato harvester.[38] These accounts, however, do not always indicate a simple inverse relationship between mechanization and reduced demand for farm labor. Instead, they point out that under certain conditions mechanization may increase total output of the commodity affected by this change, as well as substitute capital for labor. Adoption of the tractor, for example, increased total farm output in the United States by releasing land formerly needed to maintain draft animals for production of human foods. This increased cropland raised total farm output, and tended to increase the demand for farm labor. But, because adoption of the tractor also increased the number of acres that a single farmworker could cultivate, it simultaneously reduced the per acre demand for farm labor. The net effect of the tractor on total demand for farm labor, therefore, depended on the size of its "output effect" relative to its "labor substitution effect." What was true for the tractor also is true for other mechanical innovations in agriculture. In each case, their net effect on farm labor is a function of the trade-offs between their output effect and their labor substitution effect. Certain kinds of farm machinery, of

[35]Robert W. Herdt and Willard Cochrane, "Farm Land Prices and Farm Techological Advance," *Journal of Farm Economics*, **48**, May 1966, p. 262.

[36]*Technology on the Farm*, U.S. Department of Agriculture, 1940, p. 65.

[37]William Metzler, *Farm Mechanization and Labor Stabilization*, Series on Technological Change and Farm Labor Use, Part 2, Division of Agricultural Sciences, University of California, Giannini Research Report No. 277, September 1974, p. 1.

[38]Andrew Schmitz and David Seckler, "Mechanized Agriculture and Social Welfare: The Case of the Tomato Harvester," *American Journal of Agricultural Economics*, **52**, September 1970, p. 576.

course, are likely to have more significant output effects than others. Farm machines designed to permit more extensive cultivation, for example, are likely to have more important output effects than large-scale harvesting equipment. Even farm machinery having significant "output effects," however, generally precipitates major displacements in farm labor. This is because the labor substitution effects associated with these innovations consistently dwarf their output effects. In spite of the tractor's sizable output effect, for example, its adoption in the plantation South displaced approximately 500,000 farm laborers.[39]

Although mechanization has radically reduced the total demand for farm labor, it has sharply increased the productivity of the farmworkers who remained employed in agriculture. The clearest single indication of this increased productivity is provided by the fact that total farm output in the United States has increased 63% during the last thirty years, in spite of a 70% reduction in man-hours worked by farm labor.[40] Almost all major farm commodities can be produced today with only a fraction of the farm labor they required thirty years ago. In the 1945–1949 period, for example, production of 100 bushels of corn required 53 hours of farm labor. By 1975, production of the same amount of corn required only 6 hours of farm labor. Similarly, impressive reductions in farm labor requirements can be seen for most farm commodities.[41]

How much of these dramatic increases in farm labor productivity have been translated into higher wages for retained farm workers is unclear. The majority of mainstream economists have long presumed that wages are directly tied to worker productivity. This presumption leads them to suggest that increases in workers' productivity will automatically lead to increases in workers' wages.[42] A growing number of political economists, however, have begun to challenge this hypothesis. The critics' perspective on this question is perhaps best represented by the remarks of Ernest Mandel, who has argued that

the rise in real wages does not follow automatically from the rise in the productivity of labour. The latter only creates the possibility of such a rise... For this potential increase to become actual, two interlinked conditions are needed: a favorable evolution of relations of strength in the labour market... and effective organization... of the wage workers which enables them to abolish competition among themselves and so to take advantage of these favorable market conditions.[43]

[39]*Technology on the Farm*, p. 64.

[40]Carter and Johnson, *op. cit.*, p. 24.

[41]United States Department of Agriculture, *Agricultural Statistics*, 1976, cited in Carter and Johnson, *op. cit.*, p. 32.

[42]Melvin Reder, "Wage Structure Theory and Measurement," in *Aspects of Labor Economics*, New York, National Bureau of Economic Research, 1962, cited in David M. Gordon, *Theories of Poverty and Underemployment*, Lexington, Massachusetts, Lexington Books, 1972, p. 29.

[43]Ernest Mandel, *Marxist Economic Theory*, New York, Monthly Review Press, 1968, p. 145, cited in Gordon, *op. cit.*, p. 66.

Empirical support for this position has been provided, in part, by Barry Bluestone's study of the characteristics of low-wage industries. According to Bluestone,

the productivity gains in low-wage industry are not reflected in the relative wage rate changes in low-wage industry. Rather than contributing to higher wages, productivity increases are either being absorbed into broader profit margins or otherwise into lower prices due to raging competition.[44]

Although Bluestone's observations regarding productivity gains and wage increases are based primarily on manufacturing industries, they appear to explain the relationship between these two variables in agriculture as well. Between the 1945–1949 period and the 1970–1975 period the productivity of farm labor increased by 133%.[45] Between 1948 and 1975, however, the average hourly wage received by hired farmworkers increased by only 51%.[46] These figures suggest that, although an important part of the gains generated by increased farmworker productivity has gone to farm labor in the form of higher wages, an even greater part has been translated into broader farmer profit margins or passed on to consumers in the form of lower prices.

The consequences of biological and chemical innovations for farm labor are less clear-cut than those generated by mechanical technologies. Some biological and chemical innovations are designed primarily to stimulate increased yields per acre. Generally these technologies increase the total demand for farm labor. This increased demand for farm labor tends to increase the wages paid to farmworkers.[47] Other biological and chemical innovations are designed primarily to substitute capital for labor. Herbicides designed to eliminate hand weeding and plant varieties designed to make crops more amenable to machine harvest are obvious examples of biochemical technologies having major labor substitution effects. These innovations, of course, reduce the total demand for farm labor and work against wage increases for farm labor.

Technological Change and Nonfarm Labor

Technological changes in agriculture may have serious employment consequences for workers other than farmworkers. They also may affect the total demand for labor and the job structure in such agriculturally related industries as

[44]Barry Bluestone, "The Characteristics of Marginal Industries," in *Problems in Political Economy: An Urban Perspective*, David M. Gordon, ed., Lexington, Massachusetts, D.C. Heath, 1971, p. 103.

[45]Carter and Johnson, *op. cit.*, p. 24.

[46]U.S. Department of Agriculture Statistical Research Service, *Farm labor: 1948–1977*, cited in Sue E. Hayes, "Farm and Nonfarm Wages and Farm Benefits, 1948–1977." Paper presented at Workshop on Mechanization and Rural Employment sponsored by the Giannini Foundation of Agricultural Economics, Cooperative Extension and the Kellogg Program, April 1978, p. 8.

[47]Yujiro Hayami and Vernon W. Ruttan, *Agricultural Development: An International Perspective*, Baltimore, Maryland, Johns Hopkins Press, 1971, pp. 43, 50.

farm machinery manufacture, food processing, and transportation. Unfortunately, very few analyses of technological change in agriculture have focused on the employment consequences generated by new farm technologies in the nonfarm sector. An exception to this rule is a study of the employment consequences of the mechanical tomato harvester conducted by Jon Brandt, who found that this innovation eliminated approximately 325,000 weeks of work for farm laborers, but created 110,000 additional weeks of work in the food processing industry, and 5714 additional weeks of work in the farm transportation industry.[48] Brandt's estimates of the amount of additional employment created by the tomato harvester were based on the claim that this new technology allowed growers to plant an additional 51,000 acres of tomatoes in areas of California thought to be "labor shortage areas".[49] This claimed existence of farm labor shortage areas in California is, of course, very debatable, and no doubt would be seriously contested by farm labor groups in the state. Less debatable than Brandt's estimates of new employment created by the harvester in the food processing and farm transport industries is the additional employment this new technology opened up in the manufacture and marketing of the harvesters themselves. Unfortunately, Brandt did not attempt to estimate the new employment the harvester generated in this activity. Therefore he was unable to draw any generalization regarding the extent to which labor losses by farmworkers in the field were offset by labor gains to workers in the nonfarm sector.

Even if Brandt or others had successfully calculated the additional employment created by new farm technologies in agriculturally related industries, they still would have provided only a partial assessment of the nonfarm employment consequences generated by these innovations. A complete investigation of these consequences would also have to consider employment gains in wholly unrelated industries that might have experienced growth due to the reinvestment of the economic surplus generated by cost-reducing farm technologies. Economists have developed input–output models that allow them to estimate how changes in one sector of the economy effect changes generated throughout the system. As yet, however, these models have not provided any reliable generalizations regarding the total nonfarm consequences generated by the adoption of new farm technologies.

Conclusion

I began this paper by searching for major generalizations regarding the costs and benefits generated by the principal kinds of farm technologies for the main groups in society affected by them. The only certain generalization that has emerged from the multitude of studies on this issue is that there *are* no certain

---

[48]Jon Brandt, "An Economic Analysis of the Processing Tomato Industry," (Ph.D. dissertation, University of California, Davis, California, 1977, pp. 55–67.

[49]*Ibid.*

generalizations concerning the distribution of these costs and benefits. Instead, we have seen that the distribution of gains and losses varies a great deal from one technology to another, from one commodity–industry to another, and from one socioeconomic setting to another. What degree of competition exists in the various markets linking farmers to consumers? How well organized is labor at the various levels of the food chain? And how equally are capital and other technological prerequisites distributed among potential adopters of the new technology? The answers to these questions will vary tremendously from one commodity–industry to another. Because the answers to these questions will vary greatly, so also will the costs and benefits of new farm technologies adopted in different commodity–industries. Generalizations that seek to simplify by skirting this fundamental observation will only serve to deepen the confusion that presently characterizes public debate over further mechanization of agriculture in California.

25. Energy and Development: The Case of Asia

Roger Revelle

One of the crucial problems affecting the prospects of developing countries is the availability of energy. The increasing scarcity and rising costs of energy affect all countries, but they are a particularly acute problem for developing countries because energy is critically related to their rate and type of economic growth—activity in agriculture and manufacturing that is at the core of the development process. In this paper, I shall examine energy demands and potential supplies in thirteen of the largest developing countries of Asia.

Only approximate figures can be given for the fossil fuel and potential hydropower resources of Asian countries. In general, the location and the extent of these resources have been inadequately explored, and one can only guess at the fraction of the resource that can be recovered, or at the costs of recovery.

Table I shows some of the basic characteristics of thirteen of these countries with populations of more than 10 million people. Together, they contain half of the world's population, but only about 20% of the inhabited land area and a third of the earth's cultivated land. Because of their large populations and relatively small cultivated areas, the density of population on cultivated land is very high, averaging more than 5 persons per cultivated hectare.

Estimated energy resources for these countries are given in Table II in terms of a common unit—millions of metric tons of coal equivalent (1 million metric tons of coal equivalent contain 27.8 trillion BTUs, or 7×10^{12} kcal, close to 1000 megawatt-years, or 10^{-3} terawatt).*

Several striking facts are evident from Table II. Taken together, these thirteen Asian countries appear to possess adequate fossil fuel and hydropower resources—over 400 metric tons of coal equivalent per person—which could serve as a basis for significant economic development. Coal is the predominant fossil fuel, making up 88% of the total "commercial" resources, followed by petroleum

---

* One terawatt = 1 trillion watt-years, or 8760 billion kilowatt-hours.

Table I
Population, Agricultural Land, and Forests in Asian Countries[a]

Country	Population (millions)	Total area (million hectares)	Agricultural area Irrigated[b] (million hectares)	Agricultural area Total (million hectares)	Population density Forests (million hectares)	Forest area (people/ cultivated hectare)	Forest area population (hectare/caput)
Bangladesh	85	14.3	1.2	9.1	2.2	9.3	0.03
China	900	959.7	76.0	127.0	111.8	7.1	0.12
India	638	328.0	31.6	165.0	65.8	3.9	0.10
Indonesia	137	190.4	6.9	18.1	121.8	7.6	0.89
Iran	34	164.8	5.3	16.2	18.0	2.1	0.53
Japan	116	37.2	2.6	5.3	24.5	21.8	0.21
Republic of Korea	37	9.8	0.8	2.4	6.6	15.4	0.18
Malaysia	13	32.9	.3	3.5	23.5	3.7	1.81
Pakistan	77	80.4	14.0	19.4	2.6	4.0	0.03
Philippines	46	30.0	1.2	11.1	13.9	4.1	0.30
Thailand	45	54.4	2.5	13.9	29.0	3.2	0.64
Turkey	42	78.1	1.9	27.2	1.3	1.5	0.44
Vietnam	50	33.2	.6	5.3	13.5	9.4	0.27
Other Asia[c]	168	744.1	7.3	53.2	112.7	3.2	0.67
Totals	2388	2754.3	152.2	476.7	564.2	5.6	0.24

[a]Source: Population and total area: Asian Development Bank, 1979, except for China, Iran, Japan and Turkey, for which see FAO Production Yearbook, 1978. land use: FAO Production Yearbook, 1974, Table 1.
[b]Irrigated area around 1970.
[c]Afghanistan, Burma, Campuchea, Hong Kong, Iraq, Israel, Jordan, Lebanon, North Korea, Laos, Mongolia, Nepal, Saudi Arabia, Singapore, Sri Lanka, Syria, Yemen, South Yemen, and ten countries with less than 1 million population.

Table II
Estimated Fossil Fuel, Hydropower, and Forest Energy Resources in Some Asian Countries[a]
(in millions of metric tons of coal equivalent)

Country	Coal and lignite[b]	Petroleum and natural gas liquids	Natural gas	Hydropower[c]	Forests[d]	Total	Total per caput (metric tons/person) Total	Less forests
Bangladesh	1,187	–	290	10	433	1,920	23	17
China	719,000	17,300	26,500	18,000	22,000	802,650	891	867
India	85,800	3,000	9,800	730	12,960	112,290	176	156
Indonesia	2,520	4,100[e]	NA	410	24,000	31,030	226	51
Iran	385	15,000[e]	NA	445	3,550	19,380	570	466
Japan	8,760	11	150[e]	7,950	4,830	21,700	187	136
Republic of Korea	1,515	–	NA	615	1,300	3,430	93	58
Malaysia	NA	490[e]	NA	4,915	4,630	10,035	772	416
Pakistan	650	33	690	470	510	2,350	31	24
Philippines[f]	920	400[e]	NA	240	2,740	4,300	94	34
Thailand	235	1.1[e]	190	825	5,710	6,960	155	28
Turkey	6,880	370	NA	NA	3,600	10,850	258	173
Vietnam	>12	NA	NA	>335	2,660	>3,000	>60	>7
Totals	827,864	40,705	37,620	34,945	88,923	1,029,875	470	430

[a]Sources: Coal and lignites from World Energy Conference, *Coal Resources*, Guildford, England, IPC Science and Technology Press, 1978. Petroleum, natural gas, and hydropower resources from *Survey of Energy Resources 1978*, Guildford, England, IPC Science and Technology Press, 1978. Forest resources from *FAO Production Yearbook*, 1974.

[b]Total known resources of coal and lignite; economically recoverable resources may be less than these values.

[c]Hydropower resources taken as 100 times potential annual production of electric energy, converted to coal equivalents, assuming one equivalent ton of coal = 3×10^{10} joules.

[d]Forest resources taken as 100 times potential annual yield (= 22% of estimated net annual primary production of 8.07 metric tons of coal equivalent per hectare).

[e]Reserves taken as 40 times recent annual production.

[f]Solid fuel and petroleum resources from data in: Republic of the Philippines, Ministry of Energy, *Ten-Year Energy Program, 1979–1988*. The Philippines also have significant geothermal resources. The planned electric generating capacity for geothermal sources may be nearly 1900 megawatts by 1988, equivalent to an annual combustion of 6 million metric tons of coal equivalent.

and natural gas liquids (4.3%), and natural gas (4%). Potential hydropower is 3.7% of the total. Much less is known about the resources of natural gas than about those of coal and oil, and consequently the natural gas resources given in Table II may be considerably underestimated.

These thirteen countries exhibit large differences in total and per caput energy resources. The People's Republic of China has by far the largest resources, both in total and per caput. China's fossil fuel and hydropower resources are more than seven times those of India, which has the next largest total. On a per caput basis, Iran and Malaysia rank after China in "commercial" resources but before Turkey and India. The last two countries, however, possess enough fossil fuels and hydropower potential for more than 200 years, at their present or near-future rates of use. Bangladesh, Pakistan, Thailand, and the Philippines possess hydropower and fossil fuel resources that are smaller by more than an order of magnitude than those of China, Iran, or Malaysia, and less than a fifth those of Turkey and India. These resource-poor countries could, however, still be self-sufficient at their present rates of use of commercial energy for 75 years. But other sources would need to be found for desirable economic growth. The six countries with relatively high fossil fuel and hydropower resources per caput—China, India, Iran, Japan, Malaysia, and Turkey—contain nearly 80% of the populations of the countries listed in Table II.

Energy theoretically obtainable on a sustained basis from forests, providing the entire forested areas are subjected to intensive silviculture, is also shown in Table II. As will be pointed out later, we have assumed that the maximum sustainable energy available from the forests is about 22% of net primary photosynthetic production, estimated as 7 tons of carbon per hectare per year.[1] The figures should be divided by 100 to give potential annual energy production. It will be seen that the potential forest biomass energy is more than half the total energy resources for Vietnam, Indonesia, the Philippines, and Thailand, and a relatively small fraction in Bangladesh, China, India, Iran, and Pakistan. Fractions attributable to forests in Korea, Malaysia, Japan, and Turkey are intermediate between these extremes.

Present and Future Uses of Energy in Asia

Most available statistics on energy use refer only to so-called "commercial" energy—that is, energy from fossil fuels and hydroelectric installations. As is well known, both commercial energy use and national incomes per caput vary over a wide range between developed and less-developed countries. But on a worldwide basis, a fairly good correlation exists between national income per caput and use of commercial energy. Within the group of Asian countries we are discussing here, the gross national product per caput in 1965 ranged from $101 in Pakistan to $1671 in Japan, and commercial energy use varied from 75 kg of coal

[1] R.H. Whittaker, and G.E. Likens, "The Biosphere and Man," in *Primary Production of the Biosphere,* H. Lieth and R. H. Whittaker, eds., New York, Springer-Verlag, 1975.

equivalent per caput to 2000 kg. But the ratio of commercial energy used to income varied by only slightly more than a factor of 2.

For the poor countries of Asia, the figures on commercial energy use are misleading. For example, total energy use in India in 1970 was more than three times the quantity of commercial energy. The major part of the difference was made up of biomass fuels—wood, charcoal, agricultural residues, and cattle dung. About 15% was accounted for by human and animal labor. Estimated biomass fuels, plus human and animal labor, corresponded to about 285 kg of coal equivalent per caput. This figure is almost identical to the figure of 273 kg per caput arrived at in Briscoe's careful, year-long study of energy use in a rural village in Bangladesh in 1977.[2] However, the composition of biomass fuels in India and that in Bangladesh differed significantly. Of the 122 million coal equivalent tons of biomass fuels in India, 74 million tons, or 61% consisted of fuel wood, and 48 million tons, or 39%, were obtained from crop residues and cattle dung. In Bangladesh, fuel wood accounted for only 20% of biomass fuels, whereas crop residues and cattle dung made up 62%. The energy content of crop residues and cattle dung used for fuel in rural Bangladesh was nearly twice the food energy consumed by human beings, whereas in rural India it was somewhat smaller than the food energy intake.[3]

More than half the total energy use in India and over 80% of biomass fuels were devoted to cooking and other household uses. This proportion is much higher than that in the developed countries, as is the proportion of energy used in agriculture—12.5%, compared with about 3% in the United States.

The ratio of commercial energy to total energy increases with time and economic growth. In India, between 1970 and 1978, this increase resulted chiefly from growth of the industrial and transportation sections of the Indian economy. Real gross national product per caput rose by approximately 2% per year from 1970 to 1978.

Little information exists on total energy use in South and East Asia, and particularly on the nature and amount of noncommercial energy uses in countries other than India and Bangladesh. The total energy use for all Asian countries has been estimated as 1.2 billion tons of coal equivalent per year,[4] or about 500 kg of coal equivalent per caput. As in India, the ratio of commercial energy use to total energy probably varies with income, in part because industry and transportation, the economic sectors that use most commercial energy, have greater relative importance in higher-income countries, and in part because convenient commercial sources are substituted for traditional ones. If we make the simplifying assumptions that traditional energy use per caput remains

[2]John Briscoe, "The Political Economy of Energy Use in a Bangladesh Village," Division of Applied Science, Harvard University, mimeograph report, 1979.

[3]Roger Revelle, "Energy Use in Rural India," *Science,* **192,** 1976, 969–975.

[4]Harry Perry, and Hans H. Landsberg, "Projected World Energy Consumption," in *Energy and Climate,* National Research Council, Geophysics Study Committee, Washington, D.C., National Academy of Sciences, 1977.

constant, independent of income, while commercial energy use increases at 1.2 times the rate of income rise, and if we further assume that the average rates of economic and population growth in India over the next 50 years will be 4.5% and 1.25%, respectively, total energy use in India by the year 2030 would be close to 1500 million tons, of which around 1200 million tons would be "commercial" energy and 300 million tons would come from traditional sources. The population would be 1160 million people, and the income per caput would be $750 per year. Total energy use would be 6.5 times energy use in 1970. By extrapolation to the probable total population of 4000 million people for all of South and East Asia except Japan, energy use in the year 2030 would be about 8 billion metric tons of coal equivalent, or 8 terawatts.[4]

Future Energy Needs and Supplies in Rural Areas of Asia

To achieve a leveling off of population growth in the next 50 years, the incomes of most people in the Asian countries must rise substantially. Thus, demands for food, fibers, and energy should be much more than twice those of today.

As is suggested above, the population of Asia, outside of Japan and the Soviet Union, may be around 4000 million in the year 2030—half rural and half urban. With a sufficient rise in incomes, average diets could be based on 4000 or more kcal of food-grain equivalents per day, all locally produced. Total food supplies should be of the order of 1.7 billion tons of food-grain equivalents (800 million tons of carbon), nearly four times the present production, with an energy content of 0.8 terawatt. Double-cropping of 375 million hectares, using modern agricultural technology to obtain yields of 2.5 tons/hectare per crop, would provide this quantity of food, allowing 10% of production for seed and wastage.

Chemical and mechanical energy would be needed on the farms mainly for three inputs: pumped water for irrigation, chemical fertilizers, and operation of farm machinery. If we assume that 1 hectare meter of irrigation water per hectare would be pumped from an average depth of 10 meters by an internal combustion engine operating at 25% fuel efficiency, the energy for pumping would be about 0.5×10^6 kcal per gross-cropped hectare. (One gross-cropped hectare corresponds to one crop on 1 hectare.)

Application of 100 kg of nitrogen per gross-cropped hectare and of appropriate amounts of phosphate, potash, and other nutrients would require energy for fertilizer manufacture of around 2×10^6 kcal/hectare.[3] If small tractors and related farm machinery were used for seed bed preparation, cultivation, harvesting, threshing, and other on-farm work, the required fuel energy would be about 10^6 kcal per gross-cropped hectare.[3,5] The sum of these estimated on-farm energy inputs is 3.5×10^6 kcal per gross-cropped hectare, somewhat less than half the food energy utilized in human diets. The total for 375 million double-cropped hectares would be 0.35 terawatt.

[5]Arjun Makhijani, with Alan Poole, *Energy and Agriculture in the Third World,* Cambridge, Massachusetts, Ballinger, 1975.

Thermal energy would be needed in each rural household for cooking, and for water and space heating. Provided that the efficiency of household fuel use can be doubled[3] to 20% by installation of more efficient stoves,[6] the fuel energy per caput used in domestic activities of rural people might be somewhat more than half the 1.7×10^6 kcal per year now used in rural India, or 0.23 terawatt for 2 billion people.

Any estimate of the energy in liquid fuels required for transportation can be only a rough guess. Half the food produced and part of the products of rural industry would be transported to the cities. Most of the fertilizer and part of the fuels, equipment, and supplies used on farms and in rural households would be transported in the reverse direction. With the economic and social development of rural areas, transportation of human beings for many purposes would greatly increase over present levels. I estimate that between 35 and 350 liters of gasoline equivalent per person would be used annually, or between 56 million and 560 million metric tons for the entire rural population, corresponding to between 0.08 and 0.8 terawatt.

Estimates of power needed for industry in towns and small cities of rural areas are equally uncertain. Present industrial uses of energy in rural India are only about 0.5 kWh per person per day,[3] approximately 7% of the total energy used. To increase incomes and provide employment for the much larger rural populations expected 50 years from now, it will probably be necessary to raise the level of energy use in rural industries ten-to twentyfold, 5 to 10 kWh per person per day. For 2 billion rural people in the year 2025, the total industrial energy use would then be 0.4 to 0.8 terawatt. Household lighting could add 1 kWh per person per day, or 0.08 terawatt.

The maximum estimated rural energy use, not counting human energy in manual labor, or thermal energy used to generate electric power, would be 2.3 terawatts. This is 28% of the entire world energy consumption at the present time, and twice the present use of energy in Asia, outside of Japan and the USSR. Use per caput would correspond to 1.1 tons of coal equivalent, approximately three times the 1975 value in rural India.[3]

Maximum energy use for the rural population would be only 30% of the estimated total for Asia in the year 2030,[4] based on an annual growth rate in energy use of about 4%. Because of the close relationship between incomes and energy use, rural incomes would still be considerably lower than those of the urban populations.

Some of the energy demands we have described could be met in a variety of ways; others would require a more limited range of energy sources.[7] Depending on the local situation, energy for irrigation pumping could come, as it mainly does at present, from central power station thermal or hydroelectric plants, and

[6] J. Goldemberg, and R. I. Brown, "Cooking Stoves: The State of the Art," manuscript, Institute of Physics, University of São Paulo, Brazil, 1978.

[7] National Research Council, Panel on Renewable Energy Resources, *Energy for Rural Development, Renewable Resources and Alternative Technologies for Developing Countries,* Washington, D.C., National Academy of Sciences.

fossil fuel hydrocarbons. In other regions, biogas produced from agricultural wastes, locally generated hydropower, windmills, direct solar energy conversion, or local electric power generation based on energy plantations or agricultural residues could be used. Liquid fuels would be required for the vehicles used in transportation and probably would be the most economical source of energy for tractors and other farm machinery. These could be derived from fossil fuel hydrocarbons, or alcohol and possibly hydrocarbons could be obtained from sugar cane and other highly productive crops in countries with abundant sunshine and land and water resources. Production of nitrogen fertilizers would require either the consumption of fossil fuels or hydroelectric power for electrolytic production of hydrogen and manufacture of ammonia. With improved forest management, household energy supplies could come from the wood of forest trees in some Asian countries. But because of the high man-to-land ratios in most countries, it might be necessary to use agricultural residues as a major source of household energy.

One of the problems of energy supply would be provision of energy for rural industries. In regions where hydroelectric power can be developed relatively inexpensively, central power station electrification would probably be most economical.[5] But in monsoon Asia, with its occasional prolonged droughts, hydropower must be balanced with thermal electric plants to ensure a reliable power supply. Thermal electric plants might be fueled in some regions with wood grown in energy plantations. In other rural areas, agricultural residues or wood harvested from well-managed forests might be used as a source of fuel.

At least 2.7 terawatts of thermal energy would be required to meet our estimated maximum energy demand of 0.8 terawatt for rural industries, if all energy were supplied as electricity generated in thermal power plants. With our estimated total food production, the energy from agricultural residues would be about 1.6 terawatts. Probably no more than half of this amount could be used as fuel. The maximum sustained supply from forests with intensive silviculture could be 0.9 terawatt. Thus, biomass energy production from forests and agricultural residues, even if stretched to the limit, would not be sufficient to supply rural industrial needs for electricity, unless a large fraction of energy demand in these industries could be met from other sources.

The uneven distribution of forests relative to populations is evident from Table I. Equally serious is the geographic maldistribution of potential hydroelectric power and of land that could be made available for biomass energy plantations. Large capital investments and sustained efforts in all aspects of energy research and development will be necessary to ensure adequate future energy supplies for the Asian countries.

Biomass Supplies in the Future World Energy Economy

Throughout history, the major source of energy for most of mankind has been the photosynthetic conversion of solar into chemical energy. In absolute terms, biomass net energy production (photosynthetic production minus plant respiration) is very large. Most of it takes place in the world's forests The quantity of biomass energy that can be utilized in the world energy economy is constrained by several factors: (1) the low efficiency of photosynthetic conversion of solar energy; (2) the fact that the sustainable yields from forests are relatively small fractions of net primary production; (3) the existence of necessary alternative uses for biomass; (4) the necessity of using nonbiomass energy to attain an increase in sustainable biomass yields: (5) the requirement for substantial investments, if sustainable yields are to be raised; (6) the great distances between many forested areas and the regions of high energy demand, which require much energy be used in the transportation of harvested wood to the point of use; (7) the necessity to use some energy in harvesting; and (8) the growing demand for forest land for agriculture and other human uses.

The net primary production of the world's forests, which are estimated to cover over 41 million square kilometers,[8] about 25% of the earth's land area, corresponds to only about 0.3% of the energy in the solar radiation received by the forests.[1] Similarly, in most less-developed countries, the chemical energy in agricultural crops, including both the edible portion and the nonedible recoverable residues, is about 0.2% of incoming solar radiation.

Energy from Forests and Agricultural Residues

The energy that can be obtained for human use on a sustainable basis from the world's forests is much less than the net annual energy production as we have defined it. This is in contrast to cultivated agricultural areas, where about half of the net primary production can be harvested indefinitely. Forests in their natural state are virtually closed ecosystems, in which fires and the heterotrophic respiration of animals and microorganisms just about balance the net primary production.[9] The harvesting of forest products by man immediately creates an artificial ecosystem. It changes the age distribution, size, and composition of the populations of trees and other plants in the forest and, hence, the rates of net primary production of organic matter, as well as the rates of heterotrophic respiration and fires. The maximum sustainable yield from the forest will be the difference between these two rates, as each is affected by harvesting and other human actions. On a long time scale, the sustainable yield depends on the balance between the rates of replacement and removal of nitrogen, phosphorus, and other nutrients in a harvested forest.[10]

Most of the nutrients are contained in the forest canopy of leaves and needles and in the litter of dead plant materials covering the ground, which together,

[8]Food and Agricultural Organization of the United Nations, *FAO Yearbook of Forest Products,* Rome, FAO, 1979.

[9]Bert Bolin, "Global Ecology and Man," Paper presented at the World Climate Conference, Geneva (February 1979), World Meteorological Organization, Geneva, 1979.

[10]A. Baumgartner, "Climate Variability and Forestry," Paper presented at the World Climate Conference, Geneva (February 1979), World Meteorological Organization, Geneva, 1979.

according to Bolin,[9] make up at least 50% of the net primary production of organic matter in a forest. Only small amounts of nitrogen and phosphorus are contained in wood. Earl[11] estimates that wood above ground in an average forest comprises 20 to 25% of the total organic matter and a similar percentage of the net primary production. He states that only 13.5% of annual wood production is now harvested. More than half of this is used for lumber, paper pulp, and other industrial purposes, and somewhat less than half is used as fuel. Thus, only 3.1% of net primary production in the world's forests is now being utilized by human beings. A similar estimate has been given by Revelle and Munk.[12]

That the sustained yield of wood can be a much larger percentage of the net primary production is shown by experience in Europe, where scientific silviculture to increase yields has long been practiced. In 1974, the ratio of wood harvest (including bark) to estimated net biomass production in European forests was 11.4%.[3,8] In the United States, which possesses some 202 million hectares of commercial timberland, about 225 million metric tons of wood are harvested annually by forest products industries. This is somewhat less than the estimated annual growth rate of 273 million metric tons of wood. According to Spurr,[13] the potential growth would be twice this figure " if all the commercial forest areas were fully stocked with trees and balanced in age classes. With a wide spread application of intensive silviculture, including the complete use of hardwoods and the utilization of the entire tree of every harvested species, the potential productivity of the forest in the U.S. would be closer to three times its present level." If one assumes that the potential productivity is 800 million metric tons of wood from the 200 million hectares classified as commercial forest, the yield per hectare would be equivalent to 2 metric tons of carbon. This is 36% of the net primary production of 5.6 tons of carbon per hectare, estimated for 830 million hectares of North and Central American Forests.[12] At least one-third of this total forested area is not of commercial grade,[13] and hence the figure for potential production is undoubtedly an upper limit to what can be accomplished by advanced silvicultured technology. Tyers[14] estimates that the sustainable harvest from managed wood lots in rural areas of Bangladesh is of the order of 1.8 metric tons of carbon per hectare, probably about 25% of net primary production.

On the basis of these scanty data, I estimate that the sustainable yield from the world's forests, if the entire forested area were subjected to intensive silviculture, would be about midway between the present yield of European forests, and the potential maximum production from the commercial forests of the United States—say 22% of the net primary production.

[11] D.E. Earl, *Forest Energy and Economic Development,* London, Oxford University Press, 1975.

[12] Roger Revelle, and Walter Munk, "The Carbon Dioxide Cycle and the Biosphere," in *Energy and Climate,* National Research Council, Geophysics Study Committee, Washington, D.C., National Academy of Sciences, 1977.

[13] Stephen H. Spurr, "Silviculture," *Scientific American,* 240, 1979, pp. 76–91.

[14] Rodney Tyers, *Optimal Resource Allocation in Transitional Agriculture: Case Studies in Bangladesh,* Ph.D. dissertation, Cambridge, Harvard University, 1978.

Estimates of recoverable residues from agricultural production of foods, fibers, and animal feeds differ widely.[5,15] Taking the mean values of the range given by different authors for food grains (cereals and pulses), we obtain 1.75 billion metric tons of dry organic matter in wheat and rice straw, maize stover, rice husks, and other residues from the world production in 1975 of 1300 million tons of humanly edible food grains. To this figure should be added the dry weight of part of the dung produced by domestic animals, taken as one-third of the weight of cereals, pulses, and cultivated pasture grasses and fodder consumed, or 700 million tons; 100 million tons of residues from world producion of plant fibers, including cotton seed and lint; and perhaps 300 million tons of residue from other crops. (Shacklady[16] estimates 240 million tons of residues from world production of only five such crops—cassava, bananas, citrus fruits, coffee, and sugar cane.) The total of the above estimates is 2850 million metric tons of residues, corresponding to 1425 million tons of carbon, or 1640 million tons of coal equivalent (about 1.6 terawatts).

An almost identical result in terms of tons of coal equivalent is obtained by taking 1.15 times the total of world food production in food grain equivalents (the quantity of food grains having the same energy content as the crops actually harvested). In Asia, where statistics on food production are uncertain except for a few staple crops, it is perhaps better to use twice the average energy content of human diets—about 2100 kcal per person per day—multiplied by 1.1 to account for seed, feed, and wastage, as the basis for computing potential energy yield of recoverable residues from agricultural production. In this way, we obtain an energy content of 500 million tons of coal equivalent (0.47 terawatt) in agricultural residues for the estimated Asian population of 2330 million people. This result is slightly too high because the Asian countries taken together are not self-sufficient in food production.

Energy Plantations to Increase Biomass Yield

Higher energy yields from biomass might be obtained from cultivated and irrigated plantations of fast-growing trees in tropical areas of abundant sunshine. Experience on the Coromandel Littoral of the State of Tamil Nadu in India with small plantations of a fast-growing tree species, *Casuarina equisetifolia,* shows that 175 tons of dry wood per hectare can be obtained after four years, with an energy content of close to 5000 kcal/kg.[17] On the basis of an efficiency of 30% for conversion of thermal to electrical energy, a plantation with a total area of 11,400 hectares would provide fuel for an electric power plant generating 100 mW of

[15]National Research Council, Panel on Methane Generation, *Methane Generation from Human, Animal, and Agricultural Wastes,* Washington, D.C., National Academy of Sciences, 1977.

[16]Shacklady, C.A., "Bioconversion of Organic Residues for Rural Communities," work in progress, Tokyo, United Nations University, 1978.

[17]C.V. Seshadri, G. Venkataramani, and V. Vasanth, "Energy Plantations: A Case Study for the Coromandel Littoral," monograph series, Vol. 2: *Engineering of Photosynthetic Systems,* Madras, India, MCRC, 1978.

average power production, equivalent to 8.8 electrical kW/hectare. The capital cost for the plantation is estimated at $26 million, of which 80% would be allocated to road construction and the purchase of land. Annual operating costs would be $4.4 million, plus interest, and depreciation on buildings, machinery, tools, and work animals. If interest and depreciation amount to 10% of capital costs, the total annual cost of the plantation would be $7 million, corresponding to $14.70 per metric ton of dried wood. This is 10 to 15% less than the price of Indian coal delivered at power plants in Tamil Nadu, with a roughly equal energy content.

The plantation would provide employment for about 12,000 people—of whom one-third would be women—at a capital cost (not including cost of land) of about $1000 per worker. The ratio of the output of thermal energy to the input of thermal and mechanical energy, including human and animal work and diesel fuel, is estimated to be close to 100:1. The plantation would not require application of chemical fertilizer, because *Casuarina* roots have a symbiotic, nitrogen-fixing actinomycete, which fixes about 80 kg of nitrogen per hectare per year.

An energy plantation utilizing the giant "Ipil-Ipil" (*Leucaena latisiliqua*) is now operating in the Philippines, providing fuel for an average electrical power generation of 50 mW.[18] The efficiency of electrical generation is only 18% of the thermal energy in the wood, versus the 30% assumed for the proposed plantation in Tamil Nadu. This suggests the need for a research and development program in combustion engineering of wood for the generation of electric power.

In principle, approximately 100 million hectares in energy plantations would be required to produce 1,000,000 mW (1 terawatt) of steady electrical power. Scaling up from the estimates for Tamil Nadu, the capital cost would be close to $260 billion, not counting the cost of the electrical power plants. If these were included, the total cost would be about $900 billion.

Manufacture of Liquid Fuels

For transportation and many industrial uses, liquid or gaseous fuels, with as high an energy density (energy per unit volume) as possible, have clear advantages over solid fuel, particularly those of low energy density, such as wood or agricultural residues. One way to obtain liquid fuels from biomass materials is to produce ethyl alcohol (ethanol) by microbial fermentation and subsequent distillation. Brazil is embarked on a practical application of this process.[19] All 1.5 million automobiles in the city of Sao Paulo are now using a mixture of 20% ethanol and 80% gasoline, and , by 1982, 15 to 20% of all cars built in the country will be manufactured to run with 100% ethanol. It is planned that over 3 million tons of ethanol will be produced by 1980, and 26 million tons by 1990, at which

[18]Jose A. Semana, "Energy Plantations for Steam Power Plants." FORPRIDECOM, Manila, Philippines (August 1977), quoted by Seshadri, *op. cit.*

[19]Jose Goldemberg, "Global Options for Short-Range Alternative Energy Strategies," manuscript, Institute of Physics, University of São Paulo, Brazil, 1979.

time alcohol will be substituted completely for the potential uses of gasoline and diesel oil (roughly 40% of present use of petroleum). To produce this large amount of alcohol will require capital investments of $15 to $30 billion. Spread out over 15 years, the annual investment would be about 1% of Brazil's gross national product and one-fourth to one-half of the present cost of imported petroleum.

The alcohol is produced from sugar cane at a recovery rate of 70 liters/ton—about 6% by weight and 11% of chemical energy in the cane. Part of the remaining energy is used in distillation and other processing. With a yield of 60 tons of sugar cane per hectare, about 0.13% of the energy in the incident solar radiation is converted to the energy in the alcohol.

Approximately 8 million hectares would be required to produce 26 million tons of alcohol planned for 1990. Brazil is said to possess 70 million hectares of fertile actual or potential agricultural land, and at least one-third of this could be devoted to production of biomass energy without seriously interfering with desirable production of food and fibers, including export crops. Even such a large allocation of arable land to production of liquid fuel in the form of ethanol would provide less than 1 liter of gasoline equivalent per person per day for Brazil's estimated population of 150 million people in the 1990s. In this large country, where transportation over long distances is essential for socioeconomic development, very careful planning and strict control of the means of transportation would be required to eke out the meager supply of liquid fuels. Nevertheless, the scheme is well-suited to take advantage of Brazil's abundant, under-utilized resources of land, water, and sunshine. It would be much less useful in more densely populated, less-developed countries with high man-to-land ratios. But liquid fuels for rural industries might be produced on a local scale in this way. For example, 10,000 hectares planted in sugar cane for ethanol production would provide over 30,000 tons of alcohol, with an energy content close to 30 thermal megawatts.

Biomass Energy in Rural Asia

Probably 75% of Asia's 3000 million people live in rural areas, and their major source of energy comes from biomass production.[3,5] Wheat, rice, and other food grains are the principal components of their diets, and those must be cooked before they are eaten. Hence, fuel is just as essential as food for the rural people of Asia. Global totals for Asia conceal marked regional differences. Indeed, in some regions—for example, the Hills of Nepal and similar hilly regions of Pakistan and India—the amount of biomass energy now used each year is considerably greater than the sustainable yields from forests and grasslands. These natural resources are being depleted by overcutting and overgrazing.[11] In areas of abundant natural vegetation, biomass energy has been traditionally thought of as a free good. But this energy is not free when it is in short supply. The price of biomass fuel in regions of short supply will be set by the local price of commercial fuels having the same energy content, in terms of energy delivered to the food being cooked.[14] In the hills of Pakistan, the cost of wood used in cooking is even now about one-fourth of the cost of the food itself.

Like petroleum, which can be used to manufacture petrochemicals in addition to its uses as a fuel, biomass materials have several alternative uses—as construction materials, livestock feed, fertilizers, and soil conditioners, and to prevent or slow down erosion of valuable soils and lands. Consequently, only a fraction of biomass yields, which differ in magnitude in different rural societies, can be used for fuel.

With growing populations throughout the rural areas of Asia, the supply of both food and fuel must be increased at a rate at least equal to the rate of population growth. To avoid destruction of the resource base, this increase must be achieved by obtaining higher sustainable yields from forests, pastures, and cultivated fields. In general, even a modest increase in the sustainable biomass yield will require the introduction of nonbiomass energy—for example, in irrigation and in the form of chemical fertilizer.

Rural Electrification

Rural electrification has been a major development activity for many years. In general, it has been justified by the evident need for more energy in rural areas and by an implicit belief that electrification would bring modernization in its train. Recently, these justifications have been seriously questioned—among other reasons, on the grounds that the poorest classes receive little or no benefit from the typical rural uses of electricity.

More energy is certainly needed in rural South and Southeast Asia in order to raise the incomes of the people and to improve the quality of their lives. However, these increased energy supplies are required in a variety of sources and forms. Gas (including biogas), liquid, or solid fuels, th latter including wood, charcoal, and agricultural residues, are probably the most useful energy sources for cooking and water- and space-heating. Liquid or compressed-gas fuels are needed to power farm machinery and vehicles for transportation. Electric power is useful for lighting and pumping water for irrigation, although other forms of energy can be substituted. But because of its flexibility and ease of use, *electrical energy is probably more satisfactory than any other form to provide shaft-power for industries.* Many industries suitable for rural areas also require process heat or process steam, or both, as well as shaft-power. In most cases, these can probably be obtained more economically by direct combustion of fuels rather than by electric heating.

In the Asian region, agricultural modernization is—and will be—essential to provide enough food for growing populations, but it cannot provide sufficient employment for rapidly growing rural labor forces. In order to diminish rural poverty, new nonfarming jobs must be created. Considerable employment can be provided by development of small-scale rural industries, based on processing and adding value to agricultural products, and on meeting local needs for consumer and capital goods. From a strictly economic standpoint, the costs and difficulties of transporting raw materials and finished products, together with the existence of a pool of surplus labor, are two of the justifications for establishing small-scale rural industries. In many situations, these factors will overbalance the economies of scale realized by large, centralized industries in large cities.

Rural employment is not an end in itself, but rather a means to attain greater equalization of incomes; to enhance human dignity based on the freedom that comes from self-reliance; and to raise average incomes through fuller use of human resources. Empirical evidence indicates that equalization of incomes acts as a powerful force toward the stabilization of population size.

If, as the above discussion indicates, the long-term salvation of the rural poor depends both on agricultural modernization and on rural industrial development, the future of rural electrification needs to be re-examined in terms of the fundamental role of electricity as an energy source for small-scale industries. Moreover, planning for future rural electrification should be based, wherever possible, on analysis of the possibilities in a given region for creation of viable rural industries, as well as on potential uses of electricity in agriculture. Rural industries have sprung up spontaneously in some small cities of the Indian subcontinent, and these have been helped by government or private agencies. Jobs in construction, kraft paper manufacturing, diamond cutting and polishing, rug weaving, making of dry foods such as "papar," and the production of brown sugar are among the industries established by rural development agencies. Some have been created at an average capital cost of $500 per worker.

I have seen several examples of industrial applications of electricity, commonly combined with other forms of energy, in villages, market towns, and small cities:

1. In a village near Faisalibad, Pakistan, about twenty electrified power looms are being used to weave "gray goods" from purchased yarn, giving employment to perhaps fifteen villagers.

2. In Gujranwala and Sarghoda, small cities in the middle of the Pakistan Punjab, there are machine shops that make diesel engines and pumps for tube wells. Other shops manufacture coir and metal well-screens and casing for tube wells.

3. The Amul dairy in the Kaira district of Gujarat in India collects milk from farmers in the district, chills and pasteurizes it, and ships it to New Delhi for distribution and sale. This enterprise has provided considerable employment and has resulted in significant increases of incomes, particularly for landless people who own only a few head of buffalo or cows.

4. A variety of small industries has been established in Rajkot, a small city in the center of Saurashtra in the state of Gujarat, India.

5. Ludhiana is a medium-sized city in the northwestern part of the Indian Punjab. Farmers within a radius of about 20 miles commute on their bicycles to work in the factories and shops of the city, returning home to "moonlight" on their farms.

6. In Comilla in Bangladesh, the cooperatives established by Akter Hameed

Khan invested their savings in a foundry, a potato cold-storage plant, a dairy, a facility for parboiling and milling rice, and shops for the maintenance and repair of irrigation pumps and farm machinery and for the manufacture of well-drilling equipment.

7. An "industrial park" is being developed in Hetara, a relatively new town on the northern edge of the Terai in Nepal. A flour mill, a saw mill, a brewery, a seed-processing plant, an aquaculture farm, and several other small industrial establishments are in operation there.

Considerable effort has gone into rural electrification in the Philippines. Several small industries have emerged in the area of the Misamis Oriental Rural Electric Service Cooperative (MORESCO) on the island of Mindanao. The role of electrification in this development should be studied.

Although rural electrification may often be a necessary condition for the establishment of small industries in rural areas, experience shows that it is far from a sufficient condition. Other factors that are likely to be important include (1) the actual or potential existence of adequate transportation facilities to and from markets and sources of raw materials; (2) the availability of an adequate supply of other forms of energy needed for a particular industry or industrial complex; (3) sufficient investment capital, in the form of credit or savings; (4) the presence of trained technicians and of managers familiar with problems of marketing, cost accounting, personnel management, and business operation; (5) the existence of a market town or small city in which a complex of mutually supporting industries can be established; (6) the availability, skills, discipline, and comparative costs of labor in the rural areas versus the labor force in large cities; and (7) the reliability of the electric supply.

A second emerging issue in rural electrification is the size and location of electricity-generating plants. In principle, local energy sources have a considerable advantage over centralized sources in that the cost of transmission of electrical energy can be lessened. Because of very low load factors, the real cost per unit of output of electrical energy carried to rural areas over a transmission grid from a central power station is high. This is especially true when electricity is used primarily for pumping irrigation water, plus domestic or community lighting. Irrigation pumps, for example, may be operated as little as 4% of the time. For small farmers who must pump water for irrigation, local energy sources such as windmills, biogas plants, or pumps powered by photovoltaic cells may be more satisfactory than central-station electricity.

In some areas such as the Hills of Nepal and the mountainous regions of India and Pakistan, local generation of "microhydropower" from small run-of-the-river installations can be used. When sufficient water and land are available, it may be possible to develop plantations of fast-growing trees (producing up to 50 tons of wood per hectare per year) as a source of fuel for small thermal-electric plants. Continuous generation of 10 megawatts of electric power should be possible for each thousand hectares planted in suitable tree varieties under good growing conditions. Under some (probably unusual) circumstances, it may be helpful to generate small quantities of electric power from community biogas plants, using agricultural residues and human and animal wastes as the feedstock.

On the other hand, generation of electricity in central power stations feeding into a transmission grid can make effective use of the reserves of coal and natural gas in India, Pakistan, and Bangladesh, and of the hydropower potential of larger rivers in India, Pakistan, Nepal, Sri Lanka, and the countries of southeast Asia, thereby reducing dependence on imported oil supplies.

Important policy questions emerge in considering rural electrification. How much primary and secondary employment, what increases in incomes, and what multiplier effects can be expected from rural industrialization based on electrification? How can the load factors on electrical transmission grids from central station power systems be raised? What kind of public sector intervention (for example, credit for investments, technical training, road-building, construction of storage and market facilities, and subsidies for other necessary forms of energy for certain industries) should supplement electrification in order to stimulate rural industrialization? What factors need to be taken into account in appraising benefit/cost ratios? What types of data are needed for formal project evaluation?

Problems of energy supply and use in the developing countries of Asia are problems not only for their own people but also for the people inhabiting more industrially advanced countries elsewhere. The advanced countries need markets for their manufactured goods; they are competitors with developing countries for scarce fossil fuels. For these and other reasons, including humanitarian concerns, it would be prudent for the advanced countries to make more strenuous efforts to assist the developing countries to determine the most appropriate mix of energy sources and to make the best use of local sources and advanced technology. Scientists and engineers can play a major role in this effort.

26. Arms Control in the Era of Limits

Edmund G. Brown, Jr.

I am not what I would call a theologian of the arcane theology of strategic thinking, but as one of the lemmings on the way to the sea I feel that I have the obligation to make a few observations and try to set in a better perspective the kind of world we are in and are about to enter. The premises that have been programmed into our minds are often difficult to eradicate, but what was true in one period may not necessarily be true in another, and the history of evolution is the history of those species that survived longer than others did. Most people would consider that statement a tautology, and it probably is, but it is about the most abstract way of expressing the observation that some species fall into a cul de sac from which they never emerge, and others are able to adapt and flourish by changing their environment. We look at the world today, and we see the rising population curve, the depletion of our resources, and the escalating arms race. We know that the terms of the equation are changing, but much of our analysis is based on premises developed after the Second World War during the 1950s and 1960s, in a period that was quite different.

That was a period when the position of America was unchallenged throughout the world, when in our minds we divided the world between the Russian bloc and the American bloc, when resource depletion had not reached its present level, when the energy crisis was merely mentioned in a forgotten paper presented to President Truman in the early 1950s, when the notion of a world of limits existed only in the minds of the most astute thinkers. In the 1950s personal income grew at a rate of about 3.5% a year. In the 1960s it was about 3% a year. In the 1970s personal income per capita in this country was growing at about 0.5%, and in the foreseeable future this figure will probably remain at about zero or decline.

This situation means that the choices we have are limited, and that the selection of option A will foreclose option B. That is why several years ago I used the phrase "era of limits" to describe the world in which we live. Some people interpret that as an antibusiness, pessimistic, lower-your-expectations philosophy. I see it as just a sort of shorthand to describe the reality we face.

We are in an era of limits—limits that are being imposed by agriculture, by the excessive use of pesticides, by population growth, by the arms race, by the inflationary spiral, and by the arrangement of powers in the world. We know that we are in a very different world, because when we choose a new course of action we do not fund it out of the annual dividend that a growing economy yields. Rather, we fund it out of the reduced standard of living of all the people in the country, and that tends to make people grumpy, if not downright nasty. Perhaps the negative attitude toward our presidents and our governors and even our mayors is due to this phenomenon. People understand that government activity is no longer a matter of taking a percentage of the 3% annual increase in our personal wealth, but rather it is an incision into the standard of living to which we have become accustomed. If the economic stagnation of the last few years continues, as many predict, then the competition for resources will become all the more intense, and the various constituencies that have mobilized their troops, their doctrines, and their lobbyists will have to confront the unpleasant reality that there is not enough to go around.

That is why it is very important today to set forth what we described in the 1960s as a reordering of our priorities. Such a reordering is painful, because the concept of a reordering of our priorities means that one takes a priority that is higher on the list and moves it down on the list, and then takes another priority that is lower and raises it higher, and for those who are hurt by these changes, life is not pleasant. Among those who benefit, expectations rise, and therein we have created a political conflict. In my own state of California, I have tried to impose some limits on the budget, and I have tried to impose some limits on the degradation of our environment, because I recognize that the course of wisdom will be found in the philosophy of imposing limits so as to yield the maximum in opportunities.

Today we are facing a price increase by the OPEC nations that certainly will make the cost of gasoline 30 to 50% higher than it was a year ago. In the 1950s, very optimistic and excited GI's returned to start their families on a supposedly unending expansion to the suburbs, with cheap housing and cheap fuel and unquestioned mobility. Now, in Levittown, Pennsylvania, we have gas riots and truck protests because a few foreign powers have decided not to give us the oil that makes this economy operate. And so, when we think of our strategic security, we must consider the resources that make a great country what it is, and when we see the symbol of Levittown turn from one of hope to one of frustration and anger and civil disarray, we can find in that a metaphor for the 1980s, if our society does not make the correct choices. If we persist in continuing to believe the myths and dogmas of the 1950s and 1960s, certainly what is now an isolated incident will become the pervasive image of our society and of the society of many of our allies.

I propose that we look very carefully at our military budget, and that we examine it not just in terms of its specifics, but consider what the options are for the limited funds that are available. We have a certain budget, and a certain flexibility in how we spend it; we have a certain amount of unused potential, and the question is, Where do we put our fiscal resources? Today we are being offered a choice. I think that it is a modest step, but an important step, and that is President Carter's proposal on strategic arms limitation. There are critics from the left, and louder critics from the right, but it is fair to say that this treaty, negotiated after several years of intensive bargaining, will put a limit on the number of strategic launchers—a rather high limit, but nevertheless a cap. It will protect our capacity to verify the activities of the Soviet Union, and it will further set the stage for more intensive and more significant arms reductions in the world. The fact that it is not the whole loaf, that it is not the perfect treaty, does not negate the fact that it is a framework that leads us in the direction of what I would call an era of limits in the arms race itself. And that is why I strongly support the efforts of the President and commend the treaty to the Senate, which will be faced with the decision of ratifying it.

That is point one. However, there is point two. Along with the SALT treaty, President Carter has proposed a mass transit system for missiles. This mass transit system, called the MX, with its mobile basing system, is a 30-billion-, perhaps a 50-billion-, perhaps a 70-billion-dollar boondoggle that will jeopardize, not protect, the security of the United States. First, the system does not do what it is supposed to do. Second, it will force the Soviet Union to respond in kind by building its own mobile based missiles, thereby putting our security even more in jeopardy. Third, it shifts our strategic policies away from mutual deterrence to nuclear war fighting. Finally, it uses up very scarce capital that is necessary for waging the most important security battle of all, and that is attaining an energy resource base adequate to the needs and the expectations of the American people.

I want to go over a couple of these points. This proposal will not do what it says it will do because it will not even be in place until the late 1980s, and yet the vulnerability it is supposed to remedy may appear as early as 1981. How then can it protect us during the rather lengthy interval during the 1980s? Second, it will force the USSR to match our efforts, just as the moving of our Minutemen has forced them to follow in kind, and then we shall have another ratchet upward of the arms race. Far from having more security, we shall have less, and verification will not be any easier, because we cannot be sure that the Soviets will open their trenches for satellite verification, or that they will be so accommodating once the arms race advances this next notch higher. Next, it will shift the emphasis in a very decisive way from mutual assured deterrence, which has been the doctrine of Eisenhower, Kennedy, McNamara, Nixon, and Khruschev, to a new doctrine leading to nuclear war, where they throw one missile, we throw two, they throw six, and we throw eight. That whole theology, which reminds me of the scholastic debate about how many angels can dance on the head of a pin, is not only dangerous, it represents a pathological addiction to a numbers game that is not one of security but rather one of paranoia.

Suppose that 90% of the ICBMs of this nation were wiped out in a surprise attack. That would leave only 10%, but that 10% would still give use 210 nuclear warheads with the equivalent of 130 megatons of destruction. It would still leave the submarines and the bombers, which together have a 325-megaton equivalent, or more than the capacity for overkill. In fact, the July (1979) *Scientific American* contains a very interesting chart, which shows clearly both the immediate and the delayed effects of nuclear war.* It shows that even after a surprise attack we would still have the number of warheads necessary for mutual assured destruction, which is defined as the killing of 20 to 30% of the Soviet people and the devastation of two-thirds of their industrial capacity.

If we have only that amount which the graph represents, we can achieve mutual assured destruction. So we have many times overkill. The article also shows that we have various scenarios of surprise attack—when we are on alert, when we are not on alert—and the bottom line is that we have many, many times overkill. Because of these factors, I do not accept the notion that by agreeing to the SALT treaty somehow we are about to lose our strategic credibility, or that the perception of this alleged weakness will encourage the Soviets to expand their sphere of influence, or will intimidate our allies or our own leaders. This is a dangerous myth that must be refuted at every opportunity, because the choice of spending billions and billions on weapons we shall never use means that we forego opportunities of even higher moment.

If in the next five to ten years this nation does not have the energy to supply the mobility that really defines the social context in which we live, then the democratic fabric itself will come unwound. That is the critical question for the 1980s: Are we going to stress productivity or destructivity? It is hard for me even to pronounce the word: destructivity. How much of it is enough, how much is overkill? In the theology of waging nuclear war over a sustained period, we face the risk that will surely take the lemmings into the sea, whereas if we turn aside from that theology and follow the traditional doctrine of mutual assured destruction, which itself is nothing to be happy about, at least we shall have the billions of dollars necessary to invest in rapid transit, in solar energy, in synthetic fuels, in the photovoltaics, the geothermal, and other energy resources. The world is not just "Russia versus the United States." Vietnam proved that. Instead of the Vietnam War's leading to Chinese expansion, it led to a Vietnamese–Chinese conflict. The fact that the army of Iran was stronger than that of any other Moslem power did not lead to the security of the Shah but to his downfall, and not by a Marxist insurgency but by an aging religious leader exiled to France. The concepts of the planners have often been proved wrong, and they will be proved wrong in this case. The shrinking resource base, the competition between the North and the South, between the have-nots and the haves, between the starving masses and the wealthy, between assertions of nationalism and assertions of global interdependence, the cry rising from the many poor nations of the world—these specters will haunt the developed world in the eighties and nineties.

*Kevin L. Lewis, "The Prompt and Delayed Effects of Nuclear War," *Scientific American,* July 1979, Vol. 241, No. 1, pp. 27–39.

Today we are looking at a hundred thousand "boat people" seeking asylum. In the mid-eighties, we may have millions of "boat people"—if the weather changes, if the agricultural output is insufficient, if the birth rates continue to escalate. So we must readjust, rethink, and reexamine clearly the threats to our security. The unequal distribution of food, of energy, of opportunity—certainly these are great threats to our security. The industrialized world is becoming increasingly dependent on centralized power systems, on very sophisticated interconnections, the interruption of any part of which would cause a tremendous dislocation, so that in our great strength there lies a great weakness. It is said that pride comes before a fall, and we are very puffed up with our power. The Soviet Union is also very puffed up with its power, but when I look at the four billion people on this planet, when I look at our instantaneous audio and visual communication, when I look at the proliferating plutonium that is reaching the hands of people who may not be as restrained as are those who hold it now, I see a very troubled and difficult world. I see a world that will not be made more secure by increasing overkill by ten, or twenty, or a thousand times, but only by cooperation, in which some countries will assume the lead in providing an adequate energy supply, in examining in a forthright way the environmental and ecological degradation that surrounds us, and in considering the more equal distribution of world opportunities. It is not an easy agenda, and it does not fit into the diplomatic history books, but it is one that will increasingly occupy the United Nations, and will challenge the minds and hearts of young people as they develop their own political thoughts.

What should we do here at home? President Carter was right when he said that the energy crisis presents us with a moral equivalent of war, but we have not begun to mobilize the capital to overcome it. We are mortgaging our future to what I call a dinosaur technology of nuclear power. And when—not if, but when—the next Three Mile Island accident occurs, many if not all of the present light reactors will be closed down at an enormous loss. The better part of wisdom today is to invest in alternative technologies, in conservation, and to provide the kind of energy base that we need. In the short term—that is, in the next few years—conservation is the only option that will yield significant dividends. Some people estimate that we shall have to conserve the equivalent of ten million barrels of oil a day to maintain any kind of economic growth through the eighties. Along with conservation, we need to invest in the development of alternatives. The solar program is woefully underfunded; the geothermal project that is being explored at Los Alamos to develop energy from the hot rock—that is, putting water down on the hot rock and creating a recycling geothermal system—that project receives about twelve million dollars, a ridiculous sum of money in terms of the problems we face. The liquid fuels that will be needed, the photovoltaic cells, the electric car, rapid transit, the insulation of homes, these require huge capital investments. To achieve these goals, a tremendous squeezing of existing programs will take place at all levels of government in order to mobilize the needed capital, both public and private. So very high on the agenda I would place an energy program involving conservation and the development of new sources, with a major leadership role for the federal government. This will have to be an alternative energy program that does not put all our eggs in the plutonium basket, because I believe that the instability of that technology will be its downfall. Second, I see a need, at least in the early part of the eighties, to refocus our emphasis on the United States itself. We have been living in a period during which the capital of this country has been flowing to other nations. Technology transfer is a form of internationalism, which is a valuable part of our historical development; but in the process we have witnessed the growth of a South Bronx, a Compton, a Watts, parts of Providence, Rhode Island, Elizabeth, New Jersey, places like Detroit, places that are islands of devastation and despair. We need the capital to rebuild those urban areas that do not represent the American Dream. That capital will come in part from the government, but in a larger part it should be derived from the multinational capital that now leaks out of the country, is invested in cheap labor markets, and then brings back products that displace American workers. We have a situation where American capital derived from American genius and protected by American institutions is sent to other countries where there are no environmental protection laws, no laws against child labor or overtime, no minimum wage, no aggressive labor unions. The products made cheaply abroad not only divert capital that should be invested in the United States, but they return here to displace even the capital that remains in the factories and the businesses that are already facing excessive threats from foreign competition. Third, I still believe, despite political winds to the contrary, that environmental issues are a central concern. The natural systems on which we all depend are rather inevitable. If tampered with or ignored, they tend to recoil with fatal devastation. Look at the increasing desertification, not only in Africa but even in California, where in the San Joaquin Valley four hundred thousand acres of prime land are now suffering a decline in productivity because of increasing assaults from irrigated agriculture. Two-thirds of the salmon and steelhead of this state are no longer producing, are no longer alive, the salmon resources having been destroyed by the timber practices, by the debris in the streams, and by the lack of investment in that resource. I see in the waters pesticides that probably are engendering cancer, and I see the excessive toxic materials that are let loose in the society. So investment in the enhancement of natural systems is still an imperative. It is not something we shall attain the day after tomorrow or even next year, but if we look ahead ten or twenty years, we must invest in the protection of our soil, our rivers, our forests, our fisheries, and our total environment. These are the resources we draw our life from. Ultimately we are biological creatures, and we must respect and respond to the biological laws. I shall just cite the example of Mesopotamia, where a very fertile crescent became a desert, or North Africa, which once was a forest that was chopped down to supply ships, hot tubs, and aquaducts for the Romans of that period. I do not think that what has happened before will not happen again. Many a civilization has allowed its soil to erode and has itself fallen with its forests. So, as my third priority, I would certainly list investment in natural systems.

The thought that I should like to leave with you is that we are indeed entering an era of limits, but it is also an era of great possibilities. The development of technology, of space exploration, of ecological protection and sensitivity, of a greater awareness of our own identity with all the people of this planet—these are all themes that offer opportunities. But in order for those opportunities to be realized, choices must be made. Those things that we have taken for granted— and many of them are good, such as the expanding of hospitals and freeways— and many of the programs that we like, are not going to be funded at the same level, because there is only so much capital, and the system has only so much capacity. What we must achieve in the eighties is a rejuvenation of the capital of this country: the human capital in the people who lack skills and must acquire them; the ecological capital that is needed for reforestation and for the protection of our soils, waters, and fisheries; and the development of our technological capital, which includes research and development and the genius that translates itself into the expanding revelation of our own ability to manipulate machines and materials to make a better quality of life, not only for ourselves but for those who come after us. That is our choice, and I think it is as clear to us as it is to most of the countries of the world. There will still be conflicts, there will still be uprisings, there will still be great disputes and competition, by both the great powers and by the lesser powers, but certainly the era of limits comes none too soon to the strategic arms race, especially today. The choice is clear—whether we follow the course of the lemmings into the sea with the pathological addiction that leads to destruction, or the wiser, the more restrained investment in our people, our resources, and our future.

Conference Staff and Participants

CONFERENCE STAFF

Staff Roster: JEFFREY R. LEIFER, Student Coordinator; SANFORD A. LAKOFF, Faculty Coordinator; RONALD BEE, Assistant Coordinator; DANIEL E. BAXTER, Technical Supervisor; ERIC MARKUSEN, Intern: Defense Workshop/Proceedings Reviewer; LISA J. ENDLICH, Intern: Human Value Workshop; CALMAN P. PRUSSIN, Intern: Development Workshop: BRUCE REYNOLDS, Intern: Biomedical Workshop; ELIZABETH SISCO, Housing/Food/Receptions; DONNAL L. WILLS, Office Organization; DOUGLAS BOND, Evaluation Coordinator; TABER HAND, Press Liaison.

Conference Assistants: RONALD ALCALAY, DENNECE BENNETT, ERIC ENDLICH, SHERRY HEISE, DAVID LEIFER, DAVID MILLER, MICHELE MILLER, KAREN NEPVEV, DAEL SPILLMAN.
Secretarial Assistants: SUSAN EHLINGER, BETSY FAUGHT, MICHELE WENZEL, BARBARA ZIERING.

Conference Participants

Faculty: NANCY ABRAMS, Lawyer/ Science Policy Consultant, University of California, Santa Cruz; C. FRED ALFORD, Instructor of Government, University of Texas; GEORGE ANAGNOSTOPOULOS, Associate Professor of Philosophy, University of California, San Diego; JAMES ARNOLD, Professor of Chemistry, U.C.S.D.; LAWRENCE BADASH, Associate Professor of History, University of California, Santa Barbara; RONALD BERMAN, Professor of English Literature, U.C.S.D.; JOHN BRISCOE, Research Fellow, Division of Applied Science, Harvard University; ANNE CAHN, Chief, Social Impact Staff, Arms Control and Disarmament Agency; ROSEMARY CHALK, Staff Officer, Committee on Scientific Freedom and Responsibility, American Association for the Advancement of Science; WILLIAM CHAPMAN, Commander, USN, Ret.; DANIEL ELLSBERG; HAROLD FEIVESON, Assistant Professor of Public Affairs, Princeton University; MICHAEL FLOWER, Research Associate, Salk Institute; THEODORE FRIEDMANN, Associate Professor of Pediatrics, U.C.S.D.; CLIFFORD GROBSTEIN, Professor of Biological Science and Public Policy, U.C.S.D.; DONALD HELINSKI, Professor of Biology, U.C.S.D.; GEORGE H. KIEFFER, Professor of Life Sciences, University of Illinois; GEORGE KISTIAKOWSKY, Professor

313

of Chemistry, Harvard University; WILLIAM KORNHAUSER, Professor of Sociology, University of California, Berkeley; SANFORD LAKOFF, Professor of Political Science, U.C.S.D.; HERBERT LEIFER, Physicist, Rand Corporation; ALEXANDER J. MORIN, Director, Office of Science and Society, National Science Foundation; B.E. OGUAH, Professor of Philosophy, University of Ghana; JOEL PRIMACK, Associate Professor of Physics, University of California, Santa Cruz; ROGER REVELLE, Professor of Science and Public Policy, U.C.S.D.; JONAS SALK, Fellow and founding Director, Salk Institute, San Diego; JONATHAN SAVILLE, Associate Professor of Literature, U.C.S.D.; HERBERT SCOVILLE, Jr., Arms Control Association; JEREMY J. STONE, Director, Federation of American Scientists; AVRUM STROLL, Professor of Philosophy, U.C.S.D.; WILLIAM VAN CLEAVE, Professor of International Relations, University of Southern California.

Guests: EDMUND G. BROWN, Jr., Governor of California; RICHARD GARWIN, IBM Corporation; WILLIAM McELROY, Chancellor, U.C.S.D.; WALTER MUNK, Professor of Geophysics, Scripps Institution of Oceanography, U.C.S.D.; PAUL SALTMAN, Vice Chancellor for Academic Affairs, U.C.S.D.

Students, Politics Workshop: CINDY AMATNIEK, Radcliffe College; MICHAEL BERGER, U.C.S.D.; GEOFFREY BERNSTEIN, Harvard University; CURT BIREN, George Washington University; CHRISTOPHER CHYBA, Swarthmore College; JERRY COHEN, Franklin Pierce Law Center; GREGORY GROSS, University of Wisconsin, Madison; ELIZABETH HODES, University of California, Santa Barbara; JAMES KIRCHNER, Dartmouth College; STASIA MIASKIEWIEZ, Wellesley College; TERRY OLESEN, Claremont Graduate School; BARRY PRICE, Indiana University; PETER SWIRE, Princeton University; HELENA TENCH, University of Connecticut, Storrs; JONATHAN WALLACH, M.I.T.: CHRISTOPHER WRIGHT, Stanford University; *Values Workshop:* ALEXANDRA ALLEN, Carnegie Mellon University; JOHN BELLIVEAU, California Institute of Technology; MARK BENSEN, University of North Carolina; BRINA CAPLAN, Harvard University; SETH CHAIKLIN, University of Washington; JONATHAN EZEKIEL, Amherst College; CHRISTOPHER GABRIELI, Harvard University; ANNE HEISE, Swarthmore College; JOERN KROLL, University of California, Berkeley; DIANE LEFEBVRE, Michigan State University; ARTHUR MATUCK, U.C.S.D. RICHARD RICE, Michigan State University; STEPHEN ROOT, U.C.S.D.; JEFFREY SCHLOSS, Washington University; ADAM SCHULMAN, University of Chicago; LINDA STRAUSS, U.C.S.D.; *Development Workshop:* BRUCE ALLYN, Dartmouth College; H. THEODORE BERGH, Cornell University, E. ANNE CLARK, Iowa State University; CRISTOPHER COAD, U.C.S.D.; DANIEL FLEMING, Duke University; HORACE HERRING, Washington University; G. JOHN IKENBERRY, University of Chicago; JAMES KAHN, Harvard University; ANUPAM KHANA, Stanford University; KEITH LANDA, U.C.S.D.; BONAVENTURE LUISI, California Institute of Technology; ANJUM MIR, Stanford University; CALMAN PRUSSIN, U.C.S.D.; DEBRA SALKIND, Northwestern University; PATRICIA SCARLETT, University of California, Santa Barbara; JIM SLATTEN, University of Southwestern Louisiana; JEFFREY STINE,

University of California, Santa Barbara; DEAN WILKENING, Harvard University; *Arms Control Workshop:* JEFFREY DUNHAM, Stanford University; GORDON FELLER, Columbia University; BENJAMIN FIEDLER, University of Chicago; FRASER HOMER-DIXON, Carleton University; CRAIG GLIDDEN, Tulane University; HANS-GERD LOEHMANNSROEBEN, University of North Carolina, Chapel Hill; ALDO PATANIA, Università di Catania, Italy; JED SNYDER, University of Chicago; ALAN VICK, University of California, Irvine; *Biomedical Workshop:* RICHARD ANDREWS, M.I.T.; ROBERT BUCCINI, University of Chicago; SUSAN FERGUSON, University of California, Santa Cruz; MARGARET HANSEN, U.C.S.D.; MAUREEN KING, M.I.T.; MARK LEIFER, Stanford University; JAMES MURTAGH, University of Michigan; STEPHEN REISS, Princeton University; CRAIG SHAPIRO, Harvard University Medical School; CAROLY SHUMWAY, Wellesley College; IRA SINGER, University of North Carolina; LORRAINE SKACH, Oregon State University; ELLEN TAYLOR, University of California, Berkeley; JONATHAN TEPPER, U.C.S.D.; ANTHONY VITTO, U.C.S.D.; MARK WENNEKER, Harvard University.

Select Bibliography

1. *Scientists and Ethical Responsibility (General)*

   Allan, Leslie, and Kaufman, Ellen, *Paper Profits: Pollution in the Pulp and Paper Industry*, Cambridge, M.I.T. Press, 1971.

   Aron, Raymond, *et al., Scientists in Search of Their Conscience*, Berlin, New York, Springer-Verlag, 1973.

   Atkinson, Richard C., "Rights and Responsibilities in Scientific Research," *The Bulletin of the Atomic Scientists*, December 1978, pp. 10–14.

   Ben-David, Joseph, *The Scientist's Role in Society: A Comparative Study*, Englewood Cliffs, New Jersey, Prentice-Hall, 1971.

   Beyerchen, Alan D., *Scientists under Hitler: Politics in the Physics Community of the Third Reich*, New Haven, Yale University Press, 1977.

   Brown, Peter G., and Shine, Henry, *Food Policy: The Responsibility of the United States in the Life and Death Choices*, New York, The Free Press, 1977.

   Carovillano, Robert, and Skehan, James W., *Science and the Future of Man*, Cambridge, M.I.T. Press, 1970.

   Coulson, Charles Alfred, *The Scientist's Responsibility in Society*, Edinburgh, Heriot Watt University, 1970.

   Dewees, Donald, *Economics and Public Policy: The Automobile Pollution Case*, Cambridge, M.I.T. Press, 1974.

   Dupré, J.S., and Lakoff, Sanford A., *Science and the Nation: Policy and Politics*, Englewood Cliffs, New Jersey, Prentice-Hall, 1962.

   Earl, D.E., *Forest Energy and Economic Development*, London, Oxford University Press, 1975.

   Ebbin, Steven, and Kasper, Raphael, *Citizen Groups and the Nuclear Power Controversy: Uses of Scientific and Technological Information*, Cambridge, M.I.T. Press, 1974.

   Epstein, Samuel S., and Grundy, Richard D., *the Legislation of Product Safety: Consumer Health and Product Hazards*, Cambridge, M.I.T. Press, 1974. Vol. I: *Chemical, Electronic Products, Radiation;* Vol. II: *Cosmetics and Drugs, Pesticides, Food Additives.*

   Gilpin, Robert, ed., *Scientists and National Policy Making*, New York, Columbia University Press, 1964.

   Glass, Bentley, *Science and Ethical Values*, Chapel Hill, University of North Carolina Press, 1965.

   Goldsmith, Maurice, ed., *Science and Social Responsibility*, London, Macmillan, 1975.

   Haas, Ernst, Babai, Don, and Williams, Mary P., *Scientists and World Order: The Uses of Technical Knowledge in International Organization*, Berkeley, University of California Press, 1977.

   Haberer, Joseph, *Politics and the Community of Science*, New York, Van Nostrand Reinhold, 1969.

   Hatfield, Charles, ed., *The Scientist and Ethical Decision*, Downers Grove, Illinois, Intervaristy Press, 1973.

   Heidelberger, Michael, *The Scientist and Survival in an Atomic Era*, New York, New York Society for Ethical Culture, 1958.

   Holton, Gerald, and Blanpied, William A., ed., *Science and Its Public: The Changing Relationship*, Dordrecht, Holland, Boston, D. Reidel Publishing Co., 1976.

   Kaplan, Norman, ed., *Science and Society*, Chicago, Rand McNally, 1965.

   Kevles, Daniel J., *The Physicists: The History of a Scientific Community in Modern America*, New York, Knopf, 1978.

King, Alexander, *Science and Policy: The International Stimulus,* London, New York, Oxford University Press, 1974.

Kistiakowsky, George, *A Scientist in the White House: The Private Diary of President Eisenhower's Special Assistant for Science and Technology,* Cambridge, Harvard University Press, 1976.

Krier, James E., and Ursin, Edmund, *Pollution and Policy: A Case Essay on California and Federal Experience with Motor Vehicle Air Pollution, 1940–1975,* Berkeley, University of California Press, 1977.

Lakoff, Sanford A., ed., *Knowledge and Power: Essays on Science and Government,* New York, The Free Press, 1966.

Lakoff, Sanford A., "Knowledge, Power and Democratic Theory," *The Annals,* **394,** March 1971, pp. 4–12.

———, "Science and Conscience," *International Journal,* **24,** No. 4, Autumn 1970, pp. 754–765.

———, "Accountability and the Research Universities," Bruce L.R. Smith, and Joseph J. Karlesky, eds., *The State of Academic Science,* New Rochelle, New York, Change Magazine Press, 1978, pp. 163–190.

———, "The 'Galilean Imperative' and the Scientific Vocation," *Ethics,* October 1980.

Lambright, W. Henry, *Governing Science and Technology,* New York, Oxford University Press, 1976.

Lowrance, William W., *Of Acceptable Risk: Science and the Determination of Safety,* Los Altos, California, Kaufmann, 1976.

Makhijani, Arjun, with Poole, Alan, *Energy and Agriculture in the Third World,* Cambridge, Ballinger, 1975.

Matthews, William H. Kellog, William H., and Robinson, G.D., *Man's Impact on the Climate,* Cambridge, M.I.T. Press, 1970.

Mendeloff, John, *Regulating Safety: An Economic and Political Analysis of Occupational Safety and Health Policy,* Cambridge, M.I.T. Press, 1978.

Nelkin, Dorothy, ed., *Controversy: Politics of Technical Decisions,* Beverly Hills, Sage Publications, 1979.

Nelkin, Dorothy, *The University and Military Research: Moral Politics at M.I.T.,* Ithaca, Cornell University Press, 1972.

———, *Nuclear Power and Its Critics: The Cayuga Lake Controversy,* Ithaca, Cornell University Press, 1971.

———, *Jetport: The Boston Airport Controversy,* New Brunswick, New Jersey, Transaction Books, 1974.

Price, Don K., *The Scientific Estate,* Cambridge, Harvard University Press, 1965.

Primack, Joel, and von Hippel, Frank, *Advice and Dissent: Scientists in the Political Arena,* New York, Basic Books, 1974, and New American Library, 1976.

Reid, R.W., *Tongues of Conscience: Weapons Research and the Scientists' Dilemma,* New York, Walker, 1969.

Reining, Priscilla, and Tinker, Irene, *Population: Dynamics, Ethics and Policy,* Washington, D.C., The American Association for the Advancement of Science, 1975.

Rochlin, Gene I., ed., *Scientific Technology and Social Change: Readings from Scientific American,* San Francisco, W.H. Freeman, 1974.

Rosenberg, Charles E., *No Other Gods: On Science and American Social Thought,* Baltimore, Johns Hopkins University Press, 1976.

Rotblat, Joseph, *Scientists' Responsibility in the Atomic Age,* London, Lindsey Press, 1964.

Salomon, Jean Jacques, *Science and Politics,* translated from the French by Noel Lindsay, London, Macmillan, 1973.

Segerstedt, Torgny, ed., *Ethics for Science Policy;* Proceedings of a Nobel Symposium Held at Södergarn, Sweden, August 20–25, 1978, New York, Pergamon Press, 1979.

Spiegel-Rösing, Ina, and Price, Derek, eds., *Science, Technology and Society: A Cross-Disciplinary Perspective,* Beverly Hills, Sage Publications, 1979.

Strickland, Donald A., *Scientists in Politics: The Atomic Scientists Movement, 1945–46,* Lafayette, Indiana, Purdue University Studies, 1968.

Szilard, Leo, *The Voice of the Dolphins and Other Stories,* New York, Simon and Schuster, 1961.

Teich, Albert H., ed., *Scientists and Public Affairs,* Cambridge, M.I.T. Press, 1974.

U.S. Atomic Energy Commission, *Science and Society: A Bibliography,* Division of Headquarters Services, Superintendent of Documents, Washington, D.C., U.S. Government Printing Office, 1971.

Weart, Spencer, and Szilard, Gertrud Weiss, eds., *Leo Szilard: His Version of the Facts,* Cambridge, M.I.T. Press, 1978.

Zuckerman, Sir Solly, *Scientists and War: The Impact of Science on Military and Civil Affairs,* New York, Harper and Row, 1967.

2. Scientists in Dissent

Chalk, Rosemary, and von Hippel, Frank, "Due Process for Dissenting 'Whistleblowers,' " *Technology Review,* June/July 1979.

Edsall, John T., *Scientific Freedom and Responsibility,* a report of the Committee on Scientific Freedom and Responsibility, Washington, D.C. The American Association for the Advancement of Science, 1975. (Summarized in *Science,* May 16, 1977, pp. 687–693.)

Ellsberg, Daniel, *Papers on the War,* New York, Simon and Schuster, 1972.

Ewing, David W., *Freedom inside the Organization: Bringing Civil Liberties to the Workplace,* New York, Dutton, 1977.

Graham, Frank, Jr., *Since Silent Spring,* Boston, Houghton Mifflin, 1970.

Jacobs, Donald, *A Scientist and His Experience with Corruption and Treason in the U.S. Military–Industrial Establishment,* Victoria, British Columbia, Jacobs Instrument Co., 1969.

Layton, Edwin T. Jr., *The Revolt of the Engineers: Social Responsibility and the American Engineering Profession,* Cleveland, Case Western Reserve University Press, 1971.

Nader, Ralph, Petkas, Peter, and Blackwell, Kate, eds., *Whistleblowing,* New York, Grossman, 1972.

Olson, J., "Engineer Attitudes toward Professionalism, Employment, and Social Responsibility," *Professional Engineer,* August 1972.

Senate Committee on Governmental Affairs, *The Whistleblowers,* Washington, D.C., U.S. Government Printing Office, 1978.

3. Materials on the International Pugwash Conferences

Pugwash Conference on Science and World Affairs, 1st, Nova Scotia, 1957, *Statement of International Meeting of Scientists at Pugwash, Canada, July 6–10, 1957,* Library of Congress, Washington, D.C., 1957.

Pugwash Conference on Science and World Affairs, 2nd, Lac Beauport, Quebec, 1958, *Documents of the Second Pugwash Conference of Nuclear Scientists, March 13–April 11, 1958.*

Pugwash Conference on Science and World Affairs, 10th, London, 1962 "Scientists and World Affairs," *Proceedings,* London, 1963.

Pugwash Conference on Science and World Affairs, 17th, Ronneby, Sweden, 1967, "Scientists and World Affairs," *Proceedings,* London, Pugwash Continuing Committee, 1967.

Pugwash Conference on Science and World Affairs, 22nd, Oxford, 1972, "Scientists and World Affairs," Oxford, England, Pugwash Continuing Committee, 1972.

Rabinowitch, Eugene, and Rabinowitch, Victor, eds., *Views of Science, Technology, and Development* (papers presented at Pugwash conferences), Oxford, New York, Pergamon Press, 1975.

Rotblat, Joseph, *Scientists in Quest for Peace: A History of the Pugwash Conferences,* Cambridge, M.I.T. Press, 1972.

Thorin, Duane, *The Pugwash Movement and U.S. Arms Policy,* New York, Monte Cristo Press, 1965.

United States Congress, Senate Committee on the Judiciary, Subcommittee to investigate the administration of the Internal Security Act and other internal laws. *The Pugwash Conferences Revisited; The Pugwash Movement Viewed as of October 1964, a Staff Analysis,* Washington, D.C., U.S. Government Printing Office, 1964.

Wormwald, F.L., *The Pugwash Experiment: An Essay in Liberal Education,* Washington, D.C., The Association of American Colleges, 1958.

4. Arms and Arms Control (General)

Barton, John H., and Weiler, Lawrence D., *International Arms Control: Issues and Agreements,* Stanford, Stanford University Press, 1976.

Congressional Quarterly, Inc., *U.S. Defense Policy: Weapons, Strategy and Committees,* April 1978.

Endicott, John, and Stafford, Roy W., eds., *American Defense Policy,* Baltimore, Johns Hopkins University Press, 1977.

Ford, Harold P., and Winters, Francis X.S.J., *Ethics and Nuclear Strategy,* Maryknoll, New York, Orbis Books, 1977.

Kahan, Jerome H., *Security in the Nuclear Age: Developing U.S. Strategic Arms Policy,* Washington, D.C., Brookings, 1975.

Myrdal, Alva, *The Game of Disarmament: How the United States and Russia Run the Arms Race,* New York, Pantheon, 1976.

Office of Technology Assessment, *The Effects of Nuclear War,* Washington, D.C., U.S. Government Printing Office, 1979.

Pranger, Robert J., and Labrie, Roger P., eds., *Nuclear Strategy and National Security: Points of View,* Washington, D.C., American Enterprise Institute for Public Policy Research, 1977.

Russett, Bruce, and Blair, Burce, G., eds., *Progress in Arms Control?* San Francisco, W. H. Freeman, 1979.

Stockholm International Peace Research Institute, *World Armaments and Disarmament,* SIPRI Yearbook, London, Taylor and Frances, 1978.

York, Herbert F., ed., *Arms Control: A Scientific American Reader,* San Francisco, W. H. Freeman, 1973.

York, Herbert F., *Race to Oblivion: A Participant's View of the Arms Race,* New York, Simon and Schuster, 1970.

_____, *The Advisors: Oppenheimer, Teller and the Superbomb,* San Francisco, W. H. Freeman, 1976.

5. *Consequences of Nuclear War*

Drell, Sidney, and von Hippel, Frank, "Limited Nuclear War," *Scientific American,* November 1976.

Glasstone, Samuel, ed., *Effects of Nuclear Weapons,* Washington, D.C., U.S. Department of Defense and Department of Energy, 1977.

Katz, A., *Economic and Social Consequences of Nuclear Attacks on the United States,* United States Senate Committee on Banking, Housing, and Urban Affairs, 96th Congress, First Session, March 1979.

Lewis, Kevin N., "The Prompt and Delayed Effects of Nuclear War," *Scientific American,* 241, No. 1, July 1979, pp. 35–47.

National Academy of Science, *Effects of Multiple Nuclear Explosions Worldwide,* Washington, D.C., 1975.

Petty, Geraldine, Dzirkals, Lilita, and Krahenbuhl, Margaret, *Economic Recovery Following Nuclear Disaster: A Selected, Annotated Bibliography,* Santa Monica, Ca.: Rand Corporation, R-2143, December 1977.

U.S. Arms Control and Disarmament Agency, "The Effects of Nuclear War," Washington, D.C., April 1979.

United States Congress, Senate Committee on Foreign Relations, Subcommittee on Arms Control, International Organizations, and Security Agreements, Committee Print, "Analysis of Effects of Limited Nuclear Warfare."

6. *Likelihood of Nuclear War*

Bidwell, Shelford, ed., *World War 3: A Military Projection Founded on Today's Facts,* Englewood Cliffs, New Jersey, Prentice-Hall, 1978.

Builder, Carl H., "Why Not First Strike Counterforce Capabilities?" *Strategic Review,* Spring 1979, pp. 32–39.

Calder, Nigel, *Nuclear Nightmares: An Investigation into Possible Wars,* New York, Viking, 1980.

Gray, Colin S., "Nuclear Strategy: A Case for a Theory of Victory," *International Security,* Summer 1979, Vol. 4, pp. 54–87.

Hackett, General Sir John, *et al., The Third World War,* New York, Macmillan, 1978.

Kahler, Miles, "Rumors of War; The 1914 Analogy," *Foreign Affairs,* Winter 1979-1980, pp. 374–396.

Polanyi, John, "The Dangers of Nuclear War," *Bulletin of the Atomic Scientists,* January 1980, pp. 6–10.

Sigal, Leon V., "Rethinking the Unthinkable," *Foreign Policy,* 34, Spring 1979, pp. 35–51.

7. *Nuclear Weapons and Ethical Responsibility*

Anscombe, Elizabeth, "War and Murder," in Wasserstrom, Richard, ed., *War and Morality,* Belmont, California, Wadsworth, 1970.

Brandt, R.B., "Utilitarianism and the Rules of War," *War and Moral Responsibility,* Princeton, New Jersey, Princeton University Press, 1974.

Burns, Arthur L., "Ethics and Deterrence: A Nuclear Balance Without Hostage Cities?," *Adelphi Papers,* No. 69, The International Institute for Strategic Studies, July 1970.

Ford, John C., "The Morality of Obliteration Bombing," *Theological Studies,* No. 5, 1944, pp. 261–309. Reprinted in *War and Morality,* Belmont, California, Wadsworth, 1970.

Falk, Richard A., Kolko, Gabriel, and Lifton, Robert Jay, eds., *Crimes of War,* New York, Random House, 1971.

Green, Philip, *Deadly Logic: The Theory of Nuclear Deterrence,* Columbus, Ohio State University Press, 1966.

Ramsey, Paul, *War and Christian Conscience,* Durham, North Carolina, Duke University Press, 1961.

———, "The Just Revolution," *Worldview,* 16, October 1973, pp. 37–40.

Sibley, Mulford, *Unilateral Initiatives and Disarmament,* A Study and Commentary in the Beyond Deterrence Series, American Friends Service Committee, n.d.

Tucker, Robert W., *The Just War: A Study of Contemporary American Doctrine,* Baltimore, Johns Hopkins University Press, 1960.

Von Weizaecker, Carl-Friedrich, Freiherr, "The Ethical Problem of Modern Strategy," *Problems in Modern Strategy,* New York, Praeger, 1970.

Walzer, Michael, *Just and Unjust Wars: A Moral Argument With Historical Illustrations,* New York, Basic Books, 1977.

Winters, Francis, "Morality in the War Room," *America,* February 15, 1975, pp. 106–110.

———, "Ethical Considerations and National Security Policy," *Report of the Commission on the Organization of the Government for the Conduct of Foreign Policy,* Vol. 7, Appendix W, No. 2, Washington, D.C., U.S. Government Printing Office, June 1975, pp. 293–99.

8. *Cultural Dimensions of Nuclear War*

Davidson, Ann M., "Macho Obstacles to Peace," *Bulletin of the Atomic Scientists,* June 1977, pp. 22–24.

Frank, Jerome D., *Sanity and Survival; Psychological Aspects of War and Peace,* New York, Vintage Books, 1967.

Iklé, F.C., *The Social Impact of Bomb Destruction,* Norman, Oklahoma, University of Oklahoma Press, 1958.

Jones, T.K., and Thompson, W. Scott, "Central War and Civil Defense," *Orbis,* Fall 1978, pp. 681–712.

Melman, Seymour, ed., *No Place to Hide,* New York, Grove Press, 1962.

Moore, David W., "The Public Is Uncertain," *Foreign Policy,* 35, Summer 1979, pp. 68–73.

Nuttall, Jeff, *Bomb Culture,* New York, Delacorte Press, 1968.

Russett, Bruce, "No First Use of Nuclear Weapons: To Stay the Fateful Lightning," *Worldview,* 19, November 1976, pp. 9–11.

Shaheen, Jack G., *Nuclear War Films,* Carbondale and Edwardsville, Illinois, Southern University Press, 1978.

9. *U.S. Nuclear Weapons Policy and Strategic Doctrine*

Aldridge, Robert C., "First Strike: The Pentagon's Secret Strategy," *The Progressive,* May 1978, pp. 16–19.

Barnet, Richard, *Roots of War,* New York, Atheneum, 1972.

Buchan, Alastair, *Problems of Modern Strategy,* London, Praeger, 1970.

Carter, Barry, "Nuclear Strategy and Nuclear Weapons," *Scientific American,* 230, May 1974, pp. 20–31.

Congressional Budget Office, *Counterforce Issues for the U.S. Strategic Nuclear Forces,* Washington, D.C., U.S. Government Printing Office, 1979.

Federation of American Scientists, "Solution to Counterforce: Land Based Missile Disarmament," *Federation of American Scientists Public Interest Reports,* February 1974.

Gelb, Leslie H., "Debate on U.S. Nuclear Policy: Just What Is Strategic Superiority?" *New York Times,* July 30, 1974.

Gray, Colin, "The Arms Race Phenomenon," *World Politics,* 24, October 1971, pp. 39–79.

———, "The Strategic Forces Triad: End of the Road?" *Foreign Affairs,* July 1978, pp. 772–789.

Greenwood, Ted, *Making the MIRV,* Cambridge, Ballinger, 1975.

Greenwood, Ted, and Nacht, Michael L., "The New Nuclear Debate: Sense or Nonsense?" *Foreign Affairs,* 52, July 1974, pp. 761–780.

Halperin, Morton H., *Defense Strategies for the Seventies,* Boston, Little, Brown, 1971.

Halperin, Morton H., and Kanter, A., *Readings in American Foreign Policy: A Bureaucratic Perspective,* Boston, Little, Brown, 1973.

Iklé, Fred C., "Can Nuclear Deterrence Last Out the Century?" *Foreign Affairs* 51, January 1973, pp. 267–285.

Kahn, Herman, *Thinking about the Unthinkable,* New York, Horizon Press, 1962.

Lambeth, Benjamin S., *Selective Nuclear Options in American and Soviet Strategic Policy,* Santa Monica, Rand Corporation, December 1976.

———, "Deterrence in the MIRV Era," *World Politics,* 24, January 1972, pp. 221–42.

Lodal, Jan, "Assuring Strategic Stability: An Alternative View," *Foreign Affairs* 54, No. 3, April 1976, pp. 462–480.

Long, Franklin A., and Rathjens, George W., eds., *Arms, Defense Policy, and Arms Control,* New York, Norton, 1976.

Luttwak, Edward N., *The Washington Papers 38: Strategic Power: Military Capabilities and Political Unity,* Beverly Hills, Sage Publications, 1976.

———, "Nuclear Strategy: The New Debate," *Commentary,* 57, April 1974, pp. 53–59.

May, Michael M., "Some Advantages of Counterforce Deterrence," *Orbis,* 14, Summer 1970, pp. 271–283.

Nitze, Paul H., "Deterring Our Deterrent," *Foreign Policy,* Winter 1976–77, pp. 195–210.

Panofsky, Wolfgang K.H., "The Mutual Hostage Relationship between America and Russia," *Foreign Affairs,* 51, October 1973, pp. 109–118.

Quanbeck, Alton H., and Blechman, Barry M., *Strategic Forces: Issues for the Mid-Seventies,* Washington, D.C., Brookings, 1973.

Record, Jeffrey, *U.S. Nuclear Weapons in Europe: Issues and Alternatives,* Washington, D.C., Brookings, 1974.

Sallager, F.M., *The Road to Total War: Escalation in W.W. II,* Santa Monica, Rand Corporation Report, 1969.

Sampson, Anthony, *The Arms Bazaar: From Lebanon to Lockheed,* New York, Viking Press, 1977.

Scoville, Herbert, Jr., "Flexible Madness?" *Foreign Policy,* 14, Spring 1974, pp. 164–177.

Smoke, Richard, *Deterrence in American Foreign Policy: Theory and Practice,* New York, Columbia University Press, 1974.

———, *War: Controlling Escalation,* Cambridge, Harvard University Press, 1979.

Steinbruner, J., "National Security and the Concept of Strategic Stability," *Journal of Conflict Resolution,* 22, 1978, pp. 411–428.

Stern, E. P., ed., *The Limits of Military Intervention,* Beverly Hills, Sage Publications, 1977.

Trofimenko, H., "The Theology of Strategy," *Orbis,* 21, 1977, pp. 515–597.

Tsipis, Kosta, Cahn, Anne H., and Feld, Bernard T., *The Future of the Sea-Based Deterrent,* Cambridge, M.I.T. Press, 1973.

United States, 95th Congress, *United States/Soviet Strategic Options* (hearings), Washington, D.C., U.S. Government Printing Office, 1977.

10. *Soviet Strategic Doctrine*

Arbatov, G.A., "The American Strategic Debate: A Soviet View," *Survival,* 16, May/June 1974. pp. 133–134.

Frank, Allen, *Soviet Nuclear Planning: A Point of View on SALT,* Washington, D.C., American Enterprise Institute for Public Policy Research, 1976.

Lambeth, Benjamin S., "The Sources of Soviet Military Doctrine," in Frank B. Horton III, Anthony

C. Rogerson, Edward L. Warner III, eds., *Comparative Defense Policy,* Baltimore, Johns Hopkins University Press, 1974, pp. 200–216.

Pipes, Richard, "Why the Soviet Union Thinks it Could Fight a Nuclear War and Win," *Commentary,* July 1977, pp. 21–34.

Rummel, R.J., "Will the Soviet Union Soon Have a First-Strike Capability?," *Orbis,* Fall 1976, pp. 579–594.

11. *Arms Control*

Brown, Harold, "Security through Limitations," *Foreign Affairs,* **47**, No. 3, April 1980, pp. 422–432.

Cahn, Anne H., and Kruze, Joseph J., eds., *Controlling Future Arms Trade,* 1980's Project/Council on Foreign Relations, New York McGraw-Hill, 1977.

Carlton, David, and Schaerf, Carlo, *Arms Control and Technological Innovation,* New York, Halstad Press,1977.

Diplomatic and Strategic Impact of Multiple Warhead Missiles, hearings, United States Congress, 1969.

Doty, Paul, "The Race to Control Nuclear Arms," *Foreign Affairs,* October 1976, pp. 119–132.

Epstein, W., *The Last Chance: Nuclear Proliferation and Arms Control,* New York, The Free Press, 1976.

George, Alexander S., Hall, David K., Simons, William E., *et. al., The Limits of Coercive Diplomacy,* Boston, Little, Brown, 1971.

Gompert, David, Mandelbaum, Michael, Garwin, Richard L., and Barton, John J., *Nuclear Weapons and World Politics: Alternatives for the Future,* Council on Foreign Relations, New York, McGraw-Hill, 1977.

Greenwood, Ted, Feiveson, Harold A., and Taylor, Theodore B., *Nuclear Proliferation: Motivation, Capabilities, and Strategies for Control,* Council on Foreign Relations, New York, McGraw-Hill, 1977. *Alternatives for the Future,* Council on Foreign Relations, New York, McGraw-Hill, 1977.

Lee, John, "An Opening 'Window' for Arms Control," *Foreign Affairs,* Fall 1979, pp. 121–140.

Meselson, Matthew, ed., *Chemical Weapons and Arms Control,* New York, Carnegie Endownment for International Peace, 1978.

Newhouse, John, *Cold Dawn—The Story of SALT,* New York, Holt, Rinehart and Winston, 1973.

Quester, George, H., *Nuclear Diplomacy: The First Twenty-Five Years,* New York, Dunellen Publishing Company, 1970.

———, *The Politics of Nuclear Proliferation,* Baltimore, Johns Hopkins University Press, 1973.

Schelling, Thomas C., and Halperin, Morton H., *Strategy and Arms Control,* New York, The Twentieth Century Fund, 1961.

Stone, Jeremy, *Strategic Persuasion: Arms Limitation Through Dialogue,* New York, Columbia University Press, 1967.

Talbott, Strobe, *Endgame: The Inside Story of SALT II,* New York, Harper and Row, 1979.

Willrich, M., and Rhinelander, J.B., eds., *SALT: The Moscow Agreements and Beyond,* New York, The Free Press, 1974.

12. *Biomedical Ethics*

Alford, Robert R., *Health Care Politics: Ideological and Interest Group Barriers to Reform,* Chicago, University of Chicago Press, 1975.

American Association for the Advancement of Science, Office of Science Education—*EVIST Resource Directory,* a directory of programs and courses in the field of Ethics and Values in Science and Technology, Washington, D.C., American Association for the Advancement of Science, 1978.

American Academy of Arts and Sciences, "Limits of Scientific Inquiry," *Daedalus,* Spring 1978.

Baltimore, David, "Limiting Science: A Biologist's Perspective," *Daedalus,* **107**, 1978.

Bandman, E.L., and Bandman, B., *Bioethics and Human Rights,* Boston, Little, Brown, 1978.

Beauchamp, Tom L., and Walters, Leroy, *Contemporary Issues in Bioethics,* Belmont, California, Wadsworth, 1978.

Beers, R.F., and Bassett, E.G., eds., *Recombinant Molecules: Impact on Science and Society,* New York, Raven Press, 1977.

Biotechnology and the Law: Recombinant DNA and the Control of Scientific Research, A symposium. Southern California Law Reveiw, 51, No. 6, September 1978.

Brody, Howard, *Ethical Decisions in Medicine,* Boston, Little, Brown, 1976.

Cahn, Edmund, *The Moral Decision: Right and Wrong in the Light of American Law,* Bloomington, Indiana, Indiana University Press, 1955.

Childress, James F., "Who Shall Live When Not All Can Live?" *Soundings,* 53, Winter 1970, p. 4.

Cohen, Carl, "When May Research Be Stopped?," *New England Journal of Medicine,* 296, 1977, p. 21.

Engelhardt, H. Tristram Jr., and Callahan, Daniel, *Morals, Science and Sociality,* Hastings-on-Hudson, New York, 1978.

Ethics Advisory Board, *Report and Conclusions: HEW Support of Research Involving Human in vitro Fertilization and Embryo Transfer* (with separate appendix) Washington, D.C., U.S. Government Printing Office, 1979.

Freifelder, David, *Recombinant DNA: Readings from Scientific American,* San Francisco, W.H. Freeman, 1979.

Grobstein, Clifford, *A Double Image of the Double Helix, The Recombinant DNA Debate,* San Francisco, W.H. Freeman, 1979.

————, "External Human Fertilization," *Scientific American,* 240, No. 6, pp. 57–67, Scientific American Offprint §1429.

Hastings Center, *Readings in Society, Ethics and the Life Sciences,* Hastings-on-Hudson, New York, Hastings Center, 1979.

Hiatt, Howard H., "Protecting the Medical Commons: Who Is Responsible?" *New England Journal of Medicine,* 193, July 31, 1975.

Jackson, D.A., and Stich, S.P., eds., *The Recombinant DNA Debate,* Englewood Cliffs, New Jersey, Prentice-Hall, 1979.

Karp, Laurence E., *Genetic Engineering,* Chicago, Nelson-Hall, 1976.

Kay, David A., *The International Regulation of Pharmaceutical Drugs,* Washington, D.C., American Society of Interntional Law, 1975.

Kieffer, George H., *Bioethics: A Textbook of Issues,* Reading, Massachusetts, Addison-Wesley, 1979.

Krimsky, Sheldon, "A Citizen Court in the Recombinant DNA Debate," *Bulletin of the Atomic Scientists,* October 1978.

Lappe, M., and Morison, R.S., eds., *Ethical and Scientific Issues Posed by Human Uses of Molecular Genetics, Annals of New York Academy of Science,* 265, 1976.

Leach, Gerald, *The Biocrats: Implications of Medical Progress,* Baltimore, Penguin, 1978.

Lear, John, *Recombinant DNA: The Untold Story,* New York, Crown, 1978.

National Academy of Sciences, *Experiments and Research with Humans: Values in Conflict,* Academy Forum, Washington, D.C., National Academy of Sciences, 1975.

National Academy of Sciences, *Research with Recombinant DNA,* Academy Forum, Washington, D.C., National Academy of Sciences, 1977.

Pellegrino, Edmund D., "Medical Morality and Medical Economics," *Hastings Center Report,* August 1978.

Ramsey, Paul, *The Patient as a Person,* New Haven, Yale University Press, 1970.

Reiser, Stanley J., Dyck, Arthur J., and Curran, William J., *Ethics in Medicine: Historical Perspectives and Contemporary Concerns,* Cambridge, M.I.T. Press, 1977.

Rescher, Nicholas, "The Allocation of Exotic Medical Lifesaving Therapy," *Ethics,* 79, April 3, 1969.

Rettig, Richard, *Cancer Crusade: The Story of the National Cancer Act of 1971,* Princeton, New Jersey, Princeton University Press, 1977.

Richards, John, ed., *Recombinant DNA Science, Ethics, Politics,* New York, Academic Press, 1978.

Rogers, Michael, *Biohazard,* New York, Knopf, 1977.

Shatin, Leo, "Medical Care and the Social Worth of Man," *American Journal of Orthopsychiatry,* 36, 1967.

Stein, Jan J., *Making Medical Choices,* Boston, Houghton Mifflin, 1978.

United States House of Representatives, Commission on Science and Technology, *Genetic Engineering, Human Genetics and Cell Biology. DNA Recombinant Molecule Research* (Suppl. Report II) Washington, D.C., U.S. Government Printing Office, 1976.

United States Senate, Commission on Commerce, Science and Transportation, *Hearings before the Subcommittee on Science, Technology and Space on Regulation of Recombinant DNA Research,* Serial No. 95–52, Washington, D.C., U.S. Government Printing Office, 1978.

Veatch, Robert M., *Case Studies in Medical Ethics,* Cambridge, Harvard University Press, 1977.

Wade, Nicholas, *The Ultimate Experiment: Man-Made Evolution,* New York, Walker, 1977.

Wojcik, *Muted Consent,* West Lafayette, Indiana, Purdue University Press, 1978.

Wood, Madelyn, *Medicine and Health Care in Tomorrow's World,* New York, Julian Messner, 1974.

Young, Robert, "Some Criteria for Making Decisions Concerning the Distribution of Scarce Medical Resources," *Theory and Decision,* **6**, No. 4, November 1975.

Suggested Periodicals

The Bulletin of the Atomic Scientists, monthly, Chicago.

Federation of American Scientists Newsletter, (and F.A.S. Public Interest Report), monthly, Washington, D.C.

The Hastings Center Report, Hastings-on-Hudson, N.Y., Institute of Society, Ethics and the Life Sciences.

International Security, quarterly, Cambridge, Massachusetts, Harvard University.

Science, weekly, Washington, D.C., American Association for the Advancement of Science.

Science and Government Report, monthly, Washington, D.C.

Science for the People, Jamaica Plain, Massachusetts, Scientists and Engineers for Social and Political Action.

Science, Technology and Human Values Quarterly, Cambridge, Massachusetts, M.I.T. Press, co-ponsored by the John F. Kennedy School of Government, Harvard, and the Program on Science, Technology and Society, M.I.T.

Scientist and Citizen, St. Louis, Missouri, Scientists' Institute for Public Information.

Technology Review, monthly, Cambridge, Massachusetts, M.I.T.

Acronyms

AAAS	American Association for the Advancement of Science
ABM	Anti–ballistic missile
ACDA	Arms Control and Disarmament Agency
AEC	Atomic Energy Commission (U.S.)
AIRS	Advanced inertial reference sphere
API	American Petroleum Institute
ARPA	Advanced Research Projects Agency (Defense Department)
BART	Bay Area Rapid Transit
BMD	Ballistic missile defense
BTU	British thermal unit
CBM	Confidence building measures
CEP	Circular error probability
CIA	Central Intelligence Agency
CSCE	Conference on Security and Cooperation in Europe
CTB	Comprehensive nuclear test ban
CTBT	Comprehensive Nuclear Test Ban Treaty
DASA (U.S.)	Defense Atomic Support Agency [renamed the Defense Nuclear Agency in 1959]
DBCP	Dibromochloropropane
DDT	Dichloro–diphenyl–trichloro–ethane
DNA	Defense Nuclear Agency (U.S.) [see DASA]
DNA	Deoxyribonucleic acid
DOD	Department of Defense (U.S.)
DOE	Department of Energy (U.S.)
EMP	Electromagnetic pulse
EPA	Environmental Protection Agency (U.S.)
ERW	Enhanced radiation weapons (or "neutron bombs")
FAA	Federal Aviation Administration (U.S.)
FAS	Federation of American Scientists
FBS	Forward based systems
FCC	Federal Communications Commission (U.S.)
FDA	Food and Drug Administration (U.S.)
GNP	Gross National Product

HEW	Department of Health, Education, and Welfare (U.S.)
ICBM	Intercontinental ballistic missile
ICF	Inertially confined fusion
IIIS	International Institute for Strategic Studies
IRBM	Intermediate range ballistic missile
ISC	Intersystemic communications
LIA	Lead Industries Association
LTBT	Limited Test Ban Treaty (also referred to as the SALT Partial Test Ban Treaty)
MAD	Mutual assured destruction
MARV	Maneuverable re–entry vehicle
MFR	Mutual force reduction
MIRV	Multiple independently targeted re–entry vehicle
MORESCO	Misamis Oriental Rural Electric Service Cooperative
MRV	Multiple re–entry vehicle
MX	Missile X
NAS	National Academy of Sciences (U.S.)
NATO	North Atlantic Treaty Organization
NCI	National Cancer Institute (U.S.)
NIAID	National Institute of Allergy and Infectious Diseases (U.S.)
NIH	National Institutes of Health (U.S.)
NPG	Nuclear Planning Group (NATO)
NPT	Nonproliferation Treaty
NSF	National Science Foundation (U.S.)
OPEC	Organization of Petroleum Exporting Countries
OSHA	Occupational Safety and Health Administration (U.S.)
PBB	Polybrominated biphenyl
PCB	Polychlorinated biphenyl
PNE	Peaceful nuclear explosions
PSAC	President's Science Advisory Committee (U.S.) [1957–1973]
R&D	Research and development
RNA	Rybonucleic acid
SALT	Strategic Arms Limitation Talks
SIPRI	Stockholm International Peace Research Institute
SLBM	Submarine launched ballistic missile
SMR	Scarce medical resource
SST	Supersonic transport
TNW	Tactical nuclear weapons
TREE	Transient radiation effects on electronics
TSCA	Toxic Substances Control Act
TVA	Tennessee Valley Authority
WTO	Warsaw Treaty Organization

Name Index